HUMAN CELL CULTURE
Volume IV: Primary Hematopoietic Cells

Human Cell Culture

Volume 4

Human Cell Culture

Volume IV:

Primary Hematopoietic Cells

edited by

Manfred R. Koller

Oncosis, San Diego, CA, U.S.A.

Bernhard O. Palsson

University of California,
Department of Bioengineering, La Jolla, CA, U.S.A.

and

John R.W. Masters

University College London,
Institute of Urology, London, U.K.

KLUWER ACADEMIC PUBLISHERS
DORDRECHT / BOSTON / LONDON

Library of Congress Cataloging-in-Publication data is available.

ISBN 0-7923-5821-X

Published by Kluwer Academic Publishers,
P.O. Box 17, 3300 AA Dordrecht, The Netherlands.

Sold and distributed in North, Central and South America
by Kluwer Academic Publishers,
101 Philip Drive, Norwell, MA 02061, U.S.A.

In all other countries, sold and distributed
by Kluwer Academic Publishers,
P.O. Box 322, 3300 AH Dordrecht, The Netherlands.

Printed on acid-free paper

Table of Contents

Chapter 1: Hematopoietic stem and progenitor cells 1
Manfred R. Koller, Ph.D. *Oncosis*
Bernhard O. Palsson, Ph.D. *University of California - San Diego*

Chapter 2: *In vitro* T-lymphopoiesis 31
Michael Rosenzweig, Ph.D. *New England Regional Primate Center*
David T. Scadden, M.D. *Massachusetts General Hospital*

Chapter 3: T-lymphocytes: Mature polyclonal and antigen-specific cell expansion 45
Bruce L. Levine, Ph.D. *Naval Medical Research Institute*
Katia Schlienger, M.D., Ph.D. *Naval Medical Research Institute*
Carl H. June, M.D. *Naval Medical Research Institute*

Chapter 4: The culture, characterization, and triggering of B lymphocytes 101
Gerrard Teoh, M.D. *Dana-Farber Cancer Institute*
Kenneth C. Anderson, M.D. *Dana-Farber Cancer Institute*

Chapter 5: Monocytes and macrophages 125
Ivan N. Rich, Ph.D. *Richland Memorial Hospital*

Chapter 6: Isolation and cultivation of osteoclasts and osteoclast-like cells 147
Philip A. Osdoby, Ph.D. *Washington University*
Fred Anderson. *Washington University*
William Maloney, M.D. *Washington University Medical Center*
Patricia Collin-Osdoby, *Ph.D. Washington University*

Chapter 7: Isolation and culture of human dendritic cells 171
Michael A. Morse, M.D. *Duke University*
H. Kim Lyerly, M.D. *Duke University*

Chapter 8: *In vitro* proliferation and differentiation of CD34+ cells to neutrophils 193
James G. Bender, Ph.D. *Nexell Therapeutics, Inc.*

Chapter 9: Isolation and culture of eosinophils 219
Helene F. Rosenberg, M.D. *National Institutes of Health*

Chapter 10: Isolation and culture of mast cells and basophils 241
Peter Valent, M.D. *University of Vienna*

Chapter 11: Purification and culture of erythroid progenitor cells 259
Chun-Hua Dai, M.D. *VA Medical Center*
Amittha Wickrema, Ph.D. *University of Illinois - Chicago*
Sanford B. Krantz, M.D. *Vanderbilt University Medical School*

Chapter 12: *In vitro* development of megakaryocytes and platelets 287
Marcus O. Muench, Ph.D. *University of California - San Francisco*
Jae-Hung Shieh, Ph.D. *New York Blood Center*

Chapter 13: Perspectives, ethics, and clinical issues in the use of primary human cells 317
Mary Pat Moyer, Ph.D. *InCell, Inc.*

Introduction

The daily production of hundreds of billions of blood cells through the process of hematopoiesis is a remarkable feat of human physiology. Transport of oxygen to tissues, blood clotting, antibody- and cellular-mediated immunity, bone remodeling, and a host of other functions in the body are dependent on a properly functioning hematopoietic system. As a consequence, many pathological conditions are attributable to blood cell abnormalities, and a fair number of these are now clinically treatable as a direct result of hematopoietic research.

Proliferation of hematopoietic stem cells, and their differentiation into the many different lineages of functional mature cells, is highly regulated and responsive to many environmental and physiological challenges. Our relatively advanced understanding of this stem cell system provides potentially important insights into the regulation of development in other tissues, many of which are now being acknowledged as stem cell-based, perhaps even into adulthood. The recent public and scientific fanfare following announcement of human embryonic stem cell studies suggests that stem cell research will continue to be a relevant and exciting topic.

Recent advancements in primary human hematopoietic cell culture have led to remarkable progress in the study of hematopoiesis, stem cell biology, immunology, carcinogenesis, tissue engineering, and even in clinical practice for the treatment of disease. This unique comprehensive volume in the Human Cell Culture Series encompasses research methodology for the growth and differentiation of all types of primary hematopoietic cells. Over the past decade, many new techniques have been developed to propagate human cells for a number of hematopoietic lineages, utilizing specific growth factors, stroma, medium additives, perfusion culture, and other strategies. Each of twelve hematopoietic cell types is covered by a leading expert in the field, providing insightful background information along with detailed current culture and assay techniques. In addition to uses for research applications, current and future clinical applications of large-scale culture methods are also discussed. Because the procurement and processing of primary human tissues can pose a significant barrier to new researchers in this field, this subject is covered in detail within each chapter. The final chapter is intended to guide scientists through the significant regulatory and ethical implications associated with use of human and fetal tissues. A consistent format with generous inclusion of tables and figures enables readers to locate key information about each cell/tissue type covered. Additionally, numerous literature citations provide a valuable reference for students and professionals in the hematology, immunology, oncology, and bioengineering fields. It is our goal to stimulate interest in the study of human hematopoiesis, with the belief that new therapeutic solutions for a variety of diseases will result.

Manfred R. Koller

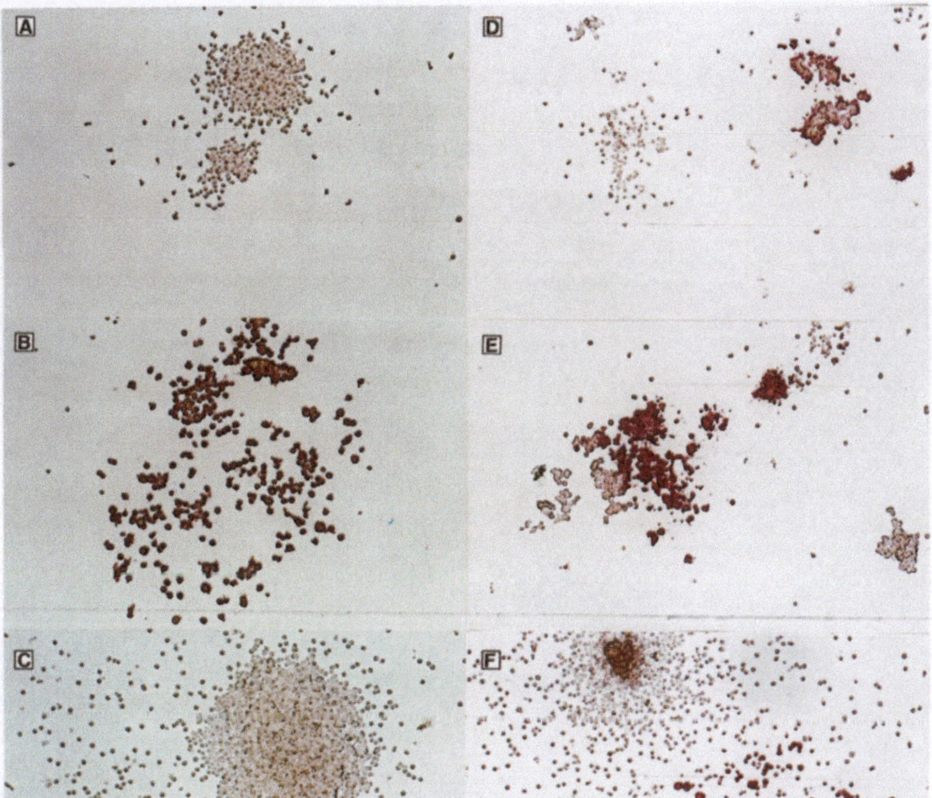
A
B
C
D
E
F

Chapter 1

Hematopoietic Stem and Progenitor Cells

[1]Manfred R Koller and [2]Bernhard O Palsson
[1]Oncosis, 6199 Cornerstone Ct., Suite 111, San Diego, CA 92121 and [2]Department of Bioengineering, UCSD, 9500 Gilman Dr, La Jolla, CA 92093-0412. Tel: 001-619-550-1770. E-mail: fredkoller@oncosis.com

1. INTRODUCTION

The human body consumes a staggering 400 billion mature blood cells every day, and this number increases dramatically under conditions of stress such as infection or bleeding. A complex scheme of multilineage proliferation and differentiation, termed hematopoiesis (Greek for blood-forming), has evolved to meet this demand. This regulated production of mature blood cells from primitive stem cells, which occurs mainly in the bone marrow (BM) of adult mammals, has been the focus of considerable research. *Ex vivo* models of human hematopoiesis now exist that have significant scientific value and promise to have an impact on clinical practice in the near future. This chapter introduces the reader to the fundamental concepts of hematopoiesis, and provides information required for the implementation of human stem and progenitor cell culture techniques. The rest of this volume contains chapters which address the isolation, culture, and utility of each of the major mature human hematopoietic cell types.

1.1 Function and Organization of the Hematopoietic System

There are approximately a dozen major types of mature blood cells which are found in the body, depending upon the subdivision nomenclature used (Fig. 1). These populations are divided into two major groups: the myeloid and lymphoid. The myeloid lineages include erythrocytes (red blood cells), monocyte lineage-derived cells (e.g. macrophages, osteoclasts, and dendritic cells), the granulocytes (e.g. neutrophils, eosinophils, basophils, and mast cells), and platelets (derived from non-circulating megakaryocytes). Thymus-derived (T)-lymphocytes, BM-derived (B)-lymphocytes, and natural killer (NK) cells constitute the lymphoid lineages. Most mature blood cells exhibit a limited lifespan *in vivo*. Although some lymphocytes are thought to survive for many years, it has been shown that erythrocytes and neutrophils have lifespans of 120 days and 8 hours, respectively [1]. As a result, hematopoiesis is a highly prolific process which occurs throughout our lives to fulfill this demand.

Mature cells are continuously produced from progenitor cells which are produced from earlier cells, which in turn originate from stem cells. At the top (Fig. 1) are the very primitive totipotent stem cells, the majority of which are in a nonproliferative state (G_O) [2]. These cells are very rare (1 in 10^5 BM cells), but collectively have enough proliferative capacity to last several lifetimes [3, 4]. Through some unknown mechanism(s), at any given time a small number of these cells are actively proliferating, differentiating, and self-renewing, thereby producing more mature progenitor cells while maintaining the size of the stem cell pool. Whereas stem cells (by definition) are not restricted to any lineage, their progenitor cell progeny do have a restricted potential and are far greater in number. Therefore, as the cells differentiate and travel from top to bottom in Fig. 1, they become more numerous, lose self-renewal ability, lose proliferative potential, become restricted to a single lineage, and finally become a mature functional cell of a particular type.

1.2 Stem Cell Self-Renewal

Although stem cells have traditionally been thought to be capable of unlimited self-renewal, new data suggest that this may not actually be the case. For example, stem cells isolated from fetal liver, neonatal umbilical cord blood (CB), and adult BM show a clear hierarchy of proliferative potential [5]. One hypothesis that has been suggested is that the length of telomeric DNA at the ends of chromosomes is shortened over time, acting as a mitotic clock that triggers replicative senescence once telomeres reach a threshold length [6]. In support of this hypothesis, longer telomeres and greater telomerase activity (which extends telomeres) have been measured in germline cells and tumor cells that do not exhibit replicative senescence [7]. Interestingly, telomerase activity is relatively high in stem cells, although the activity does not appear to be great enough to impart immortality [8]. While one study showed that introduction of telomerase

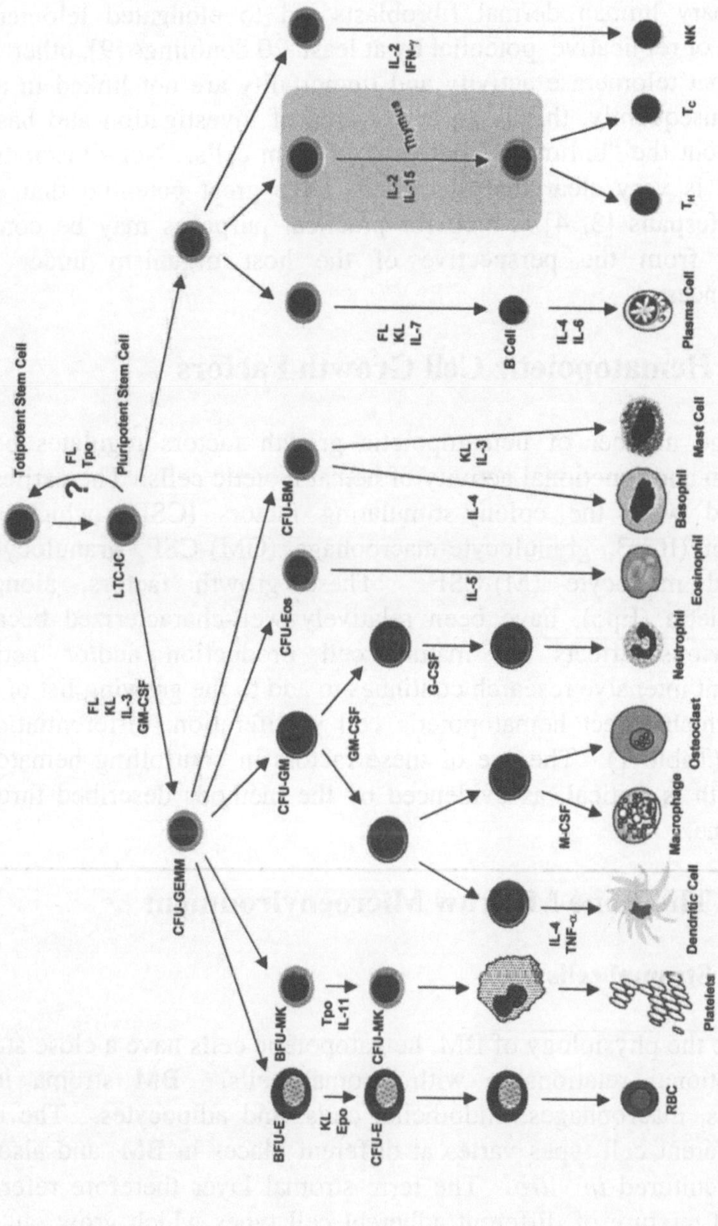

Figure 1. The hematopoietic system hierarchy. It is believed that dividing pluripotent stem cells may undergo self-renewal to form daughter stem cells without loss of potential (a matter of current debate), or may experience a concomitant differentiation to form daughter cells with more restricted potential. Continuous proliferation and differentiation along each lineage results in the production of many mature cells. This process is under the control of many growth factors (see Table 1), and the sites of action for some of these are shown. The mechanisms that determine which lineage a cell will develop into are not fully understood, although many models have been proposed.

into primary human dermal fibroblasts led to elongated telomeres and extension of replicative potential by at least 20 doublings [9], other studies suggest that telomerase activity and immortality are not linked in all cells [10]. Consequently, this is an active area of investigation and has led to debate about the "unlimited" potential of stem cells. Notwithstanding this debate, it is very clear that stem cells have great potential that exceeds several lifespans [3, 4], which for practical purposes may be considered unlimited from the perspective of the host organism under normal circumstances.

1.3 Hematopoietic Cell Growth Factors

A large number of hematopoietic growth factors regulates both the production and functional activity of hematopoietic cells. The earliest to be discovered were the colony-stimulating factors (CSF), which include interleukin (IL)-3, granulocyte-macrophage (GM)-CSF, granulocyte (G)-CSF, and monocyte (M)-CSF. These growth factors, along with erythropoietin (Epo), have been relatively well-characterized because of their obvious effects on mature cell production and/or activation. Subsequent intensive research continues to add to the growing list of growth factors which affect hematopoietic cell proliferation, differentiation, and function (Table 1). The use of these factors in controlling hematopoietic cell growth is critical, as evidenced by the methods described throughout this volume.

1.4 The Bone Marrow Microenvironment

1.4.1 Stromal cells

Due to the physiology of BM, hematopoietic cells have a close structural and functional relationship with stromal cells. BM stroma includes fibroblasts, macrophages, endothelial cells, and adipocytes. The ratio of these different cell types varies at different places in BM, and also as the cells are cultured *in vitro*. The term stromal layer therefore refers to an undefined mixture of different adherent cell types which grow out from a culture of BM cells. *In vitro*, stem cells placed on a stromal cell layer will attach to and often migrate underneath the stromal layer [11]. Under the stromal layer, some of the stem cells will proliferate, and the resulting progeny will be packed together, trapped under the stroma, forming a characteristic morphologic feature known as a cobblestone area. It is widely believed that primitive cells must be in contact with stromal cells to

Table 1. Hematopoietic growth factors

Growth Factor	Alternative names	Abbreviations	Ref.
Interleukin-1	Hemopoietin-1	IL-1	[100]
Interleukin-2		IL-2	[101]
Interleukin-3	Multi-colony-stimulating factor	IL-3, Multi-CSF	[102]
Interleukin-4	B-cell-stimulatory factor-1	IL-4, BSF-1	[103]
Interleukin-5		IL-5	[103]
Interleukin-6	B-cell-stimulatory factor-2	IL-6, BSF-2	[104]
Interleukin-7		IL-7	[105]
Interleukin-8	Neutrophil activating peptide-1	IL-8, NAP-1	[106]
Interleukin-9		IL-9	[107]
Interleukin-10	Cytokine synthesis inhibitory factor	IL-10, CSIF	[108]
Interleukin-11		IL-11	[109]
Interleukin-12	NK cell stimulatory factor	IL-12, NKSF	[110]
Interleukin-13		IL-13	[111]
Interleukin-14	High molecular weight B cell growth factor	IL-14, HMW-BCGF	[112]
Interleukin-15		IL-15	[113]
Interleukin-16	Lymphocyte chemoattractant factor	IL-16, LCF	[114]
Interleukin-17		IL-17	[115]
Interleukin-18	IFN-gamma-inducing factor	IL-18, IGIF	[116]
Erythropoietin		Epo	[117]
Monocyte-CSF	Colony-stimulating factor-1	M-CSF, CSF-1	[118]
Granulocyte-CSF		G-CSF	[118]
Granulocyte-macrophage-CSF		GM-CSF	[118]
Stem cell factor	c-*kit* ligand, Mast cell growth factor	SCF, KL, MGF	[119]
Interferon-gamma	Macrophage activating factor	IFN-γ, MAF	[120]
Macrophage inflammatory protein-1	Stem cell inhibitor	MIP-1, SCI	[121]
Leukemia inhibitory factor		LIF	[122]
Transforming growth factor-beta		TGF-β	[123]
Tumor necrosis factor-alpha	Cachectin	TNF-α	[124]
flk-2 ligand	*flt*-3 ligand	FL	[125]
Thrombopoietin	c-*mpl* ligand; Megakaryocyte growth and development factor	Tpo, ML, MGDF	[126]

maintain their primitive state. However, much of the effect of stromal cells has been attributed to the secretion of growth factors. Consequently, there have been reports of successful hematopoietic cell growth with the addition of numerous soluble growth factors in the absence of stroma [12-14]. However, this issue is quite controversial (see section 4.2), and stromal cells are still likely to be valuable because they synthesize membrane-bound growth factors [15], extracellular matrix (ECM) components [16], and probably some as yet undiscovered growth factors. In addition, stromal cells can modulate the growth factor environment in a way that would be very difficult to duplicate by simply adding soluble growth factors [17]. This modulation may be responsible for the observations that stroma can be both stimulatory and inhibitory [18].

1.4.2 Extracellular matrix

Like all other cells *in vivo*, hematopoietic cells have considerable interaction with ECM. The ECM of BM consists of collagens, laminin, fibronectin [19], vitronectin [20], hemonectin [21], and thrombospondin [22]. The heterogeneity of this system is further complicated by the presence of various proteoglycans, which are themselves complex molecules with numerous glycosaminoglycan chains linked to a protein core [23]. These glycosaminoglycans include chondroitin, heparan, dermatan, and keratan sulfates, and hyaluronic acid. The ECM is secreted by stromal cells of the BM (particularly endothelial cells and fibroblasts) and provides support and cohesiveness for the BM structure. There is a growing body of evidence indicating that ECM is important for the regulation of hematopoiesis, and these concepts have been reviewed in detail [16].

2. TISSUE PROCUREMENT AND PROCESSING

The implementation of primary human hematopoietic stem and progenitor cell culture techniques requires the use of primary human tissue. The procurement of these tissues is not straightforward, but can be accomplished through a number of mechanisms that are described below.

2.1 Human Hematopoietic Tissue Procurement and Shipping

Although hematopoiesis occurs mainly in BM of adult mammals, totipotent stem cells first arise in the yolk sac during embryonic development, are later found in fetal liver, and at birth are found in high concentrations in CB. In adults, stem cells are found in peripheral blood at very low concentrations, but the concentration increases dramatically after stem cell mobilization. Mobilization of stem cells into peripheral blood is a phenomenon which occurs in response to chemotherapy or growth factor administration *in vivo*. Therefore, hematopoietic stem cells can be collected from fetal liver, CB, BM, or mobilized peripheral blood (mPB). However, cell properties from these different tissues may vary. For example, the stem cell population within fetal liver is believed to be the most primitive and prolific, whereas adult tissue has the least proliferative potential [5]. Also, because CB stem cells are circulating *in vivo*, they have been shown to be less dependent on stroma *in vitro*, as compared with stromal-contacting BM-derived stem cells [24].

As with any human tissue, obtaining specimens for experimentation requires a donor source, a clinical collaborator, and some level of regulatory approval (see Chapter 13). The most likely source for these tissues is a hospital or clinic in which patients of interest are being treated. For example, fetal liver and CB would be available from an obstetrics/gynecology ward, whereas mPB and BM would be available from a hematology/oncology or BM transplant ward. From a practical point of view, CB is the most easily obtained tissue because it is otherwise discarded from the many deliveries that occur each day, and approximately 5×10^8 nucleated cells can be obtained from each sample. A significant amount of BM, on the order of $1\text{-}5 \times 10^8$ mononuclear cells (MNC), can be obtained by rinsing out processing sets (consisting of a bag, tubing, and stainless steel screens) that are used to filter BM during harvest in the operating room, and this source is also relatively easy to obtain because it is otherwise discarded. BM can also be obtained as a waste material from orthopedic surgeries, such as hip replacement. Unlike CB, the procurement of BM from these sources is dependent on the scheduling of operating room procedures, which occur only once or twice per week in most centers. Normal volunteer BM donors can also be recruited for small (~10 ml) iliac crest (hip) BM aspirates containing about 10^8 MNC. Somewhat less accessible is mPB because any amount taken for research comes directly from the patient, not from a waste material. Nevertheless, small mPB samples can be obtained under informed consent (see Chapter 13), or alternatively, large frozen mPB samples that were intended for transplant into deceased patients are also sometimes available. Although fetal liver is considered a waste material from spontaneous and elective abortions, there are obvious ethical issues that must be addressed when using this tissue (see Chapters 12 and 13).

Once a tissue is chosen, physicians and/or nurses from the appropriate departments should be contacted. These sources will usually be willing to provide specimens for free or at a nominal cost (e.g. $20 to $100 per sample), depending upon the source, requestor (e.g. academia vs. industry), the described use, and the interest level of the clinician. Volunteer donors should also be compensated, usually $60 to $120 for BM aspirates. Compliance at the clinical site is greatly facilitated if the proper materials are provided by the requestor. For example, provision of pre-sterilized vials containing anticoagulant and pre-labeled shipping containers usually results in greater cooperation.

Because the source of human tissue samples is usually far from the laboratory of intended use, a packing and transport protocol must be developed. For hematopoietic tissues, which are often transported for clinical transplants, protocols have been worked out in great detail. Unfortunately, almost all of these protocols are based upon the transport of cryopreserved material. For laboratory studies, it is difficult to justify the time and expense required to prepare and cryopreserve cells prior to shipment from the donor site, and cryopreserved and thawed tissue often has different properties than fresh tissue. Fortunately, overnight shipment of cells in standard wet ice containers with minimal processing has proven effective for human hematopoietic tissue sources, provided an anticoagulant such as heparin (preservative-free) or acid-citrate-dextrose (ACD) is present to prevent sample clotting. Upon arrival, samples can be maintained for an additional 24 hours at 4 °C with minimal difficulty, although viable cell number will decline with time [25].

2.2 Human Hematopoietic Tissue Processing

Except for fetal liver which must first be homogenized (see Chapter 12), human hematopoietic tissue samples essentially have the consistency and appearance of whole blood. Therefore, the cells can be counted, assayed, and cultured immediately upon arrival with little processing. Alternatively, the specimen may be processed to isolate the cells of interest for use in experimentation. Methods commonly employed for stem cell enrichment include erythrocyte lysis, density centrifugation, elutriation, adherence depletion, antibody-mediated depletion or selection, and combinations thereof (Table 2). Most protocols begin with a Ficoll-Hypaque discontinuous density gradient separation yielding the low density (<1.077 g/cm^3) MNC population at the interface while the more dense erythrocytes and granulocytes will pellet at the bottom. Perhaps the most common final processing step is CD34-selection, enriching for cells expressing the CD34 antigen (see section 3.1.2). A number of CD34-selection methods are available, and a recent study has compared the use of five of these methods [26]. As mentioned above, these cells can be cryopreserved for later use using standard techniques [27]. Although each processing step results in a more enriched population, it is important to note that a significant loss of yield is associated with each of these steps [28]. Furthermore, elimination of different cell populations results in loss of tissue function with respect to the role of the cell being eliminated.

Table 2. Human hematopoietic tissue processing methods

Method	Purpose	Reference
Erythrocyte lysis	Reduces the very large population (>99.99%) of erythrocytes, leaving the majority of the leukocyte (white cell) population	[127]
Density gradient centrifugation	Reduces the high density cells, leaving the majority of lymphocytes and more primitive stem and progenitor cells	[78]
Elutriation	Reduces cells with large size and/or density	[128]
Adherence depletion	Reduces adherent populations such as macrophages and fibroblasts	[129]
Antibody depletion	Selectively reduces cell populations targeted by specific antibodies	[129]
Antibody selection	Enriches cell populations targeted by specific antibodies, most often used with CD34	[26]

Further details on these processing methods can be found in the listed references. Also, subsequent chapters in this volume cover these methods in greater detail as they apply to the isolation of various hematopoietic cell types.

3. ASSAYS

Because hematopoietic tissues are very heterogeneous, containing many cell types, many assays have been developed in order to characterize the cell populations that are present in a sample. In fact, some of these assays are utilized in the clinic to determine the suitability of a particular cell sample for transplantation. Upon culture of these cells, the ratio of different populations may change significantly, such that the total cell number generated is not an adequate measure of outcome. Therefore, the use of appropriate assays is critical to determine the success of a particular culture technique.

3.1 Non-Biological Assays

3.1.1 Histology

The first method used to assay hematopoietic cells was histology, dating back to the late 1800's [29]. In fact, this method is still widely practiced in the clinic, utilizing spreads of Wright-Giemsa stained cells under oil-immersion microscopy, or automated instruments that have been developed to carry out these differentials (counting of different cell types). This type of analysis is most useful for assessing mature cell populations which have distinctive morphological features and are present in large numbers [30].

3.1.2 Flow cytometry

Flow cytometry has been used extensively in the study of the hematopoietic system. Antibodies to antigens on many of the cell types shown in Fig. 1 have been developed [31, 32]. Because of the close relation of many of the cell types, combinations of antigens are often required to definitively identify a particular cell, and these are described in many of the chapters that follow. Recently, much effort has been focused on the identification of primitive stem cells, and this has been accomplished by analyzing increasingly smaller subsets of cells using increasingly complex antibody combinations. By far the most utilized antigen is CD34, which appears to identify all cells from the stem through progenitor stage [33]. The CD34 antigen is stage-specific but not lineage-specific, and therefore identifies cells that lead to repopulation of all cell lineages in transplant patients [34]. However, the CD34 antigen is not restricted to hematopoietic cells because it is also found on certain stromal cells in the hematopoietic microenvironment [31]. Also, recent controversial data suggest that the most primitive stem cells are $CD34^-$ [35, 36], and that these give rise to the more mature $CD34^+$ population [36]. Although CD34 captures a small population which contains stem cells, the $CD34^+$ population is itself quite heterogeneous and can be fractionated by many other antigens. Over the past several years, many different combinations of antibodies have been used to fractionate the $CD34^+$ population. $CD34^+$ fractions which lack CD33, HLA-DR, CD38, or CD71 appear to be enriched in stem cells [31]. Conversely, $CD34^+$ populations which coexpress Thy-1 or *c-kit* appear to contain the primitive cells [31]. These studies have revealed the extreme rarity of stem cells within the heterogeneous BM population. Of the MNC subset (~40% of whole BM), only ~2% are $CD34^+$, and of those, only ~5% may be $CD38^-$. Furthermore, this extremely rare population is still heterogeneous with respect to stem cell content. Consequently, stem cells as single cells have not yet been identified, and it is important to note that stem cell phenotype does not necessarily correlate with stem cell function at the single cell level [37, 38].

3.2 Biological Assays

3.2.1 *In vivo* biological assays

The first *in vivo* assay for early hematopoietic cells was provided by Till and McCulloch in 1961 [39]. In their experiments, lethally irradiated mice were injected with BM cells from healthy donor mice. The hematopoietic system of mice receiving donor cells was reconstituted, whereas control mice died within a week. Some of the injected cells seeded in the spleen and gave rise to macroscopic hematopoietic colonies containing cells of the myeloid lineage. Cells capable of forming colonies in the spleens of ablated recipient mice are called CFU-S (spleen colony-forming unit), and although these cells were originally thought to be stem cells, subsequent work has shown them to be myeloid-committed early progenitor cells. The current definition of a stem cell therefore includes the ability to confer long-term *in vivo* repopulation of the myeloid and lymphoid lineages of an ablated host. This activity can be greatly enriched for in certain purified cell populations, supporting the hypothesis of a single pluripotent stem cell [4]. Further, genetic marking experiments in mice have demonstrated that long-term engraftment of both lymphoid and myeloid lineages can be achieved by the progeny of a single cell [40], thereby confirming the existence of true hematopoietic stem cells.

Analogous *in vivo* evidence for human stem cells is thus far lacking due to obvious experimental limitations. Several *in vivo* xenogeneic transplant models have been developed utilizing immunodeficient mice and fetal sheep as hosts for human stem cells [41-43]. A major limitation in all of these models is the very low level of human cell chimerism that is obtained in the animals. At best, only a few percent of the blood system is derived from the human donor cells, so it is difficult to state that long-term *in vivo* repopulation has been achieved. Nevertheless, these models have been used to compare *in vivo* repopulation potential of various cell populations, and the models continue to improve with time.

3.2.2 *In vitro* biological assays

In order to deal with the rarity of stem and progenitor cells, the lack of correlation between phenotype and function, and the difficulty of *in vivo* assays, many *in vitro* biological function assays have been developed. Most of these assays are performed by culturing cells under defined conditions and examining their progeny, both in number and type (Table 3) [44]. For example, the colony-forming unit (CFU) assays are performed by plating a dilute suspension of cells in semi-solid medium (most commonly 0.8% methylcellulose) containing specific growth factors. Individual progenitors are stimulated to divide and form colonies of mature cells (from 50 to several thousand) that are scored microscopically. To specify the type of progenitor, a suffix is simply added to the CFU designation. For example, granulocyte-macrophage colony-forming units (CFU-GM) proliferate and develop into mature granulocytes and macrophages. Erythrocyte colony-forming units (CFU-E) undergo growth and hemoglobinization to form mature erythrocytes. Erythrocyte burst-forming units (BFU-E) are the more commonly measured precursor of the CFU-E. Similarly, colony assays are used to study megakaryocyte progenitors (CFU-MK) which produce platelets. Multipotent cells, such as the granulocyte-erythrocyte-macrophage-megakaryocyte colony-forming unit (CFU-GEMM or CFU-Mix), can also be detected in colony assays. This technique has even been extended to include fibroblastic cells (CFU-F) which may be present in hematopoietic tissues [45].

Table 3. Human hematopoietic stem and progenitor cell assays

Method	Purpose	Reference
BFU-E	Detects outgrowth of committed erythrocytic progenitors in semi-solid culture in 2 weeks	[44]
CFU-GM	Detects outgrowth of committed granulocytic/monocytic progenitors in semi-solid culture in 2 weeks	[44]
CFU-MK	Detects outgrowth of committed megakaryocytic progenitors in semi-solid culture in 2-3 weeks	[130]
HPP-CFC	Detects outgrowth of early multi-potential progenitors in semi-solid culture in 4 weeks	[131]
LTC-IC	Detects outgrowth of primitive (perhaps pluripotent) progenitors in liquid stromal culture, followed by semi-solid culture, in 7-10 weeks	[49]
CFU-F	Detects outgrowth of fibroblast progenitors in liquid culture in 1 week	[45]
Xenogeneic transplant	Detects in vivo repopulation of human stem cells in immunodeficient animals *in weeks to years*	[41]
Flow cytometry	Detects surface antigens on individual cells in real time	[32]

Although the ability to measure stem cell potential *in vitro* is controversial, the most widely accepted primitive cell assay is the long-term culture-initiating cell (LTC-IC) assay. This assay utilizes long-term culture (five to eight weeks) of cells on a BM stromal layer, after which the cells are replated into a CFU assay [46]. This concept has been carried even further in the extended (E)LTC-IC assay, in which even more primitive cells are measured after as long as sixteen weeks *in vitro* [47]. These assays essentially rely on the fact that committed progenitor cells do not persist *in vitro* (or *in vivo*), and therefore progenitors measured after extended culture periods must be the progeny of primitive cells that were initially present in the assay sample. It is uncertain whether these *in vitro* assays are truly measuring human long-term *in vivo* repopulating cells. Fortunately, a correlation between long-term *in vitro* and long-term *in vivo* repopulating ability has been demonstrated for different murine cell populations [48], suggesting that the same may be true for human cells. Consequently, the LTC-IC assay concept has been adopted by many investigators to quantitate primitive human cells.

It is important to note that considerable care must be taken in order to ensure the quantitative fidelity of these biological assays [49]. For example, the plating of too few cells results in a low signal-to-noise ratio with weak statistics, whereas the plating of too many cells overloads the assay and underestimates the true value. Utilizing only the linear range of a biological assay can be difficult if the cell type being measured is present at a frequency that is unknown or varies widely between samples. In these cases, it is prudent to perform assays at several densities in parallel, and then utilize the results from only those densities that prove to be within the linear range [49]. Another issue to note is that some assays can behave differently when different cell populations are being assayed. For example, the LTC-IC assay is more prone to non-linearity with MNC than with CD34-enriched cells, presumably due to accessory cell effects [49].

4. CULTURE TECHNIQUES

4.1 Historical Perspective: Long-Term Bone Marrow Culture (LTBMC)

The history of hematopoietic stem and progenitor cell culture methods spans roughly the past twenty years. This history will be briefly described here, as it provides the backdrop for the current culture techniques. More complete reviews have been previously published and are recommended to the interested reader [50-52].

In the mid 1970's, Dexter and coworkers were successful in developing a culture system which maintained murine hematopoiesis for several months [53]. The key feature of this system was the establishment of a BM-derived stromal layer during the first three weeks of culture which was then recharged with fresh BM cells. One to two weeks after the cultures were recharged, active sites of hematopoiesis appeared. These sites are often described as cobblestone regions, which are the result of primitive cell proliferation (and accumulation) beneath the stromal layer. Screening of serum lots for long-term BM culture (LTBMC) support was found to be very important, and in fact, the best serum lots allowed successful one-step

LTBMC without the recharging step at week three [50]. The importance of stroma has been well documented in these Dexter cultures, because the culture outcome was often correlated with stromal development.

The adaptation of one-step LTBMC for human cells was first reported in 1980 [54]. Unfortunately, human LTBMC has never attained the productivity or longevity which is observed in cultures of other species [50]. The exponentially declining numbers of total and progenitor cells with time in human LTBMC [52] renders the cultures unsuitable for cell expansion, and indicates that primitive stem cells are lost over time. The discovery of hematopoietic growth factors was an important development in human hematopoietic cell culture, because addition of growth factors to human LTBMC greatly enhanced cell output. However, growth factors did not prolong the longevity of the cultures, indicating that primitive cell maintenance was not improved [55]. Furthermore, although the total number of progenitors obtained was increased by growth factors, it was still usually less than the number used to initiate the culture. The increased cell densities that were stimulated by growth factor addition were not well supported by the relatively static culture conditions, and a net expansion in progenitor cell numbers was not obtained.

4.2 Current Culture Techniques

Current techniques are generally based on one of two approaches to overcome the culture limitations described above; the use of CD34-enriched cells in low density static culture, or the use of high density accessory cell-containing cultures supported by continuous medium perfusion. Although both approaches have proven feasible in the laboratory and clinic, each has advantages and disadvantages depending upon the culture objectives. The salient advantages and disadvantages of these alternative culture approaches are discussed below and are summarized in Table 4.

4.2.1 Culture of CD34-enriched cells

The development of an ever increasing number of recombinant growth factors was soon joined by the discovery of the CD34 antigen [33]. As CD34-enrichment protocols became available, it was thought that the low number of CD34-enriched cells could be expanded in growth factor-supplemented cultures without the impediment of numerous mature cells. Because enrichment results in a cell population depleted of stromal cells, CD34-enriched cell cultures are often called suspension cultures, due to the lack of an adherent stromal layer. A myriad of groups have reported experiments in which $CD34^+$ cells were incubated with high doses of up to

Table 4. Advantages and disadvantages of the two major culture approaches

	Static CD34-enriched cell cultures	Perfused cultures with accessory cells
Advantages	Simple culture maintenance	Requires minimal cell processing
	Allows study of individual cells/populations	High yield of stem cells
		Mimics physiological *in vivo* tissue
	Isolates action of factors on primitive cells	Requires minimal growth factor supplementation
	Serum-free media available	Stem cells maintained/expanded in culture
		Reduced donor-to-donor variability
Disadvantages	Requires enrichment time and expense	Requires manual or automated perfusion
	Low yield of stem cells	Cannot study individual cell types
	Non-physiological culture environment	Culture obscures functional activity studies
	Requires significant growth factor supplementation	Serum currently required
	Stem cells lost in culture	
	Increased donor-to-donor variability	

seven recombinant growth factors in suspension culture carried out either in T-flasks or bags [56-58]. Although total cell numbers are expanded as much as 1000-fold, progenitor expansion is usually 10- to 50-fold, suggesting that differentiation, accompanied by depletion of primitive cells, is occurring in these systems. In fact, when LTC-IC have been measured, the numbers obtained after static culture of enriched cells have always been significantly below the input value [12, 59, 60]. Another consideration in CD34$^+$ cell culture is the loss of cells during the enrichment procedure. It is not uncommon to experience 70-80% loss of progenitors with most CD34$^+$ cell purification protocols [28, 61], and this can be very significant when trying to maximize the final cell number obtained (such as in clinical applications). Nevertheless, cultures of purified CD34$^+$ cells, and especially the smaller subsets that can be obtained by flow cytometry (e.g. Thy-1$^+$, CD38$^-$), have yielded valuable information on the biology of hematopoietic stem cells, and continue to be used by a large number of investigators. These cultures can be carried out with single cells, or more typically with low cell densities (i.e. ~5,000 per ml), but it should be noted that the inoculum density can significantly influence culture performance [62] and variability [63, 64]. Because of the low cell densities and lack of stroma, CD34-enriched cell cultures can be fed with only one or two 50% medium exchanges per week [60]. Although a number of commercial serum-free media have been developed for these cultures [65, 66], supplementation with high doses of three or more growth factors is required to obtain growth [14, 60, 67]. A typical optimized growth factor cocktail for maximum LTC-IC and CFU-

GM output from these cultures would include IL-3, GM-CSF, Epo, FL, and KL, [68] to which IL-11 and ML may also be added (see Table 1) [69].

4.2.2 Accessory cell-containing perfusion culture

An alternative approach to improve human hematopoietic cell culture has come from the realization that traditional culture protocols are highly non-physiologic, and that these deficiencies can be corrected by *in vivo* mimicry. Therefore, these techniques do not involve cell purification or high-dose growth factor stimulation, but instead attempt to grow the entire hematopoietic tissue in a high-density perfused culture. Initial studies of this hypothesis demonstrated that frequent manual medium exchange in well plates extended human LTBMC longevity from six weeks to >20 weeks [70], indicating that primitive cells were maintained for a longer period of time. Addition of low-dose growth factors to the frequent medium exchange cultures resulted in significant cell and progenitor expansion while maintaining culture longevity [71]. Co-optimization of cell inoculum densities and medium exchange rates for human BM [72] and CB [73] cultures have since been published, and it is important to note that the optimal feeding rate for a culture will depend upon the density and composition of the inoculum population. For example, a typical growth factor-supplemented BM culture should be inoculated with 5×10^5 MNC in a 24 well plate in 0.6 ml medium, and fed with 50% medium exchanges on days 4, 7, 9, 10, and 11 with harvest on day 12. The medium should contain 20% fetal bovine serum (FBS; or 10% FBS and 10% horse serum), 5 μM hydrocortisone, and antibiotics in IMDM [66]. Due to the presence of stroma, serum-free media have not been as successful in accessory cell-containing cultures as in CD34-enriched cell cultures, although progress continues to be made towards this goal [66]. A variety of growth factor supplementation strategies may be used depending upon the desired outcome, but 2 ng/ml IL-3, 5 ng/ml GM-CSF, 0.1 U/ml Epo, and 25 ng/ml FL will generally yield successful results [68]. The further addition of 10 ng/ml KL [68] or 10 ng/ml ML and 10 ng/ml IL-11 is also beneficial [69] (Table 5).

The success of this manual frequently-fed culture approach led to development of continuously perfused bioreactors for human CB [73, 74], BM [75], and mPB [76] cell culture. Human BM MNC cultures have been performed in spinner flasks in a fed-batch mode as well [77]. Slow single-pass medium perfusion and internal oxygenation have given the best results to date, yielding cell densities in excess of 10^7 per ml accompanied by significant progenitor and primitive cell expansion [78]. These systems have also been amenable to scale-up, first by a factor of ten, and then by a further factor of 7.5 [79]. When an appropriate culture substrate is provided

Table 5. Relative effects of growth factors on BM MNC expansion

Growth factor	Cells	CFU-GM	LTC-IC
IL-3/GM-CSF/Epo	1.0	1.0	1.0
+KL	1.7	2.4	1.0
+FL	1.3	4.4	2.1
+KL/FL	1.8	5.1	2.1
+FL/IL-11	1.3	4.4	3.2
+FL/ML	1.3	6.4	5.3
+FL/IL-11/ML	1.8	7.0	4.2

The effect of different growth factors on the output from BM MNC expansion perfusion cultures is shown, relative to the originally described IL-3/GM-CSF/Epo combination [71] (defined as 1.0).

[80], perfusion bioreactors support the development and maintenance of accessory cell populations, resulting in significant endogenous growth factor production which likely contributes to culture success [17, 62]. Importantly, stromal-containing cultures appear to generate greater numbers of primitive cells from a smaller initial cell sample, as measured by *in vitro* [60, 76] and *in vivo* assays [81], as compared with CD34-enriched cell cultures.

4.2.3 Donor-to-donor variability

Despite the numerous reports on human hematopoietic cell culture, there has been little analysis of inherent donor-to-donor variability. In murine studies, LTBMC longevity has been shown to vary widely among different inbred mouse strains [82], and stem cell populations from different inbred mouse strains have been found to have differential short- and long-term repopulating ability [83]. Because cells obtained from different human individuals will exhibit genetic variability, donor-to-donor variability in hematopoietic cell growth potential should be anticipated. In fact, a study examining 52 donors showed that culture outcome varied significantly from donor-to-donor, and these differences could not be correlated with donor characteristics (i.e. sex, age, weight, and height) or measured cell characteristics (i.e. $CD34^+lin^-$ cell purity, and $CD38^-$, $Thy\text{-}1^+$, and c-kit^+ subsets thereof) [84]. Interestingly, the culture of CD34-enriched cells was found to be considerably more variable than accessory cell-containing cultures of MNC from the same donors (Fig. 2). While the absolute level of most culture performance metrics varies significantly from donor-to-donor (C_v from 0.3 to 1.2), the relative responses of different samples to the same stimulus are more uniform [84]. Consequently, statistically sound experimental conclusions can be drawn, provided that the response is measured relative to a control that is included with every sample. However, it is important to note that three donor samples are often insufficient when working with primary human cells. Depending upon the variability of the response being measured, as many as 6-12 donor samples might be required in order to draw a conclusion about the effect of a particular stimulus [68, 73, 84].

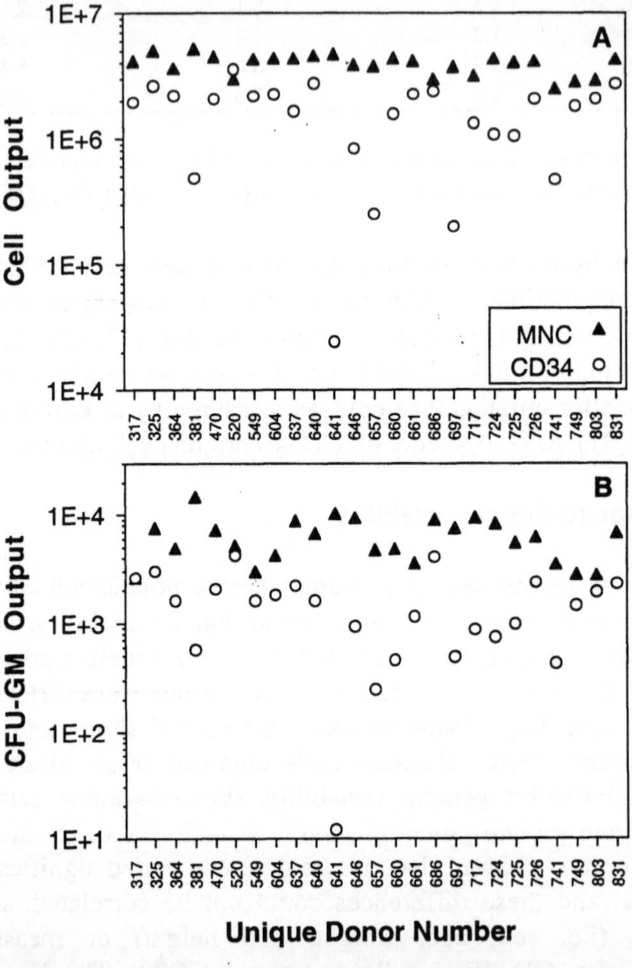

Figure 2. Donor to donor variability in BM cell expansion culture output from 23 donors. Parallel cultures of CD34-enriched cells and MNC from each donor were inoculated at a density to contain 3,000 CD34 $^{+}$lin^{-} cells each. Cultures were supplemented with IL-3, GM-CSF, Epo, and c-kit ligand, and were maintained with frequent manual medium exchange for twelve days [84]. The (A) cell and (B) CFU-GM output from each donor is shown.

5. UTILITY OF HEMATOPOIETIC STEM AND PROGENITOR CELL CULTURES

5.1 Clinical Background

In 1980, when BM transplantation (BMT) was still an experimental procedure, fewer than 200 BMT were performed worldwide [85]. Over the past decade, stem cell transplantation (SCT; includes use of mPB and CB in addition to BM) has become an established therapy for many diseases. In 1996, over 40,000 SCT were performed, primarily in the U.S. and Western Europe, for more than a dozen different clinical indications [86]. The number of SCT performed annually is increasing at a rate of 20 to 30% per year, which is expected to continue into the foreseeable future.

SCT is indicated as a treatment in a number of clinical settings because the highly prolific cells of the hematopoietic system are sensitive to many of the agents used to treat cancer patients. Chemotherapy and radiation therapy usually target rapidly cycling cells, so hematopoietic cells are ablated along with the cancer cells. Consequently, patients undergoing these therapies experience neutropenia (low neutrophil numbers, <500 per mm^3), thrombocytopenia (low platelet numbers, <20,000 per mm^3), and anemia (low red blood cell numbers), rendering them susceptible to infections and bleeding. SCT dramatically shortens the period of neutropenia and thrombocytopenia, but the patient may require repeated blood component transfusions for as long as six months. The time period during which the patient is neutropenic represents the greatest risk associated with SCT and often requires parenteral antibiotic administration. In addition, some patients never achieve engraftment (when cell numbers rise to safe levels).

SCT may be performed with patient cells (autologous) that have been removed and cryopreserved prior to administration of chemotherapy, or with donor cells (allogeneic). Autologous transplants outnumber allogeneic transplants 3:2, and the use of autologous transplants is growing more rapidly, particularly for the treatment of breast cancer. Nevertheless, there are significant advantages and disadvantages with both techniques. A major concern with autologous SCT is the possibility of reintroducing tumor cells along with the transplant. In fact, retroviral marking studies have proven that tumor cells reinfused in the transplant can contribute to relapse in the patient [87, 88]. A major obstacle in allogeneic transplantation is the high incidence of graft-versus-host disease (GVHD), in which donor T cells recognize the recipient as foreign, resulting in a strong immune response against many of the recipient's tissues.

SCT would therefore be greatly facilitated by reliable systems and procedures for *ex vivo* stem cell maintenance, expansion, and manipulation. For example, the harvest procedure, which collects one to two liters of BM, is currently a painful and involved operating room procedure. The complications and discomfort of BM donation are not trivial, and can affect donors for a month or more [89]. Through cell expansion techniques, a small BM specimen taken under local anesthesia in an out-patient setting could be expanded into the large number of cells required for transplant, thereby eliminating the large BM or mPB harvest procedures. With CB, there is a limit on the number of cells that can be collected from a single donor, and it is currently thought that this number is inadequate for an adult transplant. Consequently, CB transplants to date have been performed mainly on children and small adults, and engraftment times have been significantly delayed as compared with BM or mPB transplants [90, 91]. Expansion of CB cells may therefore enable adult transplants from the limited number of CB cells available for collection [73].

5.2 Clinical Use of Hematopoietic Cell Cultures

A number of small clinical trials have been performed to assess the safety of expanded hematopoietic cells. Thus far, all studies have been performed in the autologous setting, expanding the patient's mPB or BM cells for use after chemotherapy, and no safety concerns have appeared. Clinical trials with cells produced from CD34-enriched cell cultures have shown little effect when used to augment a standard SCT [56-58]. When used to replace a standard SCT, CD34-enriched and expanded cells have been reported to mediate engraftment in one study [56]. However, another study showed that these cells did not mediate engraftment, necessitating the use of an infusion of back-up unmanipulated cryopreserved cells [58]. The discrepancy between these studies was quite surprising given the similarity of methods used for CD34-enriched cell expansion. However, the patient treatments were very different, and this is now believed to account for the different results in these two studies. In the former study, a relatively low dose chemotherapy regimen was used [56], whereas the latter study utilized an ablative regimen including total body irradiation (TBI) [58]. In fact, it has since been acknowledged that the chemotherapy regimen in the former study was not ablative, and that patients will recover even if no cells are administered. Therefore, the currently available clinical data suggest that repopulating stem cells are not present after culture of CD34-enriched cells.

Clinical trials with cells generated from perfused culture of unpurified cells have shown modest improvements in patient course when used in conjunction with a standard SCT [92-94]. When used in place of a standard transplant in breast cancer patients receiving cytoablative chemotherapy, these cells have mediated timely and durable multi-lineage engraftment [95]. In fact, sufficient cells were generated for transplant utilizing an average of only 38 ml of BM. This low volume aspirate can be easily obtained under local anesthesia, and also significantly limits the potential number of tumor cells removed for transplant. Further, the culture procedure resulted in a significant decline in tumor cell numbers through passive purging [95], a phenomenon often observed in hematopoietic cell cultures [96-99].

5.3 Other Uses of Hematopoietic Cell Cultures

Beyond the ability to produce stem and progenitor cells for transplantation purposes lies the promise to produce large quantities of mature blood cells. Large-scale hematopoietic cultures could potentially provide several types of clinically important mature blood cells. These include red blood cells, platelets, and granulocytes. About 12 million units of red blood cells are transfused in the United States every year, the majority of them during elective surgery, and the rest in acute situations. About 4 million units of platelets are transfused every year into patients who have difficulty exhibiting normal blood clotting. Mature granulocytes, which constitute a relatively low-usage market of only a few thousand units administered each year, are involved in combating infections. All in all, the market for these blood cells totals about $1 billion to $1.5 billion in the U.S. annually, with a worldwide market that is about 3-4 times larger. Unlike *ex vivo* expansion of stem and progenitor cells for transplantation, the large-scale production of fully mature blood cells for routine clinical use is less developed and represents a more distant goal. The large market would require systems of immense size, unless major improvements in culture productivity are attained.

Functioning *ex vivo* human hematopoietic cell cultures also provide a valuable model for studying the biology of hematopoiesis. Considerable knowledge has been gained from the use of these systems, and their continued development should increase their utility. For example, testing of chemotherapeutic agents and carcinogens, currently tested in animal models, might be first evaluated in physiologically accurate human *ex vivo* systems.

6. SUMMARY

Mature blood cells, most of which exhibit a limited lifespan *in vivo*, are continuously generated from hematopoietic stem and progenitor cells. Stem cells are very rare in adult BM, but they have enough proliferative capacity to overcome stress and disease, potentially over several lifetimes. Control of stem cell growth and differentiation is a subject of intense study, and is known to be influenced by growth factors, stromal cells, ECM, and other culture conditions. These cells can be obtained from a number of primary tissue sources, and various means of processing and purification have been developed. Stem and progenitor cells are assayed through *in vivo* and *in vitro* methods, including xenogeneic transplant models, CFU assays, flow cytometry, and LTC-IC assays. Current culture methods for stem and progenitor cells are generally based on one of two approaches; the use of CD34-enriched cells in low density static culture, or the use of high density accessory cell-containing cultures supported by continuous medium perfusion. Both approaches are feasible, and each has its advantages and disadvantages for different applications. Stem cell cultures have been used in clinical studies to generate cells for SCT following cancer chemotherapy, as well as in basic scientific studies designed to better understand the complex process of hematopoiesis and carcinogenesis. Based upon the continued effort directed in this field of research and development, further advances can be expected, with the potential for considerable impact on the state of scientific knowledge and clinical practice.

LITERATURE CITED

1. Cronkite EP (1988) Analytical review of structure and regulation of hemopoiesis. *Blood Cells* 14:313-328.
2. Lajtha LG (1979) Stem cell concepts. *Differentiation* 14:23-34.
3. Boggs DR, Boggs SS, Saxe DF, Gress LA, and Canfield DR (1982) Hematopoietic stem cells with high proliferative potential. *J.Clin.Invest.* 70:242-253.
4. Spangrude GJ, Heimfeld S, and Weissman IL (1988) Purification and characterization of mouse hematopoietic stem cells. *Science* 241:58-62.
5. Lansdorp PM, Dragowska W, and Mayani H (1993) Ontogeny-related changes in proliferative potential of human hematopoietic cells. *J.Exp.Med.* 178:787-791.
6. Vaziri H, Dragowska W, Allsopp RC, Thomas TE, Harley CB, and Lansdorp PM (1994) Evidence for a mitotic clock in human hematopoietic stem cells: Loss of telomeric DNA with age. *Proc.Natl.Acad.Sci.* 91:9857-9860.
7. Broccoli D, Young JW, and De Lange T (1995) Telomerase activity in normal and malignant hematopoietic cells. *Proc.Natl.Acad.Sci.* 92:9082-9086.
8. Chiu C-P, Dragowska W, Kim NW, Vaziri H, Yui J, Thomas TE, Harley CB, and Lansdorp PM (1996) Differential expression of telomerase activity in hematopoietic progenitors from adult human bone marrow. *Stem Cells* 14:239-248.
9. Bodnar AG, Ouellette M, Frolkis M, Holt SE, Chiu C-P, Morin GB, Harley CB, Shay JW, Lichtsteiner S, and Wright WE (1998) Extension of life-span by introduction of telomerase into normal human cells. *Science* 279:349-352.
10. Holt SE, Shay JW, and Wright WE (1996) Refining the telomere-telomerase hypothesis of aging and cancer. *Nature Biotechnol.* 14:836-839.

11. Yamakazi K, Roberts RA, Spooncer E, Dexter TM, and Allen TD (1989) Cellular interactions between 3T3 cells and interleukin-3-dependent multipotent haemopoietic cells: A model system for stromal-cell-mediated haemopoiesis. *J.Cell.Physiol.* 139:301-312.
12. Verfaillie CM (1992) Direct contact between human primitive hematopoietic progenitors and bone marrow stroma is not required for long-term in vitro hematopoiesis. *Blood* 79:2821-2826.
13. Brandt JE, Srour EF, Van Besien K, Briddell RA, and Hoffman R (1990) Cytokine-dependent long-term culture of highly enriched precursors of hematopoietic progenitor cells from human bone marrow. *J.Clin.Invest.* 86:932-941.
14. Haylock DN, To LB, Dowse TL, Juttner CA, and Simmons PJ (1992) Ex vivo expansion and maturation of peripheral blood $CD34^+$ cells into the myeloid lineage. *Blood* 80:1405-1412.
15. Toksoz D, Zsebo KM, Smith KA, Hu S, Brankow D, Suggs SV, Martin FH, and Williams DA (1992) Support of human hematopoiesis in long-term bone marrow cultures by murine stromal cells selectively expressing the membrane-bound and secreted forms of the human homolog of the steel gene product, stem cell factor. *Proc.Natl.Acad.Sci.* 89:7350-7354.
16. Long MW (1992) Blood cell cytoadhesion molecules. *Exp.Hematol.* 20:288-301.
17. Koller MR, Bradley MS, and Palsson BØ (1995) Growth factor consumption and production in perfusion cultures of human bone marrow correlates with specific cell production. *Exp.Hematol.* 23:1275-1283.
18. Zipori D (1989) Stromal cells from the bone marrow: Evidence for a restrictive role in regulation of hemopoiesis. *Eur.J.Haematol.* 42:225-232.
19. Zuckerman KS and Wicha MS (1983) Extracellular matrix production by the adherent cells of long-term murine bone marrow cultures. *Blood* 61:540-547.
20. Coulombel L, Vuillet MH, Leroy C, and Tchernia G (1988) Lineage- and stage-specific adhesion of human hematopoietic progenitor cells to extracellular matrices from marrow fibroblasts. *Blood* 71:329-334.
21. Campbell AD, Long MW, and Wicha MS (1987) Haemonectin, a bone marrow adhesion protein specific for cells of granulocyte lineage. *Nature* 329:744-746.
22. Long MW and Dixit VM (1990) Thrombospondin functions as a cytoadhesion molecule for human hematopoietic progenitor cells. *Blood* 75:2311-2318.
23. Wight TN, Kinsella MG, Keating A, and Singer JW (1986) Proteoglycans in human long-term bone marrow cultures: Biochemical and ultrastructural analyses. *Blood* 67:1333-1343.
24. Hao Q-L, Shah AJ, Thiemann FT, Smogorzewska EM, and Crooks GM (1995) A functional comparison of $CD34^+CD38^-$ cells in cord blood and bone marrow. *Blood* 86:3745-3753.
25. Preti RA, Razis E, Ciavarella D, Fan Y, Kuhns RE, Cook P, Wong G, Wuest DL, and Ahmed T (1994) Clinical and laboratory comparison study of refrigerated and cryopreserved bone marrow for transplantation. *Bone Marrow Transplant.* 13:253-260.
26. de Wynter EA, Coutinho LH, Pei X, Marsh JCW, Hows JM, Luft T, and Testa NG (1995) Comparison of purity and enrichment of $CD34^+$ cells from bone marrow, umbilical cord and peripheral blood (primed for apheresis) using five separation systems. *Stem Cells* 13:524-532.
27. Rowley SD (1992) Techniques of bone marrow and stem cell cryopreservation and storage, in RA Sacher and JP AuBuchon (eds.), *Marrow transplantation: Practical and technical aspects of stem cell reconstitution*, American Association of Blood Banks, Bethesda, pp. 105-127.

28. Koller MR, Manchel I, Newsom BS, Palsson MA, and Palsson BØ (1995) Bioreactor expansion of human bone marrow: Comparison of unprocessed, density-separated, and CD34-enriched cells. *J.Hematotherapy* 4:159-169.
29. Pappenheim A (1898) Abstammung und entstehung der rotten blutzelle. *Virchows Arch.* 151:89-80.
30. Naeim F (1992) *Pathology of bone marrow*, Igaku-Shoin, New York, NY.
31. Civin CI and Gore SD (1993) Antigenic analysis of hematopoesis: A review. *J.Hematotherapy* 2:137-144.
32. Brott DA, Koller MR, Rummel SA, and Palsson BØ (1995) Flow cytometric analysis of cells obtained from human bone marrow cultures, in M Al-Rubeai and AN Emery (eds.), *Flow cytometry applications in cell culture*, Marcel Dekker, New York, pp. 121-146.
33. Civin CI, Strauss LC, Brovall C, Fackler MJ, Schwartz JF, and Shaper JH (1984) Antigenic analysis of hematopoiesis. III. A hematopoietic progenitor cell surface antigen defined by a monoclonal antibody raised against KG-Ia cells. *J.Immunol.* 133:157-165.
34. Berenson RJ, Bensinger WI, Hill RS, Andrews RG, Garcia-Lopez J, Kalamasz DF, Still BJ, Spitzer G, Buckner CD, Bernstein ID, and Thomas ED (1991) Engraftment after infusion of $CD34^+$ marrow cells in patients with breast cancer or neuroblastoma. *Blood* 77:1717-1722.
35. Osawa M, Hanada K, Hamada H, and Nakauchi H (1996) Long-term lymphohematopoietic reconstitution by a single CD34-low/negative hematopoietic stem cell. *Science* 273:242-245.
36. Goodall MA, Rosenzweig M, Kim H, Marks DF, DeMaria M, Paradis G, Grupp SA, Sieff CA, Mulligan RC, and Johnson RP (1997) Dye efflux studies suggest that hematopoietic stem cells expressing low or undetectable levels of CD34 antigen exist in multiple species. *Nature Medicine* 3:1337-1345.
37. Rebel VI, Dragowska W, Eaves CJ, Humphries RK, and Lansdorp PM (1994) Amplification of $Sca\text{-}1^+$ Lin^- WGA^+ cells in serum-free cultures containing steel factor, interleukin-6, and erythropoietin with maintenance of cells with long-term in vivo reconstituting potential. *Blood* 83:128-136.
38. Spangrude GJ, Brooks DM, and Tumas DB (1995) Long-term repopulation of irradiated mice with limiting numbers of purified hematopoietic stem cells: In vivo expansion of stem cell phenotype but not function. *Blood* 85:1006-1016.
39. Till JE and McCulloch EA (1961) A direct measurement of the radiation sensitivity of normal mouse bone marrow cells. *Radiation Research* 14:213-222.
40. Szilvassy SJ, Fraser CC, Eaves CJ, Lansdorp PM, Eaves AC, and Humphries RK (1989) Retrovirus-mediated gene transfer to purified hemopoietic stem cells with long-term lympho-myelopoietic repopulating ability. *Proc.Natl.Acad.Sci.* 86:8798-8802.
41. Zanjani ED, Ascensao JL, Harrison MR, and Tavassoli M (1992) Ex vivo incubation with growth factors enhances the engraftment of fetal hematopoietic cells transplanted into sheep fetuses. *Blood* 79:3045-3049.
42. Kyoizumi S, Baum CM, Kaneshima H, McCune JM, Yee EJ, and Namikawa R (1992) Implantation and maintenance of functional human bone marrow in SCID-hu mice. *Blood* 79:1704-1711.
43. Larochelle A, Vormoor J, Lapidot T, Sher G, Furukawa T, Li Q, Schultz LD, Olivieri NF, Stamatoyannopoulos G, and Dick JE (1995) Engraftment of immune-deficient mice with primitive hematopoietic cells from β-thalassemia and sickle cell anemia patients: Implications for evaluating human gene therapy protocols. *Human Molec.Genetics* 4:163-172.

44. Sutherland HJ, Eaves AC, and Eaves CJ (1991) Quantitative assays for human hemopoietic progenitor cells, in AP Gee (ed.), *Bone marrow processing and purging*, CRC Press, Ann Arbor, pp. 155-167.
45. Friedenstein AJ, Chailakhjan RK, and Lalykina KS (1970) The development of fibroblast colonies in monolayer cultures of guinea-pig bone marrow and spleen cells. *Cell Tissue Kinet.* 3:393-403.
46. Sutherland HJ, Lansdorp PM, Henkelman DH, Eaves AC, and Eaves CJ (1990) Functional characterization of individual human hematopoietic stem cells cultured at limiting dilution on supportive marrow stromal layers. *Proc.Natl.Acad.Sci.* 87:3584-3588.
47. Hao Q-L, Thiemann FT, Petersen D, Smogorzewska EM, and Crooks GM (1996) Extended long-term culture reveals a highly quiescent and primitive human hematopoietic progenitor population. *Blood* 88:3306-3313.
48. van der Sluijs JP, de Jong JP, Brons NHC, and Ploemacher RE (1990) Marrow repopulating cells, but not CFU-S, establish long-term in vitro hemopoiesis on a marrow-derived stromal layer. *Exp.Hematol.* 18:893-896.
49. Koller MR, Manchel I, and Smith AK (1998) Quantitative long-term culture-initiating cell assays require accessory cell depletion that can be achieved by CD34-enrichment or 5-fluorouracil exposure. *Blood* 91:4056-4064.
50. Dexter TM, Spooncer E, Simmons P, and Allen TD (1984) Long-term marrow culture: An overview of techniques and experience, in DG Wright and JS Greenberger (eds.), *Long term bone marrow culture*, Alan R. Liss, New York, pp. 57-96.
51. Greenberger JS (1984) Long-term hematopoietic cultures, in DW Golde (ed.), *Hematopoiesis*, Churchill Livingstone, New York, pp. 203-242.
52. Eaves CJ, Cashman JD, and Eaves AC (1991) Methodology of long-term culture of human hemopoietic cells. *J.Tiss.Cult.Meth.* 13:55-62.
53. Dexter TM, Allen TD, and Lajtha LG (1977) Conditions controlling the proliferation of haemopoietic stem cells in vitro. *J.Cell.Physiol.* 91:335-344.
54. Gartner S and Kaplan HS (1980) Long-term culture of human bone marrow cells. *Proc.Natl.Acad.Sci.* 77:4756-4759.
55. Coutinho LH, Will A, Radford J, Schiró R, Testa NG, and Dexter TM (1990) Effects of recombinant human granulocyte colony-stimulating factor (CSF), human granulocyte macrophage-CSF, and gibbon interleukin-3 on hematopoiesis in human long-term bone marrow culture. *Blood* 75:2118-2129.
56. Brugger W, Heimfeld S, Berenson RJ, Mertelsmann R, and Kanz L (1995) Reconstitution of hematopoiesis after high-dose chemotherapy by autologous progenitor cells generated ex vivo. *N.E.J.Med.* 333:283-287.
57. Williams SF, Lee WJ, Bender JG, Zimmerman TM, Swinney P, Blake M, Carreon J, Schilling M, Smith S, Williams DE, Oldham F, and Van Epps DE (1996) Selection and expansion of peripheral blood $CD34^+$ cells in autologous stem cell transplantation for breast cancer. *Blood* 87:1687-1691.
58. Alcorn MJ, Holyoake TL, Richmond L, Pearson C, Farrel E, Kyle B, Dunlop DJ, Fitzsimons E, Steward WP, Pragnell IB, and Franklin IM (1996) CD34-positive cells isolated from cryopreserved peripheral-blood progenitor cells can be expanded ex vivo and used for transplantation with little or no toxicity. *J.Clin.Oncol.* 14:1839-1847.
59. Sutherland HJ, Hogge DE, Cook D, and Eaves CJ (1993) Alternative mechanisms with and without steel factor support primitive human hematopoiesis. *Blood* 81:1465-1470.
60. Koller MR, Palsson MA, Manchel I, and Palsson BØ (1995) LTC-IC expansion is dependent on frequent medium exchange combined with stromal and other accessory cell effects. *Blood* 86:1784-1793.

61. Traycoff CM, Abboud MR, Laver J, Brandt JE, Hoffman R, Law P, Ishizawa L, and Srour EF (1994) Evaluation of the in vitro behavior of phenotypically defined populations of umbilical cord blood hematopoietic progenitor cells. *Exp.Hematol.* 22:215-222.
62. Koller MR, Manchel I, and Palsson BØ (1997) Importance of parenchymal:stromal cell ratio for the ex vivo reconstitution of human hematopoiesis. *Stem Cells* 15:305-313.
63. Mayani H, Dragowska W, and Lansdorp PM (1993) Characterization of functionally distinct subpopulations of $CD34^+$ cord blood cells in serum-free long-term cultures supplemented with hematopoietic cytokines. *Blood* 82:2664-2672.
64. Young JC, DiGiusto DL, and Backer MP (1996) In vitro characterization of fetal hematopoietic stem cells: Range and kinetics of cell production from individual stem cells. *Biotechnol.Bioeng.* 50:465-478.
65. Lebkowski JS, Schain LR, and Okarma TB (1995) Serum-free culture of hematopoietic stem cells: A review. *Stem Cells* 13:607-612.
66. Koller MR, Maher RJ, Manchel I, Oxender M, and Smith AK (1998) Alternatives to animal sera for human bone marrow cell expansion: Human serum and serum-free media. *J.Hematotherapy* 7:413-421.
67. Brugger W, Mšcklin W, Heimfeld S, Berenson RJ, Mertelsmann R, and Kanz L (1993) Ex vivo expansion of enriched peripheral blood $CD34^+$ progenitor cells by stem cell factor, interleukin-1β (IL-1β), IL-6, IL-3, interferon-γ, and erythropoietin. *Blood* 81:2579-2584.
68. Koller MR, Oxender M, Brott DA, and Palsson BØ (1996) flt-3 ligand is more potent than c-kit ligand for the synergistic stimulation of ex vivo hematopoietic cell expansion. *J.Hematotherapy* 5:449-459.
69. Koller MR, Jensen TC, Goltry KL, Brott DA, and Smith AK (1997) The effects of Tpo and IL-11 on human bone marrow cell expansion in a small-scale perfused culture system. *Blood* 90:537a.
70. Schwartz RM, Palsson BØ, and Emerson SG (1991) Rapid medium perfusion rate significantly increases the productivity and longevity of human bone marrow cultures. *Proc.Natl.Acad.Sci.* 88:6760-6764.
71. Schwartz RM, Emerson SG, Clarke MF, and Palsson BØ (1991) In vitro myelopoiesis stimulated by rapid medium exchange and supplementation with hematopoietic growth factors. *Blood* 78:3155-3161.
72. Koller MR, Manchel I, Palsson MA, Maher RJ, and Palsson BØ (1996) Different measures of human hematopoietic cell culture performance are optimized under vastly different conditions. *Biotechnol.Bioeng.* 50:505-513.
73. Koller MR, Manchel I, Maher RJ, Goltry KL, Armstrong RD, and Smith AK (1998) Clinical-scale human umbilical cord blood cell expansion in a novel automated perfusion culture system. *Bone Marrow Transplant.* 21: 653-663.
74. Koller MR, Bender JG, Miller WM, and Papoutsakis ET (1993) Expansion of human hematopoietic progenitors in a perfusion bioreactor system with IL-3, IL-6, and stem cell factor. *Biotechnol* 11:358-363.
75. Palsson BØ, Paek S-H, Schwartz RM, Palsson M, Lee G-M, Silver SM, and Emerson SG (1993) Expansion of human bone marrow progenitor cells in a high cell density continuous perfusion system. *Biotechnol* 11:368-371.
76. Sandstrom CE, Bender JG, Papoutsakis ET, and Miller WM (1995) Effects of $CD34^+$ cell selection and perfusion on ex vivo expansion of peripheral blood mononuclear cells. *Blood* 86:958-970.
77. Zandstra PW, Eaves CJ, and Piret JM (1994) Expansion of hematopoietic progenitor cell populations in stirred suspension bioreactors of normal human bone marrow cells. *Biotechnol* 12:909-914.

78. Koller MR, Emerson SG, and Palsson BØ (1993) Large-scale expansion of human stem and progenitor cells from bone marrow mononuclear cells in continuous perfusion culture. *Blood* 82:378-384.
79. Armstrong RD, Koller MR, Maluta J, and Ogier WC (1999) Clinical systems for the production of cells and tissues for human therapy. In Novel therapeutics from modern biotechnology: from laboratory to human testing, eds. DL Oxender and LE Post, Springer Verlag, Berlin, pp. 221-241.
80. Koller MR, Palsson MA, Manchel I, Maher RJ, and Palsson BØ (1998) Tissue culture surface characterisitcs influence the expansion of human bone marrow cells. *Biomaterials* 19:1963-1972.
81. Knobel KM, McNally MA, Berson AE, Rood D, Chen K, Kilinski L, Tran K, Okarma TB, and Lebkowski JS (1994) Long-term reconstitution of mice after ex vivo expansion of bone marrow cells: Differential activity of cultured bone marrow and enriched stem cell populations. *Exp.Hematol.* 22:1227-1235.
82. Sakakeeny MA and Greenberger JS (1982) Granulopoiesis longevity in continuous bone marrow cultures and factor-dependent cell line generation: Significant variation among 28 inbred mouse strains and outbred stocks. *J.Natl.Cancer Inst.* 68:305-317.
83. Van Zant G, Scott-Micus K, Thompson BP, Fleischman RA, and Perkins S (1992) Stem cell quiescence/activation is reversible by serial transplantation and is independent of stromal genotype in mouse aggregation chimeras. *Exp.Hematol.* 20:470-475.
84. Koller MR, Manchel I, Brott DA, and Palsson BØ (1996) Donor-to-donor variability in the expansion potential of human bone marrow cells is reduced by accessory cells but not by soluble growth factors. *Exp.Hematol.* 24:1484-1493.
85. Gratwohl A (1991) Bone marrow transplantation activity in Europe 1990. *Bone Marrow Transplant.* 8:197-201.
86. Horowitz MM and Rowlings PA (1997) An update from the International Bone Marrow Transplant Registry and the Autologous Blood and Marrow Transplant Registry on current activity in hematopoietic stem cell transplantation. *Curr.Opin.Hematol.* 4:395-400.
87. Rill DR, Santana VM, Roberts WM, Nilson T, Bowman LC, Krance RA, Heslop HE, Moen RC, Ihle JN, and Brenner MK (1994) Direct demonstration that autologous bone marrow transplantation for solid tumors can return a multiplicity of tumorigenic cells. *Blood* 84:380-383.
88. Deisseroth AB, Zu Z, Claxton D, Hanania EG, Fu S, Ellerson D, Goldberg L, Thomas M, Janicek K, Anderson WF, Hester J, Korbling M, Durett A, Moen R, Berenson R, Heimfeld S, Hamer J, Calvert L, Tibbits P, Talpaz M, Kantarjian H, Champlin R and Reading C (1994) Genetic marking shows that Ph^+ cells present in autologous transplants of chronic myelogenous leukemia (CML) contribute to relapse after autologous bone marrow transplantation in CML. *Blood* 83:3068-3076.
89. Stroncek DF, Holland PV, Bartch G, Bixby T, Simmons RG, Antin JH, Anderson KC, Ash RC, Bolwell BJ, Hansen JA, Heal JM, Henslee-Downey PJ, Jaffé ER, Klein HG, Lau PM, Perkins HA, Popovsky MA, Price TH, Rowley SD, Stehling LC, Weiden PL, Wissel ME, and McCullough J (1993) Experiences of the first 493 unrelated marrow donors in the national marrow donor program. *Blood* 81:1940-1946.
90. Wagner JE, Kernan NA, Steinbuch M, Broxmeyer HE, and Gluckman E (1995) Allogeneic sibling umbilical-cord-blood transplantation in children with malignant and non-malignant disease. *Lancet* 346:214-219.
91. Kurtzberg J, Laughlin M, Graham ML, Smith C, Olson JF, Halperin EC, Ciocci G, Carrier C, Stevens CE, and Rubinstein P (1996) Placental blood as a source of hematopoietic stem cells for transplantation into unrelated recipients. *N.E.J.Med.* 335:157-166.

92. Silver SM, Adams PT, Hutchinson RJ, Douville JW, Paul LA, Clarke MF, Palsson B, and Emerson SG (1993) Phase I evaluation of ex vivo expanded hematopoietic cells produced by perfusion cultures in autologous bone marrow transplantation. *Blood* 82:296a.
93. Champlin R, Mehra R, Gajewski J, Khouri I, Geisler D, Davis M, Oba K, Thomas M, Armstrong RD, Douville JW, Weber S, Silver SM, Muller TE, and Deisseroth AB (1995) Ex vivo expanded progenitor cell transplantation in patients with breast cancer. *Blood* 86:295a.
94. Mandalam R, Koller MR, and Smith AK (1999) Ex vivo hematopoietic cell expansion for bone marrow transplantation, in R Nordon (ed.), *Ex vivo cell therapy*, Landes Bioscience, Austin, TX (in press).
95. Stiff PJ, Oldenberg D, Hsi E, Chen B, Douville JW, Burhop S, Bayer R, Peace D, Malhotra D, Kerger C, Armstrong RD, and Muller TE (1997) Successful hematopoietic engraftment following high dose chemotherapy using only ex-vivo expanded bone marrow grown in Aastrom (stromal-based) bioreactors. *Proc.Amer.Soc.Clin.Oncol.* 16:88a.
96. Coulombel L, Eaves CJ, Kalousek D, Gupta C, and Eaves AC (1985) Long-term marrow culture of cells from patients with acute myelogenous leukemia. *J.Clin.Invest.* 75:961-969.
97. Barnett MJ, Eaves CJ, Phillips GL, Kalousek DK, Klingemann H-G, Lansdorp PM, Reece DE, Shepherd JD, Shaw GJ, and Eaves AC (1989) Successful autografting in chronic myeloid leukaemia after maintenance of marrow in culture. *Bone Marrow Transplant.* 4:345-351.
98. Chang J, Morgenstern GR, Coutinho LH, Scarffe JH, Carr T, Deakin DP, Testa NG, and Dexter TM (1989) The use of bone marrow cells grown in long-term culture for autologous bone marrow transplantation in acute myeloid leukaemia: An update. *Bone Marrow Transplant.* 4:5-9.
99. Da WM, Douay L, Barbu V, Fabrega S, Allieri M-A, Drouet X, Deloux J, Giarratana M-C, Oszahin H, Van den Akker J, Vanaeke D, and Gorin N-C (1991) Serum-free liquid marrow culture in patients with acute lymphoblastic leukaemia: a potential application to purge marrow for autologous transplantation. *Br.J.Haematol.* 78:42-47.
100. Dinarello CA, Rosenwasser LJ, and Wolff SM (1981) Demonstration of a circulating suppressor factor of thymocyte proliferation during endotoxin fever in humans. *J.Immunol.* 127:2517-2519.
101. Smith KA (1988) Interleukin-2: Inception, impact, and implications. *Science* 240:1169-1160.
102. Ihle JN, Pepersack L, and Rebar L (1981) Regulation of T cell differentiation: In vitro induction of 20 alpha-hydroxysteroid dehydrogenase in splenic lymphocytes is mediated by a unique lymphokine. *J.Immunol.* 126:2184-2180.
103. Yokota T, Arai N, de Vries JE, Spits H, Banchereau J, Zlobnik A, Rennick D, Howard M, Takebe Y, Miyatake S, Lee F, and Arai KI (1988) Molecular biology of interleukin-4 and interleukin-5 genes and biology of their products that stimulate B cells, T cells and hemopoietic cells. *Immunol.Rev.* 102:137-130.
104. Kishimoto T (1989) The biology of interleukin-6. *Blood* 74:1-10.
105. Tushinski RJ, McAlister IB, Williams DE, and Namen AE (1991) The effects of interleukin 7 (IL-7) on human bone marrow in vitro. *Exp.Hematol.* 19:749-754.
106. Herbert CA and Baker JB (1993) Interleukin-8: A review. *Cancer Invest.* 11:743-750.
107. Donahue RE, Yang Y-C, and Clark SC (1990) Human P40 T-cell growth factor (interleukin 9) supports erythroid colony formation. *Blood* 75:2271-2275.
108. Zlotnik A and Moore KW (1991) Interleukin 10. *Cytokine* 3:366-371.
109. Du XX and Williams DA (1994) Interleukin-11: A multifunctional growth factor derived from the hematopoietic microenvironment. *Blood* 83:2023-2030.

110.Wolf SF, Temple PA, Kobayashi M, Young D, Dicig M, Lowe L, Dzialo R, Fitz L, Ferenz C, and Hewick RM (1991) Cloning of cDNA for natural killer cell stimulatory factor, a heterodimeric cytokine with multiple biologic effects on T and natural killer cells. *J.Immunol.* 146:3074-3081.
111.Minty A, Chalon P, Derocq JM, Dumont X, Guillemot JC, Kaghad M, Labit C, Leplatois P, Liauzun P, and Miloux B (1993) Interleukin-13 is a new human lymphokine regulating inflammatory and immune responses. *Nature* 362:248-240.
112.Ambrus Jr. JL, Pippin J, Joseph A, Xu C, Blumenthal D, Tamayo A, Claypool K, McCourt D, Srikiatchatochorn A, and Ford RJ (1993) Identification of a cDNA for a human high-molecular-weight B-cell growth factor. *Proc.Natl.Acad.Sci.* 90:6330-6334.
113.Grabstein KH, Eisenman J, Shanebeck K, Rauch C, Srinivasan S, Fung V, Beers C, Richardson J, Schoenborn MA, and Ahdieh M (1994) Cloning of a T cell growth factor that interacts with the beta chain of the interleukin-2 receptor. *Science* 264:965-968.
114.Center DM, Kornfeld H, and Cruikshand WW (1997) Interleukin-16. *Int.J.Biochem.Cell Biol.* 29:1231-1234.
115.Yao Z, Painter SL, Fanslow WC, Ulrich D, Macduff BM, Spriggs MK, and Armitage RJ (1995) Human IL-17: A novel cytokine derived from T cells. *J.Immunol.* 155:5483-5486.
116.Ushio S, Namba M, Okura T, Hattori K, Nukada Y, Akita K, Tanabe F, Konishi K, Micallef M, Fujii M, Torigoe K, Tanimoto T, Fukuda S, Ikeda M, Okamura H, and Kurimoto M (1996) Cloning of the cDNA for human IFN-gamma-inducing factor, espression in Escherichia coli, and studies on the biologic activities of the protein. *J.Immunol.* 156:4274-4279.
117.Krantz SB (1991) Erythropoietin. *Blood* 77:419-434.
118.Metcalf D (1985) The granulocyte-macrophage colony-stimulating factors. *Science* 229:16-22.
119.Zsebo KM, Wypych J, McNiece IK, Lu HS, Smith KA, Karkare SB, Sachdev RK, Yuschenkoff VN, Birkett NC, Williams LR, Satyagal VN, Tung W, Bosselman RA, Mendiaz EA, and Langley KE (1990) Identification, purification, and biological characterization of hematopoietic stem cell factor from buffalo rat liver-conditioned medium. *Cell* 63:195-201.
120.Virelizier JL and Arenzana-Seisdedos F (1985) Immunological functions of macrophages and their regulation by interferons. *Med.Biol.* 63:149-159.
121.Graham GJ, Wright EG, Hewick R, Wolpe SD, Wilkie NM, Donaldson D, Lorimore S, and Pragnell IB (1990) Identification and characterization of an inhibitor of haemopoietic stem cell proliferation. *Nature* 344:442-444.
122.Metcalf D (1991) The leukemia inhibitory factor (LIF). *Int.J.Cell Cloning* 9:95-90.
123.Sporn MB and Roberts AB (1989) Transforming growth factor-β: Multiple actions and potential clinical applications. *JAMA* 262:938-930.
124.Pennica D, Nedwin GE, Hayflick JS, Seeburg PH, Derynck R, Palladino MA, Kohr WJ, Aggarwal BB, and Goeddel DV (1984) Human tumour necrosis factor: Precursor structure, expression and homology to lymphotoxin. *Nature* 312:724-720.
125.Lyman SD, James L, Johnson L, Brasel K, de Vries P, Escobar SS, Downey H, Splett RR, Beckmann MP, and McKenna HJ (1994) Cloning of the human homologue of the murine flt3 ligand: A growth factor for early hematopoietic progenitor cells. *Blood* 83:2795-2801.
126.de Sauvage FJ, Hass PE, Spencer SD, Malloy BE, Gurney AL, Spencer SA, Darbonne WC, Henzel WJ, Wong SC, Kuang W-J, Oles KJ, Hultgren B, Solberg J, L.A., Goeddel DV, and Eaton DL (1994) Stimulation of megakaryocytopoiesis and thrombopoiesis by the c-Mpl ligand. *Nature* 369:533-538.

127.Terstappen LWMM, Meiners H, and Loken MR (1989) A rapid sample preparation technique for flow cytometric analysis of immunofluorescence allowing absolute enumeration of cell subpopulations. *J.Immunol.Meth.* 123:103-100.
128.Wagner JE, Collins D, Fuller S, Schain LR, Berson AE, Almici C, Hall MA, Chen KE, Okarma TB, and Lebkowski JS (1995) Isolation of small, primitive human hematopoietic stem cells: Distribution of cell surface cytokine receptors and growth in SCID-Hu mice. *Blood* 86:512-523.
129.Sato N, Sawada K, Kannonji M, Tarumi T, Sakai N, Ieko M, Sakurama S, Nakagawa S, Yasukouchi T, and Krantz SB (1991) Purification of human marrow progenitor cells and demonstration of the direct action of macrophage colony-stimulating factor on colony-forming unit-macrophage. *Blood* 78:967-974.
130.Nishihira H, Toyoda Y, Miyazaki H, Kigasawa H, and Ohsaki E (1996) Growth of macroscopic human megakaryocyte colonies from cord blood in culture with recombinant human thrombopoietin (c-mpl ligand) and the effects of gestational age on frequency of colonies. *Br.J.Haematol.* 92:23-28.
131.Lu L, Xiao M, Shen R-N, Grigsby S, and Broxmeyer HE (1993) Enrichment, characterization, and responsiveness of single primitive $CD34^{+++}$ human umbilical cord blood hematopoietic progenitors with high proliferative and replating potential. *Blood* 81:41-48.

Chapter 2

In Vitro T-Lymphopoiesis

[1]Michael Rosenzweig and [2]David T Scadden
[1]New England Regional Primate Center, Southborough, MA and [2]Massachusetts General Hospital, Harvard Medical School, Partners AIDS Research Center, Room 5212, 149 13th St, Boston, MA 02129. Tel: 001-617-726-5615; Fax: 001-617-726-4691; E-mail: scadden.david@mgh.harvard.edu

1. INTRODUCTION

T-lymphocyte generation can occur in a number of different tissues *in vivo*, but none compares with the extraordinary efficiency of the thymus, where greater than 90% of the cell content is maturing T-cells. The capacity to form T-lymphocytes was first ascribed to the thymus in the 1960s when thymectomized newborn mice were noted to be lymphopenic and immunosuppressed (14). During development, the thymus forms from outgrowths of pharyngeal pouches. The pouches extend tissue derived from endodermal as well as ectodermal primordium, with both sources essential for the function of the mature organ (3). Budding epithelium from the pharyngeal pouches rapidly becomes infiltrated with mesenchymal cells originating in the neural crest (11). This infiltration is followed by mesenchymal cells from mesoderm which results in the generation of communicating vascular spaces. By 8 weeks of gestation, the thymic rudiment in the developing human begins to be populated by T-cell progenitors migrating from the fetal liver (8). Fetal liver hematopoietic stem cells migrate to bone marrow (BM) by week 16, and after approximately week 22 of gestation, all subsequent progenitor cell immigration to the thymus is exclusively from the BM (5). Cells entering the thymus have multilineage capability including the ability to form myeloid, dendritic and natural killer (NK) cells in addition to T-cells (1, 10, 12, 17). Whether these cells represent true stem cells has been controversial, though immunophenotypic data would suggest that the cells are distinct from stem cells (17).

The sequential differentiation of progenitor cells into mature T-lymphocytes occurs within the confines of the thymus with details in the human largely inferred from detailed studies of the mouse. $CD34^+$ cells migrating to the thymus acquire CD7 on the cell surface and CD2 while resident in the thymic cortex (Fig. 1). There is then expression of CD1 and CD5, low levels of CD4, and eventually, the signal transducing components of the T-cell receptor, CD3. T-cell receptor (TCR) rearrangement occurs with recombinase activating gene (RAG1 and RAG2) expression leading to complexes of either rearranged beta chains with invariant pre-alpha chains (pre-TCR) or rearranged gamma/delta complexes. The latter express neither CD4 nor CD8 while pre-TCR expressing cells become $CD4^+CD8^+$ as alpha chain rearrangement is completed and a mature TCR is expressed. The interaction of TCR with either major histocompatibility complex (MHC) class I or II leads to generation of mature $CD4^-CD8^+$ or $CD4^+CD8^-$ cells, respectively, in the thymic medulla prior to exiting the thymus. Emigration from the thymus to the periphery is associated with a $CD3^+CD45RA^+CD62L^+$ immunophenotype.

The recapitulation of thymic maturation in model systems has received considerable attention and has taken a number of different forms. Human T-cell differentiation has most thoroughly been accomplished in the laboratory setting by reconstitution of congenitally immunodeficient mice. Specifically, McCune and colleagues transplanted human fetal thymus and a source of stem cells under the kidney capsule of severe combined immunodeficiency (SCID) mice and noted successful maturation of human T-lymphocytes (13). This model has subsequently been shown to demonstrate the ability of cells to be fully immunocompetent, tolerant to antigens of donor stem cells, donor thymus and host while retaining reactivity to allogeneic cells (15). This system is limited primarily by complexity and the difficulty of sequentially isolating cells which are not abundant outside the confines of the transplanted thymus.

Other models have utilized fetal thymic lobes to create an organ culture system. This approach has been demonstrated to be successful when either human or murine sources of thymus are used (16, 18, 23). The output from these systems has been variably reported, but there is full lineage maturation and they can be applied to the analysis of lineage-differentiating events or gene manipulation of primitive cells to assess transgene expression during T-cell differentiation.

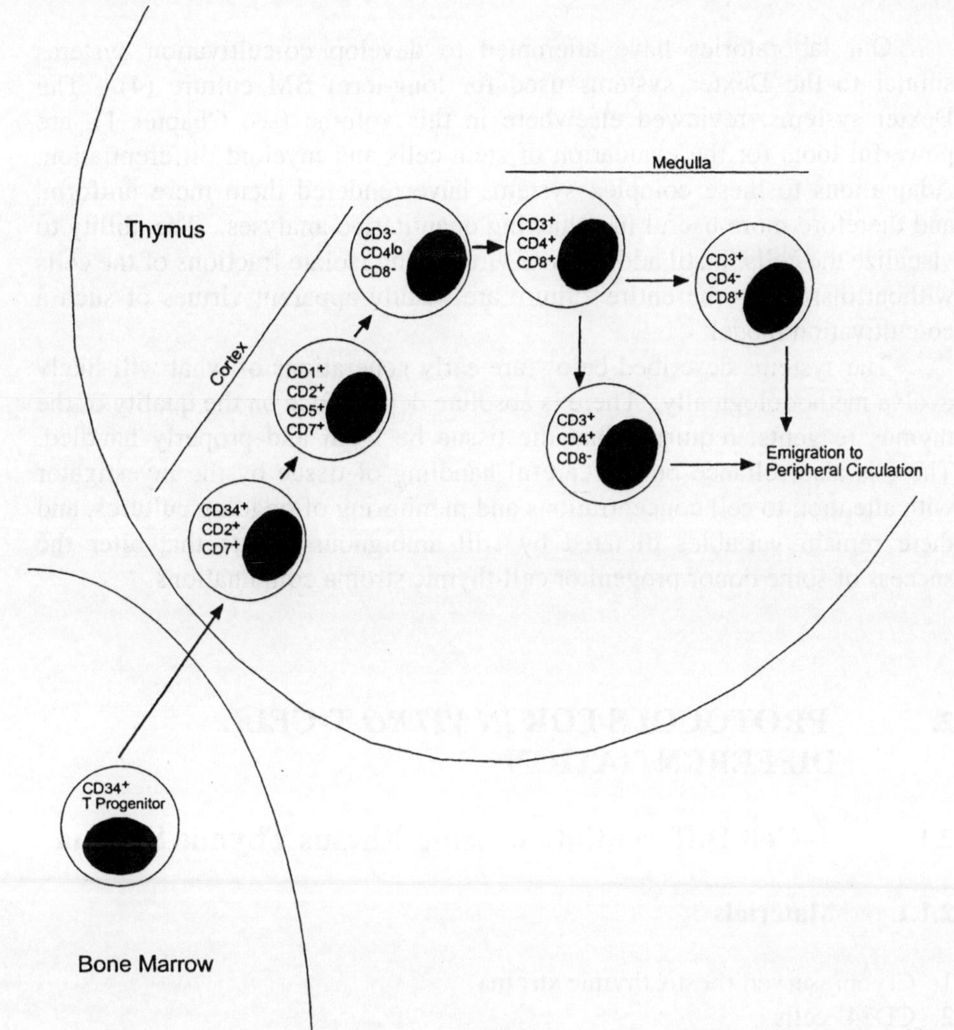

Figure 1. Model of thymocyte development. Migration of T progenitors from the bone marrow to the thymus is shown, followed by sequential maturation into $CD4^+CD8^+$ (double positive) thymocytes. Subsequent to both positive and negative selection events, these cells continue to mature into single positive $CD4^+$ or $CD8^+$ thymocytes, which migrate to the peripheral circulation.

Our laboratories have attempted to develop co-cultivation systems similar to the Dexter systems used for long-term BM culture (4). The Dexter systems, reviewed elsewhere in this volume (see Chapter 1), are powerful tools for the evaluation of stem cells and myeloid differentiation. Adaptations to these complex systems have rendered them more uniform, and therefore more useful in achieving quantitative analyses. The ability to visualize the cells, instil additives or viruses and isolate fractions of the cells without disrupting the entire culture are readily apparent virtues of such a co-cultivation model.

The systems described below are early generations of what will likely evolve methodologically. There is absolute dependence on the quality of the thymus reagents, requiring that the tissue be fresh and properly handled. There is also reliance on the careful handling of tissue by the investigator with attention to cell concentrations and monitoring of ongoing cultures, and there remain variables dictated by still ambiguous events that alter the success of some donor progenitor cell-thymic stroma combinations.

2. PROTOCOLS FOR *IN VITRO* T-CELL DIFFERENTIATION

2.1 T-Cell Differentiation Using Rhesus Thymic Stroma

2.1.1 Materials

1. Cryopreserved rhesus thymic stroma
2. $CD34^+$ cells
3. RPMI with 10% fetal bovine serum (FBS), supplemented with 10 IU/ml penicillin, 10 μg/ml streptomycin, and 1 mM L-glutamine (R-10)

2.1.2 Tissue procurement and processing

Thymic tissue is harvested under sterile conditions from third trimester (120 days gestation) or full term (165 days) neonates after delivery by cesarean section. The thymus is minced into small fragments using sterile blunt-nosed scissors. The fragments are subsequently digested into a single cell suspension by incubation in phosphate-buffered saline (PBS) with collagenase (0.5 mg/ml) and DNAse (2 U/ml) at 37 °C for 30 minutes with frequent agitation. The tissue is washed twice using R10, filtered through a

70 μm mesh and counted. In general, rhesus thymic tissue is then frozen in aliquots of 50 x 10^6 viable cells using 90% FBS and 10% DMSO for subsequent use. The tissue may also be used fresh to establish stromal cultures as described below.

2.1.3 Detailed protocol

1. Thaw a vial of rhesus thymic stroma using a general protocol for cryopreserved cells (e.g. thaw rapidly in 37 °C water bath until only small crystals of ice are present, place on wet ice for a couple of minutes, and then wash once with RPMI). Resuspend cells in 5-10 ml of R10, count, and dilute to the appropriate plating volume (1-2 x 10^6 viable cells per ml). Note that cell viability is often quite low at this stage, as there is a selective loss of thymocytes during cryopreservation.
2. Thymic stromal cultures are established by plating the cell suspension in 24-well plates at a concentration of 4 x 10^6 viable cells per well in a volume of 2 ml R10. The optimal cell concentration for establishing thymic stromal cultures will vary from one batch of thymus to another. Incubate cultures in a standard humidified tissue culture incubator at 37 °C with 5% CO_2.
3. After 3-4 days, nonadherent cells are removed by washing three times with R10. The stroma requires an additional 7-10 days to become confluent. Wells should not be used unless a confluent layer of cells develops during this time. The appearance of nodules which contain heterogeneous cell populations is considered a sign of a healthy culture. The monolayer is maintained in R10 which is changed at least twice per week.
4. After 7-10 days in culture, $CD34^+$ cells in R10 are added to the monolayer at a concentration of 1-3 x 10^5 cells per well. Cultures should be fed bi-weekly using partial medium exchanges with R10. No exogenous cytokines should be added to these cultures. As a control, include an additional well not inoculated with $CD34^+$ cells. Note that with the rhesus stromal system, both rhesus and human progenitor cells differentiate successfully into T-cells, and either may thus be used as a source of progenitors.
5. After 14-21 days, nonadherent cells should be removed from the monolayer. On average, each well will yield about 3-6 x 10^5 cells, although this varies. Aliquots can then be processed for flow cytometry, PCR, and other assays.

In order to maintain the resulting T-cells in culture, they should be stimulated with lectin. For rhesus cells, use concanavalin A (Con A) at 5 μg/ml. For human cells, use phytohemagglutanin (PHA) at 5 μg/ml unpurified (or 0.25 μg/ml purified), IL-2 (20 U/ml) and irradiated (3000 rads) human peripheral blood (PB) mononuclear cells (MNC) at approximately a 1:1 ratio with the T-cells to be stimulated. The initial restimulation is performed on the thymic monolayer at 21 days, and subsequently at approximately 14-day intervals.

2.2 T-Cell Differentiation Using Human Thymic Stroma

2.2.1 2.2.1. Materials

1. Human fetal thymic tissue
2. $CD34^+$ cells
3. Iscove's-Modified Dulbecco's Medium (IMDM) with 20% FBS, supplemented with 10 IU/ml penicillin, 10 mg/ml streptomycin, and 1mM L-glutamine (I-20)

2.2.2 Tissue procurement and processing

Human thymus glands are obtained from electively aborted fetuses of 18-20 weeks gestation after informed maternal consent. The thymus tissue is carefully manually disaggregated, passed through a mesh sieve and washed twice in PBS. In general, 1-10 x 10^8 viable cells are recovered per gland. Cells are then resuspended in I-20 medium.

2.2.3 Detailed protocol

1. Fresh human thymic tissue is plated into 24-well plates at a density of 2 x 10^6 cells/well and incubated at 37 °C with 5% CO_2.
2. After 3-4 days, nonadherent cells are removed by washing three times with I-20. The stroma requires an additional 7-10 days to achieve confluence and develop nodules.
3. Once the thymic stroma is confluent, 1-5 x 10^4 $CD34^+$ cells are added per well and cultured for three weeks at 37 °C with 5% CO_2 in I-20 supplemented with IL-12 (10 ng/ml) and flt-3 ligand (100 ng/ml).
4. After three weeks, nonadherent cells can be removed and evaluated via FACS, PCR, etc.
5. To stimulate further expansion of T-cells obtained from these cultures, cells can be stimulated with PHA (0.25 mg/ml) in the presence of IL-2

(30 U/ml). For extended maintenance, cells require this treatment every two weeks.

2.3 Thymic Organ Culture

2.3.1 Materials

1. Murine or human thymus
2. 13 mm round filters (Nucleopore)
3. 20 x 60 x 7 mm gelfoam sponges cut into 10 x 10 mm pieces
4. Media: human utilizes RPMI plus 10% pooled human AB serum supplemented with 10 IU/ml penicillin, 10 μg/ml streptomycin, and 1mM L-glutamine, whereas murine utilizes DMEM plus 10% FBS.

2.3.2 Tissue procurement and processing

2.3.2.1 Human

Intact thymic lobes are obtained from routine thoracic surgeries on neonates (1-14 days) or from elective abortions (18-20 weeks) following maternal consent. Thymic tissue is manually sectioned into 1 mm^3 fragments for organ culture using blunt-nosed scissors.

2.3.2.2 Murine

Thymic lobes are isolated from day 14 fetuses. Pregnant female mice are sacrificed at day 14 of gestation and the fetuses are removed. It may be helpful to remove the thymic lobes from the embryonal neck region using a dissecting microscope. Carefully cut the head off the fetus ensuring the forelimbs remain attached. Place the fetus in a petri dish under the dissecting microscope. Hold the fetus on its back using blunt-nosed tweezers placed under the armpits. Using sharp-nosed scissors, open the thorax along the sternum beginning at the neck. Carefully open the thorax. The thymic lobes will be located anterior to the heart. Using forceps, gently pluck the thymic lobes and place in ice cold medium.

2.3.3 Detailed protocol

Both human and murine thymus can be cultured successfully using this technique.

1. Boil Nucleopore filters for 30 minutes in distilled water, followed by autoclaving.

2. Place sterile gelfoam sponges into a 24-well tissue culture plate, and add 2 ml medium. Incubate at 37 °C for 2-3 hours prior to use.
3. The sponge will be saturated, so a Nucleopore filter can be added to float on top of the gelfoam sponge to act as a support scaffold for the thymic lobes.
4. Place lobes/fragments on top of the filter (2-4 lobes/fragments per well).
5. At the conclusion of the culture, remove the Nucleopore filter to retrieve the thymic lobes.

2.4 Reconstituted Human-SCID Mice Thymic Cultures

Fetal SCID mouse thymus lobes can be used to support differentiation of human T-cells from hematopoietic progenitors.

2.4.1 Materials

1. Fetal SCID mouse thymus lobes (day 14-15 of gestation)
2. $CD34^+$ cells
3. Medium

2.4.2 Detailed protocol

1. Fetal thymic lobes are harvested from SCID mice at day 14-15 of gestation as described above (section 2.3.2.2).
2. Using a Terasaki plate, a hanging drop culture is established with 25 µl of medium containing $CD34^+$ progenitor cells (1-2 x 10^4 $CD34^+$ cells per well). One SCID fetal thymic lobe is added to each drop. The plate is then inverted to create a hanging drop culture, and incubated for 48-72 hours.
3. Thymic lobes are removed from the Terasaki plates and added to gelfoam cultures as described above (section 2.3.3).

3. ANALYSIS OF CELL OUTPUT

3.1 Immunophenotyping of Cultured Cells by Flow Cytometry

Cells are harvested by gentle aspiration, washed twice in PBS, counted and 2-5 x 10^5 cells are stained in a final volume of 100 µl with the following monoclonal antibodies in the presence of 2% mouse serum (Dako, Carpentiera, CA). For human cells, the antibodies we use (but are by no means the only acceptable antibodies) are: anti-CD34 (Caltag, Burlingame, CA), anti-CD1a, anti-CD2, anti-CD3 (Becton Dickinson, San Jose, CA); anti-CD4, and anti-CD8 (Exalpha, Boston, MA). Antibodies used for immunophenotyping of rhesus cells include anti-CD3 (6G12), anti-CD4 (OKT4; Ortho Diagnostics), anti-CD8 (Leu-2a; Becton Dickinson), anti-CD33 (Dako), anti-CD71 (Becton Dickinson), anti-CD34 (QBend-10; Immunotech), anti-CD61 (Becton Dickinson), anti-CD64 (Immunotech) and anti-CD38 (American Type Culture Collection, Manassas, VA). FITC- and PE-conjugated isotype control antibodies are used for each culture. Stained samples are then washed three times in PBS and flow cytometric analysis performed using a FACScan (Becton Dickinson, San Jose, CA) or other cytometer immediately, or after fixation with fresh paraformaldehyde (2%).

3.2 RT-PCR Analysis

RNA is prepared from between 10^4-10^5 cells using guanidium thiocyanate and oligo dT spun columns (QuickPrep mRNA purification kit, Pharmacia) prior to DNAseI (Promega, Madison, WI) digestion. Random hexanucleotide primers and Moloney reverse transcriptase (Gibco-BRL, Grand Island, NY) are used to prepare cDNA according to standard protocols, which are then stored at –20 °C. Amplification is performed using one-quarter of the cDNA product in each PCR reaction (50 µl) with 2.5 U Taq DNA polymerase (Pharmacia), and 0.4 µM of each of the following oligonucleotide sequences: constant region of TCRδ chain: 3' primer: CAA ACG TTT CTG ACA ACG GCA G (840 - 861), 5' primer: GAA CAA ATG TCG CTT GTC TGG T (479 - 500); Gs_α (a constitutively expressed G protein): 3' primer: GCT GCT GGC CAC CAC GAA GAT GAT; 5' primer: GTG ATC AAG CAG GCT GAC TAT GTG; RAG2: 3' primer: CCT CCC ACA CGC TTG CAG T; 5' primer: GTC CCG GGC GCT GCA; T-cell receptor Vβ chains in numerical order: 1. AAG AGA GAG CAA AAG GAA ACA TTC TTG AAC, 2: GCT CCA AGG CCA CAT ACG AGC AAG GCG TCG, 3: AAA ATG AAA GAA AAA GGA GAT ATT CCT GAG, 4: CTG AGG CCA CAT ATG AGA GTG GAT TTG TCA, 5: CAG AGA AAC AAA GGA AAC TTC CCT GGT CGA, 5b: TTC CCT AAC TAT AGC TCT GAG CTG, 6: GGG TGC GGC AGA TGA CTC AGG GCT GCC CAA, 7: ATA AAT GAA AGT GTG CCA AGT CGC TTC TCA, 8: AAC GTT CCG ATA GAT GAT TCA GGG ATG CCC, 9: CAT TAT AAA TGA AAC AGT TCC AAA TCG CTT, 10: CTT ATT CAG AAA GCA GAA ATA ATC AAT GAG, 11: TCC ACA GAG AAG GGA GAT CTT TCC TCT GAG, 12: CTG AGA TGT CAC CAG ACT GAG AAC CAC CGC, 13: CAA GGA GAA GTC CCC AAT, 14:

GTG ACT GAT AAG GGA GAT GTT CCT GAA GGG, 15: GAT ATA AAC AAA GGA GAG ATC TCT GAT GGA, 16: CAT GAT AAT CTT TAT CGA CGT GTT ATG GGA, 17: GCA CAA GAA GCG ATT CTC ATC TCA ATG CCC, 18: CAT CTG TCT TCT GGG GGC AGG TCT CTC AAA, 19: ATA GCT GAA GGG TAC AGC GTC TCT CGG GAG, 20: TCT AAT ATT CAT CAA TGG CCA GCG ACC CT, 21: GCA GTA GAC GAT TCA CAG TT, 22: ATG CAG AGC GAT AAA GGA AG, 23: ATC TCA GAG AAG TCT GAA AT, 24: GAT TTT AAC AAT GAA GCA GA; 5' constant: 26-49: CCG AGG TCG CTG TGT TTG AGC CAT, 529-510: AAT CCT TTC TCT TGA CCA TG; 491-468: CTG ACC AGC ACI GCA TAC AIG GTG.

PCR amplification may be performed in a Gene Amp 9600 thermal cycler (Perkin Elmer, Norwalk, CT) with an initial denaturation (95 °C for 5 minutes) followed by 35 cycles of denaturation at 94 °C for 30 sec, annealing at 55 °C for 30 sec and extension at 72 °C for 60 sec for TCRδ amplification; 30 cycles of 94 °C, 55 °C and 72 °C for 1 minute each is used for Gs_{α} amplification. A single 10 min extension following amplification is performed at 72 °C for all reactions. PCR products (10 µl) are then electrophoreticaly separated on ethidium bromide-stained agarose (1.5%) gels and photographed under UV light.

TCR Vβ subset analyses are performed as previously described using a panel of primers specific for individual Vβ chains (6, 9). Specificity may be confirmed by Southern blot hybridization with digoxigenin-labeled internal probe at 55 °C for 2 hours using Expresshyb hybridization solution (Clontech, Palo Alto, CA). Blots should be washed under high stringency, followed by incubation with an anti-digoxigenin antibody alkaline phosphatase conjugate, and detection by chemiluminescence. Parallel extractions should be performed from samples without RT or water alone to exclude DNA contamination.

3.3 Proliferation Assays

T-cells derived from $CD34^+$ cultures are washed and resuspended in RPMI with 10% FBS at a concentration of 10^6 cell/ml, and 100 µl added to each well of a 96-well plate. Cells are stimulated with either Con A at 5 µg/ml or B-cells in the presence of IL-2 (20 IU/ml) and irradiated human PB MNC (10^5 cells per well). After 7 days in culture at 37 °C, each well is pulsed with 1 µCi of methyl-^{3}H-thymidine (DuPont, NEN, Boston, MA) for 16 hours and harvested for liquid scintillation counting. Six replicate wells should be assayed for each condition.

4. UTILITY OF THE SYSTEM

The use of co-culture systems in which events of T-lymphoid differentiation can be followed *in vitro* has been limited to date. The applications to which our laboratories have applied these systems is in the assessment of gene transfer vectors and specific anti-human immunodeficiency virus (HIV) transgenes (7, 19, 20). Through the use of these systems, the likelihood of specific constructs to either be maintained or expressed could be evaluated. For example, it was demonstrated that while recombinant adeno-associated virus could infect $CD34^+$ progenitors with high efficiency and result in transgene expression early in the culture system, the construct was eventually lost, likely due to a failure to integrate (7). A retroviral vector containing an anti-HIV tat gene was noted to be expressed in both myeloid and lymphoid progeny of $CD34^+$ cells and was able to protect differentiation of T-cells from HIV-1 infection (20). Similarly, the system was useful in demonstrating that a hairpin ribozyme was capable of protecting transduced T-cells differentiated in the co-culture system (21). Therefore, these systems do have a demonstrated ability to address questions of differentiation stage-specific expression of gene transfer vectors, recognizing that such information has not yet been correlated with *in vivo* events.

The systems may also be used to determine the impact of transgene expression. We have noted the lack of effect of the green fluorescent protein or the retroviral anti-tat construct on immunophenotypic profiles of maturing cells in these systems (7). Influences on T-cell differentiation, either as untoward effects of a transgene with another purpose or as an intentional effort to manipulate T-cell differentiation, can be monitored in these systems.

The examination of T-cell differentiation-specific effects of pathogens, such as HIV, or of gene modifications, can also potentially be applied to other additives in the *in vitro* system. Specifically, the testing of pharmacologic agents (e.g. anti-HIV-1 drugs) may be conducted in the system, analyzing for immunophenotype and cell number. Alternatively, the possibility of cell augmentation can be addressed through addition of biologicals such as cytokine analogues or peptides.

The molecular events which accompany specific differentiation events may be evaluated in the co-culture context, particularly with the availability of gene amplification methods such as semi-quantitative RT-PCR. In other culture models, the combined use of micromanipulator-guided cell isolation and single cell RT-PCR has permitted the characterization of gene expression profiles during myeloid differentiation (2). Similar approaches may be envisioned in the context of T-cell differentiation.

The necessary role of thymic stroma in the induction of T-cell differentiation leads to the central question of which stromal elements provide the nurturing milieu for a multipotent cell to enter T-lineage commitment. An *ex vivo* culture system provides a possible starting point for addressing that question. It may be possible to seek the necessary gene product(s) from the feeder layers, although the complexity of the systems presently defy any simple read-outs for such a cloning enterprise.

Finally, the development of feeder layer-based *ex vivo* cell generation systems has led to the pursuit of clinical-scale cell production in other contexts. Specifically, BM stromal feeder layers have entered into the arena of bioreactors for the expansion of hematopoietic progenitors (see Chapter 1) (22). Whether such adaptations of *in vitro* systems for T-cell differentiation will permit *ex vivo* generation and education of T-cell populations is an issue that remains to be explored.

5. CONCLUDING REMARKS

The development of co-culture systems for the maturation of T-cells from primitive precursor populations offers manipulable models to dissect the biology of human T-cell commitment. From a more utilitarian point of view, these models may be useful for testing agents which may alter T-cell differentiation, or for assessing transgene expression during that process, offering assays, and possibly, scalable techniques for T-cell generation. However, the current state of the art in these systems continues to require much art. The systems are temperamental and tend to perform only for those exacting in method and nurture. Realization of the full potential of these systems will require better understanding of the variables at play, and harnessing those variables to provide uniformity of output. Achieving that goal will provide an exceptional tool for immunobiology.

LITERATURE CITED

1. Barcena, A., A. H. M. Galy, J. Punnonen, M. O. Muench, D. Schols, M. G. Roncarola, J. E. deVries, and H. Spits. 1994. Lymphoid and myeloid differentiation of fetal liver $CD34^+$ lineage cells in human thymic organ culture. *J Exp Med.* 180:123.
2. Cheng, T., H. Shen, D. Giokas, J. Gere, D. G. Tenen, and D. T. Scadden. 1996. Temporal mapping of gene expression levels during the differentiation of individual primary hematopoietic cells. *Proc Natl Acad Sci.* 93:13158-13163.
3. Cordier, A. C., and S. M. Haumont. 1980. Development of thymus, parathyroids and ultimo-branchial bodies in NMRI and nude mice. *Am J Anat.* 157:227-263.
4. Dexter, T. M., T. D. Allen, and L. G. Lajtha. 1977. Conditions controlling the proliferation of hemopoietic stem cells in vitro. *J Cell Physiol.* 91:335-344.

5. Ford, C. E., H. S. Micklem, E. P. Evans, J. S. Gray, and D. A. Ogden. 1966. The inflow of bone marrow cells to the thymus. *Ann NY Acad Sci*:283-296.
6. Freedman, A. R., H. Zhu, J. D. Levine, S. Kalams, and D. T. Scadden. 1996. Generation of human T-lymphocytes from bone marrow CD34^{+} cells *in vitro*. *Nature Medicine*. 2:46-51.
7. Gardner, J. P., H. Zhu, P. C. Colosi, G. J. Kurtzman, and D. T. Scadden. 1997. Robust, but transient expression of adeno-associated virus-transduced genes during human T-lymphopoiesis. *Blood*. 90:4854-4864.
8. Haynes, B. F., and C. S. Heinly. 1995. Early human T-cell development: analysis of the human thymus at the time of initial entry of hematopoietic stem cells into the fetal thymic microenvironment. *J Exper Med*. 181:1445.
9. Kalams, S. A., R. P. Johnson, A. K. Trocha, M. Dynan, H. N. Ngo, R. T. D'Aquila, J. T. Kurnick, and B. D. Walker. 1994. Longitudinal analysis of T-cell receptor (TCR) gene usage by human immunodeficiency virus 1 envelope specific cytotoxic T-lymphocyte clones reveals a limited TCR repertoire. *J. Exp. Med.* 179:1261.
10. Kurtzberg, J., S. M. Denning, L. M. Nycum, K. H. Singer, and B. F. Haynes. 1989. Immature human thymocytes can be driven to differentiate into nonlymphoid lineages by cytokines from thymic epithelial cells. *Proc Natl Acad Sci*. 86:7575.
11. LeLievre, C. S., and N. M. LeDouarin. 1975. Mesenchyme derivatives of the neural crest: analysis of chimeric quail and chick embroyos. *J Embryol, Exp Morphol*. 34:125-136.
12. Marquez, C., C. Trigueros, E. Fernandez, and J. L. Toribio. 1995. The development of T and non-T-cell lineages from CD34^{+} human thymic precursors can be traced by the differential expression of CD44. *J Exp Med*. 181:475.
13. McCune, J. M., R. Namikawa, H. Kaneshima, L. D. Shultz, M. Lieberman, and I. L. Weissman. 1988. The SCID-hu mouse: Murine model for the analysis of human hematolymphoid differentiation and function. *Science*. 241:1632-1639.
14. Miller, J. F. A. P. 1994. The thymus: maestro of the immune system. *BioEssays*. 16:509-512.
15. Nikolic, B., J. P. Gardner, D. T. Scadden, and M. Sykes. 1999. In vivo development of a human T-cell repertoire that is tolerant to porcine xenoantigens. *J. Immunol.*, in press).
16. Plum, J., M. DeSmedt, M. P. Defresne, G. Leclereq, and B. Fandekerckhove. 1994. Human CD34^{+} fetal liver stem cells differentiate to T-cells in a mouse thymic microenvironment. *Blood*. 84:1587-1593.
17. Res, P., E. Martinez-Caceres, A. Jaleco Cristina, F. Staal, E. Noteboom, K. Weijer, and H. Spits. 1996. CD34^{+}, CD38^{dim} cells in the human thymus can differentiate into T natural killer and dendritic cells, but are distinct from pluripotent stem cells. *Blood*. 87:5196-5206.
18. Rosenzweig, M., E. M. Bunting, R. L. Damico, and G. N. Gaulton. 1995. Human neonatal thymic organ culture: An ex vivo model of thymocyte ontogency and HIV-1 infection. *Pathobiology*. 62:245.
19. Rosenzweig, M., D. F. Marks, D. Hempel, M. Heusch, G. Kraus, F. Wong-Staal, and R. P. Johnson. 1997. Intracellular immunization of rhesus CD34^{+} hematopoietic progenitor cells with a hairpin ribozyme protects T-cells and macrophages from simian immunodeficiency virus infection. *Blood*. 90:4822-4831.
20. Rosenzweig, M., D. F. Marks, D. Hempel, J. Lisziewicz, and R. P. Johnson. 1997. Transduction of CD34^{+} hematopoietic progenitor cells with an anti-tat gene protects T-cell and macrophage progeny from AIDS virus infection. *J Virol*. 71:2740-2746.
21. Rosenzweig, M., D. F. Marks, H. Zhu, D. Hempel, K. G. Mansfield, P. K. Sehgal, S. Kalams, D. T. Scadden, and R. P. Johnson. 1996. In vitro T-lymphopoiesis of human and rhesus CD34^{+} progenitor cells. *Blood*. 87:4040-4048.

22. Koller, M. R., S. G. Emerson, B. O. Palsson. 1993. Large-scale expansion of human stem and progenitor cells from bone marrow mononuclear cells in continuous perfusion cultures. *Blood* 82: 378-384.

23. Yeoman, H., R. E. Gress, C. V. Bare, A. G. Leary, E. A. Boyse, J. Bard, L. D. Shultz, D. T. Harris, and D. DeLuca. 1993. Human bone marrow and umbilical cord blood cells generate $CD4^+$ and $CD8^+$ single-positive T-cells in murine fetal thymus organ culture. *Proc Natl Acad Sci USA*. 90:10778.

Chapter 3

T-Lymphocytes: Mature Polyclonal and Antigen-Specific Cell Culture

Bruce L Levine, Katia Schlienger and Carl H June
Abramson Family Cancer Research Institute at the University of Pennsylvania Cancer Center, 1225 Penn Tower, 3400 Spruce Street, Philadelphia, PA 19104-4283. Tel: 001-215-349-5323; Fax: 001-215-662-6809; E-mail: levineb@mail.med.upenn.edu

1. INTRODUCTION

T- and B-lymphocytes are both designed to recognize specific foreign or dangerous molecules (antigens) and mount a response. While B-lymphocytes can recognize soluble antigen (Ag), T-lymphocytes recognize cell-associated Ag presented in the context of major histocompatibility complex (MHC) Class I (CD8 cells) or Class II (CD4 cells). Immune responses coordinated by T-lymphocytes are initiated by processed Ag fragments in the context of MHC binding to the T-cell receptor (TCR), the primary means of activation, and a supplementary signal which allows activation to progress. TCR stimulation is Ag-specific and is frequently referred to as signal 1, while co-stimulation is Ag non-specific and is commonly referred to as signal 2. This 2 signal model of lymphocyte activation was first proposed by Bretscher and Cohn [1] and later modified by Lafferty and Cunningham [2] for T-cells. Stimulation of $CD4^+$ T-cells via Ag in the context of MHC Class II leads to T-cell proliferation and clonal expansion and, through the action of released cytokines, can induce the activation of not just other T-cells, but also B-cells and cells of the macrophage/monocyte lineage. $CD4^+$ T-cells which secrete cytokines that activate other T-cells and $CD8^+$ T-cells to differentiate into cytotoxic lymphocytes (CTL) are termed T-helper type 1 (Th1), while $CD4^+$ T-cells that secrete cytokines that enable B-cell differentiation and immunoglobulin

(Ig)-type switching are termed T-helper type 2 (Th2) [3,4]. Recent evidence indicates that $CD8^+$ T-cells may also be divided into subsets based on patterns of cytokine secretion [5]. Thus, the activation of T-cells is central to the mobilization of an effective immune response because T-cells can exert regulatory influence over other cells of the immune system. This chapter will describe some of the established and newer methods of polyclonal and antigen-specific T-cell culture and *in vitro* measures of function with an emphasis on those methods suitable for longer term (>7 days) culture of T-cells.

1.1 Anatomy *In Vivo*

T-lymphocytes, like all other hematopoietic cells, arise from a pluripotent stem cell in the bone marrow (BM). Following maturation into committed lymphoid progenitor. cells, T-cell precursors migrate to the thymus where they differentiate into mature T-lymphocytes. In the thymus, T-cells rearrange their gene for the antigen receptor, which consists either of an alpha and beta chain (90-95% of T-cells) or a gamma and delta chain. These cells also express the CD3 complex and both CD4 and CD8. The first step in thymic selection, which occurs in the cortex of the thymus, is the positive selection of cells that bind with a certain affinity for MHC expressed by thymic stromal cells. Cells that are positively selected can then recognize foreign Ag as presented in the context of MHC and migrate next to the medulla. Here, negative selection occurs and cells that react to self-Ag are removed or inactivated. The last step in the maturation of T-cells in the thymus is that $CD4^+CD8^+$ cells become single positive for either CD4 or CD8 and exit the thymus. Only 5-10% of T-cell progenitors survive thymic selection. T-lymphopoiesis *in vivo* and *in vitro* is discussed in detail in Chapter 2 of this volume.

Following maturation and exodus from the thymus, mature T-lymphocytes seed the spleen, lymph nodes and lymphoid tissue in the gut (Peyer's patches), tonsils and skin. These sites are commonly referred to as the secondary lymphoid organs, where lymphoid maturation takes place, as opposed to the BM and thymus, which are referred to as the primary lymphoid organs. The secondary lymphoid organs are the main sites of antigen trapping and clonal expansion of antigen-specific T-cells and B-cells. Naive T-cells which have never encountered Ag express CD45RA and CD62L (L-selectin) and home to the peripheral lymph nodes. This occurs via a specific interation between CD62L on T-cells and an adhesion molecule termed addressin on the high endothelial venules (HEV) entering the lymph nodes. Memory T-cells, or cells that have previously encountered Ag or have been activated in some way, migrate to the skin or mucosa where they may act as sentries to a repeat assault from a foreign invader. Memory T-cells express CD45RO and may or may not express CD62L. The transition from $CD45RA^+$ to $CD45RO^+$ is not immediate and some cells may express both isoforms of CD45 [6,7]. Additional markers such as CD62L and CD27 can serve to aid the discrimination between naive and memory cells [8]. In the peripheral blood, approximately 50% of T-cells are $CD45RA^+$ and 50% are $CD45RO^+$, although this varies with the age of the donor, with elderly people tending to have a greater percent of $CD45RO^+$ T-cells [9]. The peripheral blood ratio of $CD4^+$ to $CD8^+$ T-cells is 2:1, with slight variation. In human immunodeficiency virus (HIV) disease, where the absolute $CD4^+$ T-cell count declines, the CD4/CD8 ratio declines as well [10]. In a healthy adult human, the total number of T-cells is ~10^{12}, with only ~2% circulating in the peripheral blood [11]. This means that any sampling of peripheral blood can be only an approximation of the total body T-cell phenotype or functional status.

1.2 T-Cell Activation

Our understanding of T-cell activation through cell surface receptors and proteins is now recognized to be a complex multistage process of recognition, adhesion and stimulation [12,13]. As mentioned above, Ag presented in the context of MHC binds to the TCR. Adhesion molecules present on both the T-cell and the antigen presenting cells (APC) strengthen this initial interaction. Besides CD8 and CD4, which interact with MHC I and II, respectively, CD2 on T-cells interacts with LFA-3 (CD58), and LFA-1 (CD11a/CD18) on T-cells interacts with ICAM-1 (CD54). An additional surface interaction between T-cells and APC occurs between CD40 on the surface of an APC and CD40 ligand (CD154) on the T-cell. This results in the upregulation of co-stimulatory receptors CD80 (B7-1) and CD86 (B7-2) on the APC, and finally, full activation of the T-cell via the co-stimulatory receptor CD28 [14]. The signal delivered through the TCR/CD3 complex results in activation of phospholipase C (PLC) which cleaves phosphatidyl inositol bisphosphate (PIP2) into inositol trisphosphate (IP3) and diacylaglycerol (DAG) [15]. IP3 is believed to induce an early and transient release of $[Ca^{++}]$ from intracellular stores [15,16]. DAG and $[Ca^{++}]$ are cofactors for the activation of protein kinase C (PKC), and this can be mimicked by pharmacological agents [17] (see section 3.3.2). TCR engagement also results in the activation of several tyrosine kinases that are critical to T-cell activation, including p56lck, p60fyn, and ZAP kinase [18].

As discussed above, T-cells require a second co-stimulatory signal in order for clonal expansion to take place. This supplementary signal can be delivered by the engagement of cell surface co-stimulatory receptors such as

CD28, whereas in memory or effector cells cytokines such as IL-2 may be sufficient [19]. In naive T-cells, the same APC on which the T-cell recognizes its specific antigen must deliver this signal. CD28 stimulation drives cyclosporine-independent T-cell proliferation and cytotoxic activity through de novo synthesis and mRNA stabilization of cytokines [20]. CTLA-4, which is structurally related to CD28, is upregulated on activated T-cells [21]. Recent evidence indicates that CD28 and CTLA-4 have opposing effects on T-cell activation, with CTLA-4, which binds with greater affinity to CD80 and CD86 than does CD28, serving to inhibit or down-regulate a response [22]. Thus, the activation state of a T-cell and the regulation of an immune response may be determined by the relative strength of the signal delivered by either CD28 or CTLA-4, which is dependent on the absolute amount, as well as the relative levels, of the co-stimulatory ligands CD80 and CD86. An artificial APC may be constructed by coating beads with monoclonal antibody (mAb) directed against CD3 and CD28 to bypass the negative signal delivered by CTLA4 (Fig. 1). Other surface receptors including CD40L (CD154) [23] and 4-1BB (CD137) [24,25] may provide co-stimulation independent of, or supplementary to, CD28.

An activated T-cell expresses higher levels of many cell surface molecules including adhesion receptors, cytokine receptors, co-stimulatory receptors, and general markers of activation such as CD69 and the transferrin receptor (CD71). $CD4^+$ T-cells have been shown to secrete a large number of cytokines upon activation including IL-2, IL-3, IL-4, IL-5, IL-6, IL-10, IL-13, IFNγ, TNFα, TNFβ, GM-CSF, MIP-1α, MIP-1β, and RANTES [26-29]. $CD8^+$ T-cells may secrete IFNγ and TNF and lower levels of IL-2, IL-4, IL-5 and IL-10 [26]. The spectrum of cytokines secreted by an individual cell or population of cells varies widely as a function of the activation and differentiation status of the cell. Multi-parameter intracellular cytokine staining, a powerful technique for which commercial grade reagents have recently become available, should aid in the characterization of T-cell subsets and the cytokines secreted. Other useful assays for characterizing cultured T-cells include allo-specific and Ag-specific proliferation assays, the use of cell tracking dyes to measure the

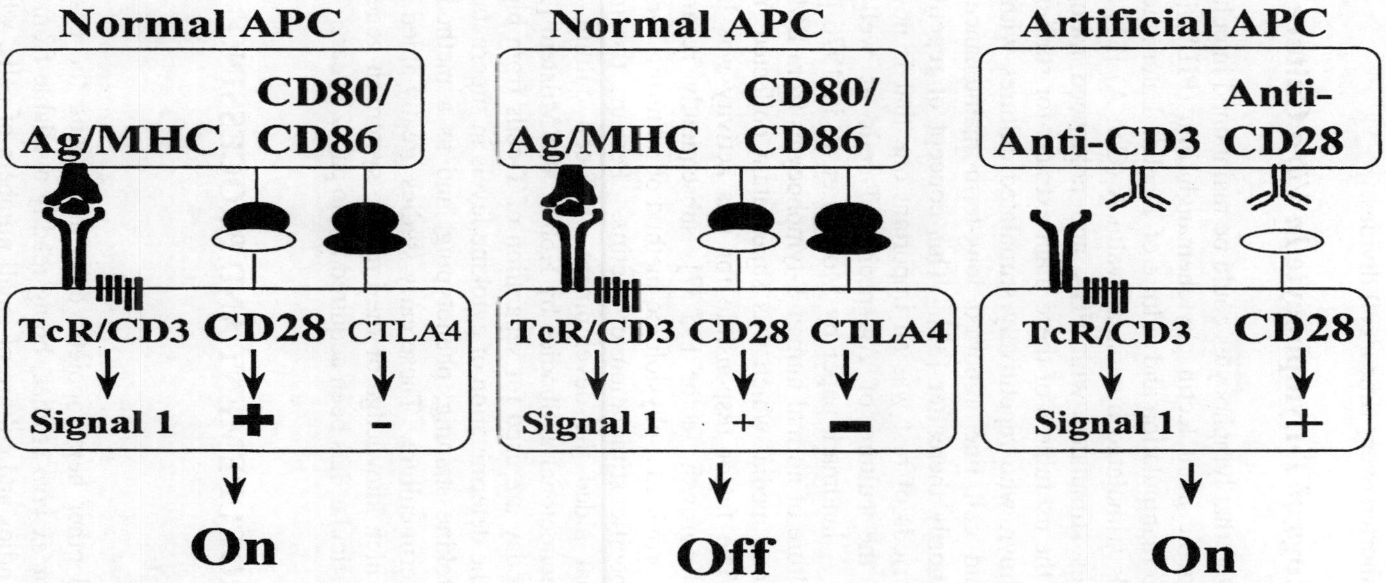

Figure 1. The 3 signal model of T-lymphocyte activation. Antigen/MHC interactions with T-cells generate signal 1. In resting T-cells (left), there is a much higher level of CD28 present than CTLA4 (CD152) so that a positive signal (signal 2) is delivered through CD28. In activated T-cells (center), CTLA4 is upregulated and, because of its higher affinity for CD80 and CD86, may deliver a negative signal (signal 3) to the T-cell. The negative regulatory effects of CTLA4 may be overcome by constructing an artificial APC using mAb directed against CD3 and CD28, and thus bypassing negative signals delivered by CTLA4.

number of times a given cell population divides, and assays for cytotoxic T-lymphocyte activity. Another measure of the replicative history of a population of cells is the measurement of telomeres, structures on the ends of chromosomes, which shorten every time a cell divides.

1.3 The Beginnings of T-Lymphocyte *In Vitro* Culture

The discovery in 1960 that lymphocytes could be maintained in culture by addition of the mitogenic plant lectin phytohemagglutinin (PHA) [30] paved the way for *ex vivo* manipulation and culture of T-cells. Later studies on murine lymphoid cells stimulated by concanavalin A (Con A) [31] and on human T-lymphocytes stimulated with PHA and pokeweed mitogen (PWM) [32] established the usefulness of these plant extracts for studies of immune function. Svedmyr, who repetitively stimulated cultures with an allogeneic lymphoblastoid cell line, reported long-term maintenance of human T-lymphocytes shortly thereafter [33]. This concept of repetitive stimulation following periods of rest was an important one that is now a standard technique for the culture of Ag-specific T-cells as well as polyclonal T-cell lines. A landmark paper by Morgan et al. in 1976 [34] reported the selective culture of normal human T-lymphocytes from BM by maintenance in conditioned media, which was supernatant collected from PHA-stimulated cells. The factor responsible for this activity was later determined to be T-cell growth factor [35-37], subsequently renamed interleukin-2 (IL-2), and provided a physiologic method of culturing T-cells compared to the non-specific stimulation of lectins. Perhaps the most significant discovery that aided the development of T-cell culture and separation was that of monoclonal antibodies by Kohler and Milstein [38]. Monoclonal antibodies today are used for separation of T-cells from blood and BM, in ELISAs for the determination of cytokine levels in supernatants, and for intracellular cytokine staining, phenotyping, and as a method to stimulate cells for long term culture. These early studies greatly aided and accelerated the rate at which knowledge of the immune system in general, and T-lymphocytes in particular, has been acquired in the past 25 years.

2. TISSUE PROCUREMENT AND PROCESSING

2.1 Procurement

T-lymphocytes and other hematopoietic cells are among the easiest human cells to obtain for *ex vivo* studies, being readily available from the peripheral blood. For studies and culture of small numbers of cells, whole blood may be obtained by venipuncture and collected in a syringe or tube containing an anti-coagulant. Heparin may be used at 1 ml of 1000 U/ml preservative-free heparin per 100 ml of whole blood. Whole blood collected in this manner may sit for several hours or even overnight prior to processing, however, it is best to start the isolation as soon as possible [39]. The absolute T-cell count in a healthy adult ranges from 1-1.5 x 10^6 cells/ml of blood [40]. Accommodations in the amount of blood drawn should be made to account for a loss of cells during processing. Guidelines may vary among institutions, but in general, the amount of whole blood that may be donated is limited to 400 ml in a given 8-week period. T-cells can also be obtained from cord blood (CB) collected from the umbilical vein immediately after delivery [41]. The advantage of working with T-cells isolated from CB is that 90% of the T-cells are naive as defined by CD45RA expression, compared to 30-50% in healthy adults [40]. In the tonsil, although B-cells predominate, a minority of lymphocytes are T-cells and may be obtained for *in vitro* culture using separation methods described below. Most tonsil specimens are obtained from children, which means that more of the T-cells will be $CD45RA^+$ than in adults [40], however, nearly all will be from donors with chronic infections that may affect the activation/differentiation state of the T-cells present.

When large numbers of cells are required for *in vitro* culture, the sources of T-cells mentioned above are insufficient. Large numbers of mononuclear cells (MNC) may be obtained by apheresis in a blood bank or collection center. The apheresis machine is a self-contained continuous-flow centrifuge that separates whole blood into components and returns unneeded cells or components to the donor. An experienced apheresis technician using a Baxter CS3000 (Baxter Healthcare Corporation, Fenwal Division, Deerfield, IL) or COBE Spectra (COBE, Lakewood, CO) may process 5-10 liters of peripheral blood in 2-3 hrs with a yield of 5-15 x 10^9 MNC with slight modification to established protocols [42,43] (H. Cullis, unpublished). Other instruments, such as the Haemonetics MCS 3P (Haemonetics, Braintree, MA) and the Fresenius AS104 (Fresenius, Walnut Creek, CA), have been used for peripheral blood progenitor cell collection [44] and may be adapted to collect MNC for further enrichment of T-cells. An apheresed product may be further enriched for lymphocytes by the use of another large-scale processing device, a counterflow centrifugal elutriator [45-47]. The most common use of an elutriator is to deplete or adjust the T-cell fraction in BM or mobilized peripheral blood (mPB) prior to a BM transplant. As with an apheresis machine, one drawback of this device is its cost and the requirement for extensive training of a technician for its operation.

2.2 Enrichment/Purification from Whole Blood or an Apheresis Product

Once whole blood or apheresis-isolated MNC have been obtained, $CD4^+$ or $CD8^+$ T-cells within the bulk population may need to be separated before *in vitro* culture. If autologous human serum is desired for *in vitro* studies from the same sample of peripheral blood, the blood can be collected without anti-coagulants and then immediately defibrinated by the use of glass beads or other objects [48]. The defibrinated blood can then be spun to isolate serum. A common first step when starting with heparinized whole blood is the separation of erythrocytes and granulocytes from MNC (monocytes and lymphocytes) by a ficoll gradient [49]. Ficoll is a sucrose polymer with a density of 1.077 gm/ml that forms a discontinuous density gradient upon centrifugation. Heparinized blood is first diluted with Ca^{++}/Mg^{++} free Hank's or Dulbecco's phosphate buffered saline (PBS), and layered over the ficoll in a centrifuge tube which is spun at 800*g* for 20 min with no brake at room temperature. The erythrocytes and granulocytes will be at the bottom of the tube, followed by the ficoll, the MNC layer, and on top will be plasma and platelets (Fig. 2A). Cells should be washed out of the ficoll as soon as possible. This MNC layer can be further processed to obtain T-cells or T-cell subsets as described below.

An alternative protocol for the first step of purification from whole blood or an apheresis unit is to first lyse the red cells using ACK lysis buffer, which is readily available from a variety of commercial suppliers. The leukocytes can be separated into a lymphocyte fraction and monocyte fraction by centrifugation with percoll at a density of 1.099 mg/ml. Percoll (Pharmacia) is a non-toxic colloidal silica solution which forms a density gradient during centrifugation due to heterogeneity of particle sizes. Shown in Fig. 2B is a percoll gradient with a distinct monocyte layer at the top and a lymphocyte layer near the bottom. Any red cells remaining after ACK lysis and the neutrophils form a pellet at the bottom of the tube. An

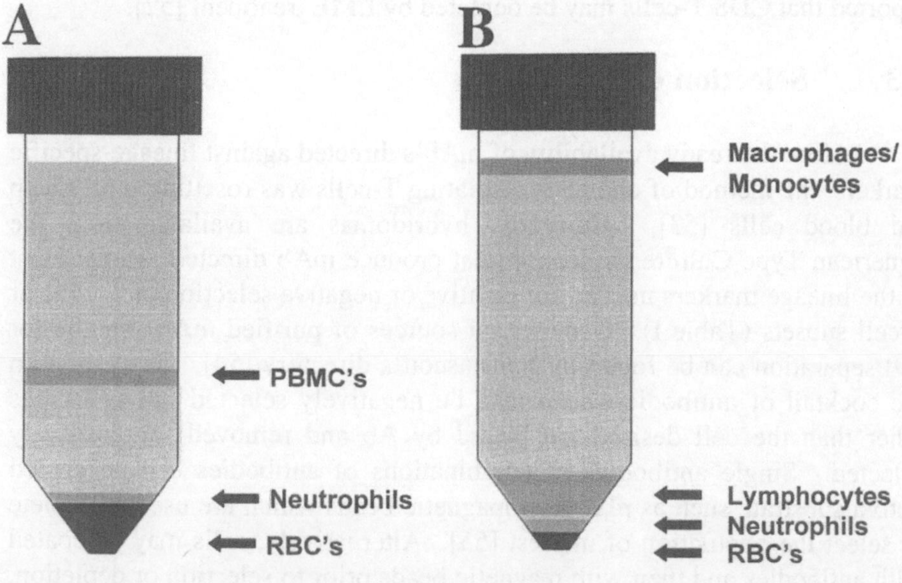

Figure 2. Two examples of gradient centrifugation useful in the enrichment of T-cells from whole blood or an apheresis leukopack. A ficoll gradient (A) is able to separate PB MNC, which contain lymphocytes, monocytes, macrophages and granulocytes, from neutrophils and RBC's. A percoll gradient (B) is able to separate B- and T-lymphocytes from macrophages/monocytes and neutrophils and RBC's.

advantage of percoll over ficoll for separating T-cells is that the monocytes are removed. This can save time by eliminating the need for adhesion steps to remove monocytes, and it can save reagents when beads or particles are being used for purification of T-cells via negative selection. ACK lysis can be avoided by centrifuging heparinized blood and removing the buffy coat of leukocytes with a small bore pipette [39]. Platelets can be removed from lymphocyte preparations by several centrifugation washes at slow speed.

If ACK lysis of red cells and percoll separation is not feasible, monocytes and macrophages can be removed from a post-ficoll layer preparation by plastic adherence. Cells are suspended in medium in a plastic tissue culture flask and incubated at 37 °C for >1 hr. The efficiency of monocyte/macrophage removal will depend on the length of incubation and the surface area available for adherence, and should be confirmed by FACS analysis. When large numbers of T-cells are being processed, plastic adherence may not be practical. An additional non-specific method of T-cell enrichment is the use of leucine methyl ester (LME) to eliminate granulocytes and large granular lymphocytes (NK cells) from a population of leukocytes [50,51]. LME is taken up by phagocytic and cytotoxic cells and converted to leucyl leucine methyl ester, which is toxic. It has also been reported that CD8 T-cells may be depleted by LME treatment [52].

2.3 Selection of T-Cells

Before the ready availability of mAb's directed against lineage-specific markers, the method of choice for isolating T-cells was rosetting with sheep red blood cells [53]. Currently, hybridomas are available from the American Type Culture Collection that produce mAb directed against most of the lineage markers needed for positive or negative selection of T-cells or T-cell subsets (Table I). Commercial sources of purified mAb suitable for cell separation can be found in the Linscott's directory [54]. Depending on the cocktail of antibodies, cells may be negatively selected (all cell types other than the cell desired are bound by Ab and removed) or positively selected. Single antibodies or combinations of antibodies can be coated onto a substrate such as plastic or magnetic beads which are used to deplete or select the population of interest [55]. Alternatively, cells may be coated with antibodies and then with magnetic beads prior to selection or depletion. One concern with positive selection is that Ab bound to a substrate and used to separate cells may also as an unintended consequence activate or inhibit the cell. Negative selection results in a cell population that is not affected by Ab bound to the cell surface, but additional beads or substrate may be required to remove unwanted cells when the population of interest is scarce. This may drive up the cost of separation depending on the method used. Plastic petri dishes have been used as a substrate to bind Abs to select cells or deplete cells [56]. The efficiency with which different Abs bind to plastic can vary, and thus will affect the cell yield. This may be improved by first coating the plastic surface with an anti-mouse Ab. Another disadvantage of this panning technique is that the separation/adherence of large numbers of cells becomes cumbersome, as the surface area for these separations can be quite large. For cell separation from an entire apheresis product, the use of beads or particles as a substrate for Ab to deplete or select cells is more readily applicable to large-and small-scale culture.

A number of commercial suppliers sell reagents for the purification of T-cells. StemCell Technologies (Vancouver, B.C.) provides reagents for negative selection of T-cells and T-cell subsets. Undesired cells can be removed from cell preparations or from whole blood. The unwanted cells are bound with antibody conjugated with dextran, which binds the magnetic colloid added after incubation. The cell mixture with antibodies and colloid is then loaded on a column packed with iron mesh that is bracketed with a

Table 1. Hybridomas producing mAb directed at human immune cell surface markers, available from the ATCC

	Target cell	Clone[a]
CD4	T-helper	OKT4
CD8	T-cytotoxic	51.1
CD11b	Granulocytes, monocytes, NK cells, macrophages	OKM1
CD14	Monocytes, granulocytes, macrophages	63D3
CD16	NK cells, granulocytes, macrophages, monocytes	N/A[b]
CD20	B-cells	IF5
CD21	B-cells	THB-5
HLA-DR	Macrophages, monocytes, activated T-cells	2.06

[a]additional clones may be available, check the ATCC web page (http://www.atcc.org) for updated catalogue information
[b]currently not available

magnet, and cells unbound by antibody pass through the column [57]. This system can also be used for positive selection. Miltenyi Biotech (Auburn, CA) also sells positive and negative T-cell selection reagents. Antibodies are either directly conjugated to paramagnetic microbeads 100-150 nm in diameter for positive selection, or unconjugated antibodies are bound to cells targeted for negative selection and removed using a secondary mAb directly conjugated to microbeads [58]. Both of these systems require purchase of antibodies, cell separation columns filled with iron mesh and magnets designed specifically for the cell separation columns. If positive selection is desired, particles or beads may remain attached to cells. These manufacturers state that this should not affect cell function, however investigators should determine this in their own laboratory for their specific application. Another consideration for the clinical researcher is that positively selected cells with beads or colloid bound to the cell surface may not be suitable for adoptive immunotherapy. PerSeptive Diagnostics (Cambridge, MA) sells silanized iron oxide particles 0.5 – 1.5 μM in diameter conjugated with CD2, CD3, CD4 or CD8 for positive selection of T-cells. Dynal (Dynal, Lake Success, NY) sells beads coated with goat or sheep anti-mouse antibodies or cell lineage markers that enable negative and positive selection of T-cells. These beads are paramagnetic, round, 4.5 μM in diameter and consist of iron coated with polystyrene. The sheep anti-mouse beads are also available in preparations suitable for use in clinical trials. Because Dynal beads are much larger than particles sold by StemCell Technologies, Miltenyi, and PerSeptive Diagnostics, their magnetic susceptibility is also much larger and the beads move to magnets in seconds to minutes. Rare earth cobalt magnets available from a variety of scientific suppliers may be used for cell separation as well as the Dynal magnets. Dynal beads conjugated with mAb for positive selection can be detached from the selected cells by use of a reagent containing Ab directed against the Ag binding region of the Ab (Fab) conjugated to the magnetic bead. This is a competitive reaction between the Ag on the surface of the target cells and the anti-Fab Ab, which has been reported to preserve phenotype and function in positively selected $CD8^+$ T-cells [59].

Our laboratory has developed a negative selection protocol for the separation of T-cells and T-cell subsets from MNC following ficoll or lymphocytes following percoll. This method utilizes mAb directed against CD11b, CD14, CD16, CD20 and HLA-DR to remove all cells except $CD28^+$ T-cells. $CD4^+$, $CD8^+$, $CD45RA^+$, or $CD45RO^+$ T-cells can also be obtained by adding anti-CD8, anti-CD4, anti-CD45RO, or anti-CD45RA respectively to the cocktail of mAb [60]. Each Ab must be titered before use as there may be lot-to-lot variability, particularly with mAb purified from ascites fluid or culture supernatants. Once titered, the Ab cocktail can be aliquotted and frozen. The cells targeted for removal are coated with Ab cocktail at 4 °C for 30 min followed by centrifugation washes to remove excess Ab. Dynal goat anti-mouse coated magnetic beads are then added to the cells and allowed to bind to the targeted cells coated with mouse Ab. The percent of cells targeted for removal is based upon an estimation of the frequency of T-cells or T-cell subsets in the starting cell preparation. The number of beads needed per target cell should also be titrated. One disadvantage of this technique for negative selection of large numbers of cells is that the cost of Dynal beads may be prohibitive.

An alternative to magnetic selection or depletion is the use of dense particles conjugated to mAb that can then be separated by gravity or centrifugation over a gradient medium (1.077 gm/ml). Zwerner et al. have published a cell separation method using mAb conjugated to nickel particles which settle by gravity within 4 min [61]. These particles are irregular and have a large surface area that improves the efficiency of separation. As of this writing, these particles are not yet commercially available. Patel et al. [62] have used antibody-coated dense polystyrene beads to obtain T-cells. Porous polystyrene beads were modified with colloidal gold and then coated with antibody. These modified beads could then be incubated with a MNC population during which the selected cells bound to the beads. Bead-bound cells could be isolated by centrifugation with a density gradient medium. Either of these methods could be a less expensive option compared to magnetic beads. Polystyrene or magnetic beads can be purchased from a number of scientific suppliers (see http://biomednet.com/). For all of the methods described above, the efficiency and purity of the separation should be monitored by flow cytometry. This is of particular importance because small subsets of cells other than the desired population which contaminate a cell culture can cloud the interpretation of studies designed to measure small changes in function. In addition, interaction of T-cell subsets may alter the function of the population as a whole. Thus, quality control during separation is a critical element of successful T-cell culture.

2.4 Cryopreservation

Lymphocytes can be maintained for long periods of time at less than -120 °C in liquid nitrogen freezers or newer mechanical freezers. If the temperature is maintained, cells can be thawed and successfully cultured many years after the initial freezing. To prevent irreversible cell damage, it is necessary to first freeze cells in a medium containing a cryoprotectant to prevent the formation of ice crystals that will damage the cell. The current cryprotectant of choice is 10% dimethyl sulfoxide (DMSO) [63]. The freezing medium should also contain at least 10% protein, which can be provided by fetal bovine serum (FBS), human serum, or albumin formulations. Secondly, since the phase change from liquid to solid is an exothermic reaction, a freezing method designed to prevent a freeze, thaw and freeze of cells is needed to minimize cell damage. In a comparison between bulk freezing of samples in a mechanical freezer to a controlled rate freezing method, the controlled rate freezing of cells at 1 °C per minute resulted in a higher post-thaw viability [63]. For an excellent review of cryopreservation techniques, see [64]. One can expect a certain amount of cell loss in viability and recovery following thawing of lymphocytes. T-lymphocytes from patients who have undergone extensive chemotherapy and HIV^+ donors are more sensitive to the stresses of cryopreservation and thawing, perhaps due to a lowered threshold for apoptosis. Another concern is the loss of cell function or phenotype. The expression of CD2, CD3, CD4, and CD8 has been shown to remain stable up to three months following cryopreservation [65]. Other T-cell phenotypic and functional markers such as CD45RA and CD62L [66] are not as stable. The investigator should take care to evaluate cell phenotype and function on fresh T-cells compared to cryopreserved T-cells.

3. CULTURE TECHNIQUES

Primary T-lymphocytes isolated from healthy individuals are quiescent and do not spontaneously proliferate *in vitro*. Therefore, second to the basic constituents of culture media, T-cells require some sort of stimulation to enter the cell cycle and proliferate. For a more detailed description of the many ways to activate T-cells, see [67-69]. In this chapter, traditional methods of T-cell activation and those newer methods that show promise for wide application will be emphasized. This section will describe common culture media and vessels followed by methods for stimulation and maintenance of polyclonal T-cell populations, and the development and culture of Ag-specific T-cell lines. For an excellent reference on the classical techniques of tissue culture, see [70].

3.1 Media and Serum

There is a wide variety of defined culture media available [71,72] (see also supplier catalogues), however, the medium most favored for culture of T-cells is RPMI (Roswell Park Memorial Institute) 1640 [73]. Some laboratories supplement media with glutamine in order to replace glutamine that degrades over time. RPMI 1640 is sold with or without 2 mM glutamine and formulations supplemented with glutamine are guaranteed for up to a year, so addition of glutamine after the initial preparation of complete medium may not be necessary. Rapidly dividing T-cell cultures may benefit from the addition of a supplementary buffer, either 20 mM HEPES or 0.2% sodium bicarbonate, to maintain pH at 7.4. Phenol red in media should appear reddish orange in well-maintained cultures. A reddish purple indicates a more basic pH, perhaps due to an incubator with less than 5% CO_2, while a yellow color usually indicates acidic conditions due to acid metabolites produced by cells and indicates the need to feed and split the culture. All culture media should be protected from light when not in use, as riboflavin and tryptophan yield toxic byproducts when exposed to light [74]. Some refrigerators require modification to minimize exposure of media to light during storage.

RPMI 1640 must be supplemented with 5-10% serum, either autologous, pooled human or fetal bovine. Serum is the undefined aspect of tissue culture media, in that the precise formulation of specific proteins, hormones, or lipids necessary for cell growth remains unidentified. In one study, RPMI 1640 supplemented with IL-2 and transferrin was sufficient for delayed optimal proliferation of T-cells [75]. We have found that for polyclonal T-cells grown in RPMI 1640 and stimulated via CD3 and CD28, autologous serum provides the most robust growth, followed by pooled human serum and FBS (Levine et al., unpublished). Autologous serum may not always be available and, particularly in donors with an infectious disease such as HIV, the time needed for preparation and storage may make the use of pooled human serum more practical. Serum from patients with LGL leukemias or NK cell lymphomas may contain elevated levels of Fas ligand [76], which may induce cell death in activated T-cells expressing CD95 (Fas). For culture of Ag-specific cell lines, 1-3% human serum is preferred. Even though this is more expensive, it avoids presentation of bovine proteins as Ags by APC. If using pooled human or FBS, it is important for the investigator to establish a benchmark of performance such as proliferation or cytokine secretion, as the serum components vary from lot to lot. Lipopolysaccharide and xenogeneic protein present in FBS may cause background proliferation in mixed lymphocyte proliferation assays. Media supplemented with FBS should not be used for immunotherapy protocols, as recipients of cells grown in supplemented media developed anti-FBS Abs and exhibited arthus-like reactions upon subsequent infusions of cells [77,78]. Serum can also be a source of culture contamination by mycoplasma and viruses and interfere or enhance the maintenance of antigenic specificity [79] depending on the source and type of serum. In order to remove the variability inherent in different lots of serum, defined medium supplemented with insulin, transferrin and other hormones and lipids has been formulated in recent years which allows the culture of T-cells under serum-free conditions. Of the pre-made serum free media, X-VIVO 15 (BioWhittaker, Walkersville, MD), AIM-V (Gibco Life Technologies, Gaithersburg, MD) and EX-CELL 300 (JRH Biosciences, Lenexa, KS) have been developed for human T-cell culture. Some laboratories find it necessary to supplement serum-free media formulations with 1-3% human serum albumin or pooled human serum.

3.2 Culture Vessels and Large-Scale Culture

T-cells differ from most mammalian cells in that they do not require a surface to adhere to *in vitro*, but instead grow in suspension. For long-term culture, a density of 0.5 - 2 x 10^6 cells per ml is preferable. T-cells may be cultured at a lower density during the development of Ag-specific T-cells (see below) or at a slightly higher density if the culture medium is buffered appropriately or changed often. As for density per cm^2 of growth surface, see Table 2 for rough guidelines for seeding polyclonal T-cells. For Ag-specific cells or T-cell clones, round bottom wells are frequently used to increase cell-to-cell contact. A higher cell concentration may be maintained for Ag-specific T-cells if the cultures are refed often. Evaporation from plates is a concern with long-term cultures and care should be taken to maintain sufficient humidity levels. To avoid large fluctuations in pH and cell density, T-cells should be counted and fed with additional medium every 1 to 3 days, depending on the rate of cell growth.

Table 2. Recommended number of T-cells to seed in tissue culture plates or flasks

Culture vessel	T-cell number to seed
48-well plate (1 well)	$0.5\text{-}1 \times 10^6$
24-well plate (1 well)	$1\text{-}2 \times 10^6$
12-well plate (1 well)	$2\text{-}4 \times 10^6$
6-well plate (1 well)	$5\text{-}10 \times 10^6$
T-25 (standing up)	$5\text{-}10 \times 10^6$
T-75 (standing up)	$10\text{-}25 \times 10^6$
T-75 (lying down)	$25\text{-}50 \times 10^6$
T-175 (standing up)	$50\text{-}100 \times 10^6$
T-175 (lying down)	$100\text{-}200 \times 10^6$

In addition to the plates and flasks listed in Table 2, large numbers of T-cells can be cultured in roller bottles. The disadvantage of roller bottles is that they take up a lot of space. Designed to conserve incubator space, the Nunc cell factory is composed of multiple stacking chambers to increase surface area. However, chamber dimensions limit the amount of medium per cm^2 and for high densities of T-cells, medium should be changed often. A novel culture system developed by Knazek et al. [80] is the hollow fiber bioreactor. The device consists of a medium reservoir bottle, a peristaltic pump and gas permeable tubing connected to hollow fibers encased in a cartridge. Cells are seeded into the cartridge and medium is constantly pumped through the hollow fibers, which allows for very high cell densities (5×10^8 cells/ml). The monitoring of glucose or lactate concentration determines when fresh medium must be added or when cells must be harvested from the device. Tumor infiltrating lymphocytes (TIL) have been reported to expand 100-fold in 20 days and maintain a viability of 91% [80]. This system, currently available from Spectrum Laboratories (Laguna Hills, CA), may provide a way to culture very high numbers of T-cells ($>3 \times 10^{10}$) suitable for adoptive immunotherapy [81]. However, hollow fiber cartridges may not be suitable for Ag-specific T-cells where contact with dendritic cells (DC) or other APC is necessary. In addition, cell viability may become suboptimal if medium exhaustion occurs or if the population of T-cells is not amenable to culture at such high densities. Aastrom Biosciences (Ann Arbor, MI) is testing a self-contained cartridge system that employs continuous radial medium flow for large-scale, automated T-cell and progenitor cell growth (see Chapter 1). Finally, for investigators with a larger budget, a self-contained lymphocyte cell culture device including incubator and filtration system for harvesting supernatant has been developed for deployment aboard spacecraft to study the effects of microgravity [82].

An alternative culture vessel suitable for large numbers of T-cells and utilized in many immunotherapy protocols is the gas permeable bag. Polyolefin bags were first used to culture lymphokine activated killer (LAK) cells in a closed system [83]. A closed system provides benefits over flasks or roller bottles because of the reduced risk of contamination of the culture, as well as, with T-cell cultures from HIV^+ donors [84], reduced biohazard risk to the culture technician. Gas permeable bags are available from Nexell Therapeutics (formerly Baxter Immunotherapy, Irvine, CA), Ethox (Buffalo, NY), American Fluoroseal (Columbia, MD) and TC Tech (Minneapolis, MN). In order to maintain a closed sterile fluid path when feeding or harvesting cells, these bags have a combination of luer lock connectors, or other type of locking connectors, and spike receptacle ports. In addition, the tubing connected to the ports on the bags can be sterile welded to tubing connected to a medium bag or an empty flask by the use of an FDA-approved sterile connect device (Terumo Medical Corporation, Somerset, NJ). One variation on these bags is a dual bag system where cells are loaded into an inner bag, while medium is loaded into an outer bag [85]. To date, gas permeable bags have been used in a number of immunotherapy protocols [86-89]. The disadvantage of these bags for large-scale culture is the amount of medium necessary to maintain the cells. With the expansion in immunotherapy and gene therapy protocols predicted in the future, improved large-scale culture systems in development may provide advantages over current systems.

3.3 Polyclonal Activation and Culture of T-Cells

The growth of polyclonal cells as defined here is accomplished by stimulation with any reagent that does not result in the selective stimulation or survival of an Ag-specific clone of cells. Some reagents, such as lectins, pharmacologic activators and anti-CD3 Abs plus anti-co-stimulatory receptor Abs, are able to stimulate T-cells and maintain a population in culture representative of the starting population. Superantigens bind to one or more TCR Vβ families to initiate the expansion of those T-cells. For most of the culture methods discussed below, stimulation of peripheral blood MNC (PB MNC) or peripheral blood lymphocytes (PBL) will eventually favor the numerical expansion of T-cells over B-cells or monocytes. However, the greater the proportion of contaminating cells, the greater the decline in total cell number during the initial days of culture until a relatively pure population of T-cells is achieved. The advantage of purifying T-cells or their subsets before the initiation of culture is that monocytes may produce cytokines such as IL-10 that are inhibitory under certain culture conditions [90], even while having a stimulatory effect under other conditions [91]. Regardless of the starting population, it is important to phenotype the cells prior to the initiation of culture and at intervals following stimulation.

3.3.1 Lectins

Lectins are proteins or glycoproteins that bind to specific glycoproteins or glycolipids on the cell surface and have been used to stimulate lymphocytes for 40 years [30]. While lectins can be isolated from a variety of organisms, the lectins mitogenic for T-cells, Con A, PHA, and PWM are all plant extracts [92]. PHA and PWM can stimulate both B- and T-cells, while Con A is specific for T-cells. PHA has been found to bind the TCR while Con A binds to the CD3 complex. Signaling of T-cells by lectins is recognized to mimic many of the same pathways initiated by TCR engagement of the Ag-MHC complex [93]. Lectins can induce the transit from G_0 to G_1 of the cell cycle in T-cells, but a second signal is needed to induce proliferation. Accessory cell interaction or exogenous IL-2 can provide this second signal [15,94]. Very pure populations of T-cells (>95%) may require the addition of IL-2 to lectins in order to proliferate [93]. Lectins must be added only at the initiation of a T-cell culture, while IL-2 should be added with each subsequent addition of fresh medium. Irradiated autologous or allogeneic feeder cells may also be added at intervals to maintain a long-term culture. While long-term culture of T-cells is possible with mixtures of feeder cells providing co-stimulation, lectins and IL-2, this approach does not yield polyclonal populations of cells [95]. Con A, PHA and PWM will stimulate T-cells at 1-10 μg/ml, although this may vary depending on the population of cells and other culture conditions. Lectins can be reconstituted in PBS or medium, aliquotted and frozen for future use. Some lectins, in particular Con A, are not stable after reconstitution and freezing.

3.3.2 Membrane bypass or pharmacologic activation

While lectins depend on interaction at the cell membrane to transmit mitogenic signals, pharmacologic activation by phorbol esters, bryostatins and calcium ionophores bypass the membrane to directly activate intracellular signaling pathways [96]. Both phorbol esters [97,98] and bryostatins [99] bind to and activate protein kinase C (PKC), a cyclic nucleotide-independent protein kinase. A combination of diacylglycerol and the release of Ca^{++} from intracellular stores activate PKC following TCR stimulation [15,100]. Phorbol esters or bryostatins by themselves upregulate IL-2R, but require a second signal, either exogenous IL-2 or calcium ionophore, for T-cell proliferation [101,102]. PMA (12-*O*-tetradecanoylphorbol-13-acetate) and the water soluble PDBu (phorbol 12,13 dibutyrate) are the most commonly used phorbol esters to stimulate T-cells, and bryostatin-1 is the most commonly used isoform of bryostatin. Interestingly, while phorbol esters are tumor promoters, bryostatin possesses anti-neoplastic activities [103,104]. PMA and bryostatin are stimulatory at very low doses, in the range of 5-50 nM with the addition of exogenous IL-2. Alternatively, phorbol esters or bryostatin may be combined simultaneously with calcium ionophore (A23817 or ionomycin). As with phorbol ester, in the presence of accessory cells, ionomycin triggers T-cells to proliferate [105]. For purified (>95%) T-cells or CD4$^+$ T-cells, 1.9 nM PMA plus 0.08 μg/ml ionomycin may be used as effective stimulatory concentrations [29]. While these pharmacologic activators can be used to culture T-cells for at least two weeks, the potential for extended polyclonal expansion is limited. In addition, while pharmacologic activation can mimic the signaling pathways following TCR stimulation, the availability of purified recombinant cytokines and Ab directed against stimulatory cell surface receptors allows a more (but not completely) physiologic T-cell activation and growth.

3.3.3 Cytokines

As discussed briefly in section 1.3 earlier, the addition of exogenous IL-2 can enable T-cell growth in concert with other stimuli. Since IL-2R are not expressed on resting cells [106], an additional stimulus must be added with IL-2 that upregulates IL-2R. IL-2 can be used in concert with lectins, phorbol esters, other cytokines as discussed in this section, or anti-CD3 Abs as discussed below, to culture T-cells for several weeks. With these reagents, IL-2 added at 100 IU/ml is a sufficient dose. In some systems, IL-2 alone may be used to culture T-cells. For example, when TIL are cultured in very high dose (1000 U/ml) IL-2, a costimulus is not necessary for CD8$^+$ T cell expansion [107]. Even low levels of IL-2R present on TIL are sufficient to initiate proliferation and autocrine upregulation of IL-2R. Proliferation can be induced in resting lymphocytes for several days after the addition of very high doses of IL-2 [108], but this generally cannot be sustained in long-term culture when starting with resting T-cells. Combinations of the cytokines IL-2 and IL-4 have been used to stimulate neonatal CD4$^+$ T-cells and these cells can be restimulated with PMA and ionomycin [109]. IL-2, IL-4 and IL-6 supplementation can be used to culture CTL's in concert with irradiated tumor cells [110]. CTL activity was maintained for over a month in this system. In a similar system, PB MNC cultured with irradiated melanoma cells and mixtures of IL-2, IL-4 and IL-6 could be cultured for over 3 months and maintain CTL activity [111]. Finally, combinations of cytokines may be used to induce proliferation of naive T-cells [112]. A limitation of this approach is that only about 3% of the input T-cells proliferate in the cytokine only approach.

IL-7, produced by BM stromal cells, supports the growth of T-lymphocyte progenitors as well as mature T-cells. IL-7 alone at 30 ng/ml has recently been shown to support vigorous growth of $CD4^+CD45RA^+$ cells from CB, while adult $CD4^+CD45RA^+$ cells were marginally responsive [113,114]. IL-7 may serve to complement other cytokines or other stimuli in culturing T-cells. IL-7 added to peptide-generated CTL enhanced specific lysis and increased the percentage of specific CTL in the total population [115]. IL-12 can also act as a growth factor for T-cells [116] and, in concert with IL-7 and anti-CD3, could augment the development of CTL activity [117]. IL-12 can synergize with other methods of stimulation, including CD28 stimulation and mitogen stimulation at doses much lower than IL-2 [118]. IL-15 and IL-16 can also supplement IL-2 to induce T-cell proliferation. IL-2 plus IL-16 can induce long-term proliferation of $CD4^+$ T-cells [119]. Allospecific $CD8^+$ CTLs can be generated by cytokine co-culture with irradiated stimulator cells. If the $CD8^+$ cells were co-cultured with either IL-12 or TGF-β, CTLs secreting IL-2 and IFN-γ (Tc1) could be obtained, while if $CD8^+$ cells were co-cultured with IL-4, CTLs secreting IL-4, IL-5, and IL-10 (Tc2) could be obtained [5]. Cytokines alone, or in concert with Ag stimulation, should prove useful in T-cell culture and hold promise for the development of improved culture methods with clinical applications.

3.3.4 Surface receptor co-stimulation

Receptors and counter-receptors present on the surface of T-cells which are involved in intercellular communication can provide another method for stimulation and culture of T-cells, as Abs directed against stimulatory receptors can substitute for natural ligand. The signal delivered by Ag-MHC to the TCR can be mimicked by anti-TCR or anti-CD3 complex Abs. This type of stimulation is sufficient to induce the expression of the IL-2R, CD69, and low levels of proliferation. However, anti-CD3 stimulation without the addition of IL-2 or other costimulus is not sufficient for full activation of the T-cell and long-term growth [106]. IL-2 added as a costimulus to anti-CD3 mAb results in an effective method for culture of T-cells, although in a mixed population of $CD4^+$ and $CD8^+$ T-cells, high dose IL-2 will tend to favor the predominance of $CD8^+$ T-cells [120-122]. The level of proliferation induced by anti-CD3 and other stimulatory Abs and the amount of cytokines produced varies depending on the method of presentation. Soluble Ab may elicit T-cell proliferation if there are Fc-receptor bearing cells that can bind and present the Ab to T-cells. More effective than soluble Ab is the immobilization of Ab by coating on plastic or beads.

If anti-CD3 immobilized on beads is combined with co-stimulation of CD28 delivered by B7-1 or B7-2, long-term autocrine growth of $CD4^+$ T-cells is possible [27]. When mAb directed against the co-stimulatory molecule CD28 provides signal 2, an even more vigorous and durable polyclonal proliferation and cytokine secretion is seen (see Fig. 3A) [29]. However, the method of presentation of anti-CD3 and anti-CD28 is critically important. Our laboratory has uncovered striking differences when 'cis' vs. 'trans' presentation of signal 1 and signal 2 using mAb directed against CD3 and CD28 coupled to magnetic beads is employed. We have observed higher levels of cytokine and chemokine secretion delivered by anti-CD3 and anti-CD28 immobilized on the same bead (cis) compared to anti-CD3 and anti-CD28 immobilized on separate beads or soluble mAb (trans) (Table 3). This confirms previous cell-based studies of co-stimulation showing that co-stimulation in cis is more efficient than in trans [123]. Cis presentation of anti-CD3 and anti-CD28 mAb on beads in the absence of exogenous cytokines or feeder cells enables a 10^9- to 10^{11}-fold polyclonal expansion of adult $CD4^+$ T-cells [29]. It is important to note that cell growth will continue past this threshold level of numerical expansion. However, with further culture the antigen receptor repertoire will become restricted with oligoclonal outgrowth. Fresh beads can be added to restimulate cells. Unseparated T-cells can be cultured in this manner as well as purified $CD4^+$ or $CD8^+$ T-cells. This system, in contrast to methods in which feeder cells provide co-stimulation, bypasses the negative signal delivered by CTLA4 through specific anti-CD28 mAb. CD3/CD28 stimulation by immobilized mAb enables the growth of $CD4^+$ T-cells from HIV^+ donors without the addition of anti-retroviral agents to the culture medium [124]. The viral load in these HIV^+ $CD4^+$ T-cells declines during culture due to a CD28-induced down-regulation of CCR5, a coreceptor for HIV-1 [125], and provides a method for adoptive immunotherapy and gene therapy for HIV [126]. Bispecific Ab for CD3 or CD28 and tumor antigens has been explored as a method for immunotherapy of Hodgkin's disease and ovarian cancer [127,128].

Anti-CD2 mAb has also been used in concert with anti-CD28 mAb to stimulate T-cell clones and enable autocrine proliferation [129]. In fact, combinations of Ab directed at various cell surface receptors can propagate T-cells, although most require the addition of exogenous IL-2 for longer-term expansions. Anti-CD3 plus anti-CD2 or anti-CD4 or anti-CD5

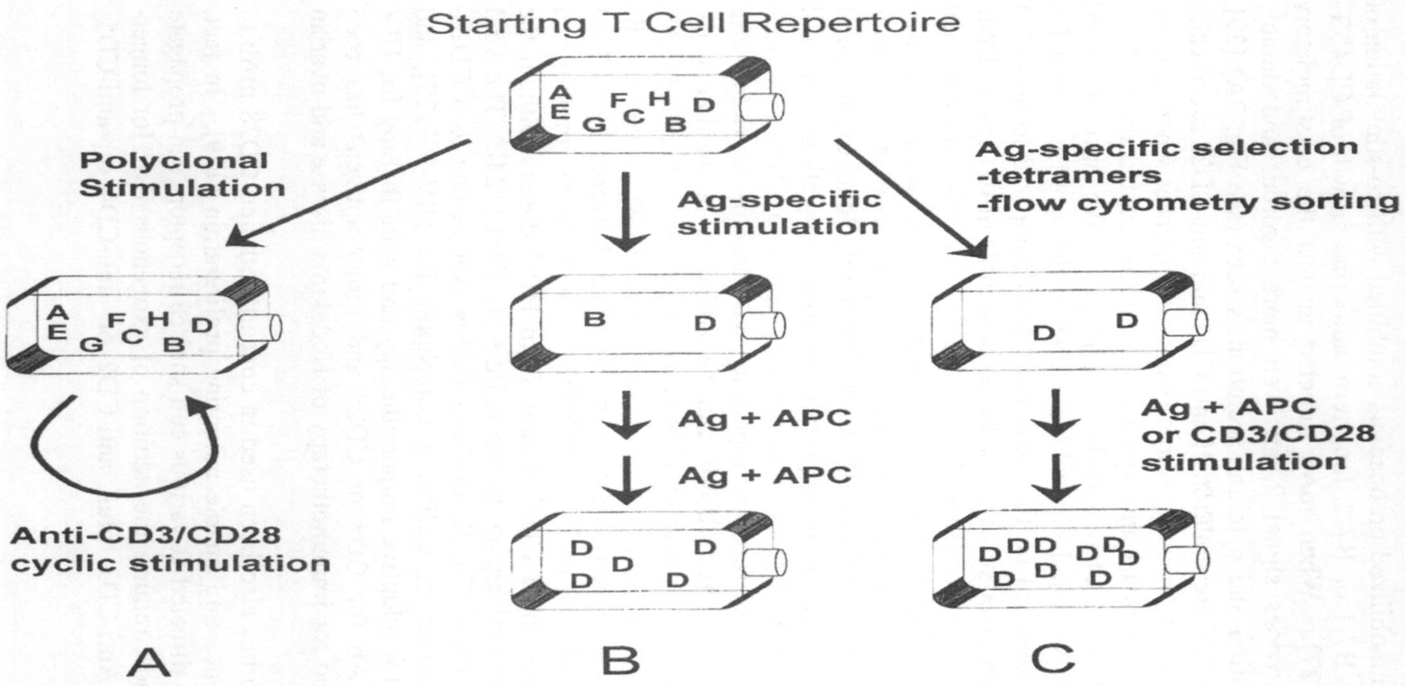

Figure 3. Methods of polyclonal and antigen-specific T-cell propagation. A) Polyclonal stimulation via CD3/CD28 stimulation or other methods maintains the TCR repertoire of the starting population. B) Ag-specific T-cell culture without prior selection or enrichment requires several rounds of stimulation. C) Selection of Ag-specific cells via methods discussed in section 4.4 can improve the efficiency of Ag-specific T-cell generation and several methods, such as cyclic anti-CD3/CD28 stimulation, can be used for *ex vivo* expansion of these Ag-specific cells.

immobilized on beads can effectively be used to culture CD4$^+$ T-cells [28], as can anti-CD3 plus anti-CD7 (Riley et al., unpublished). However, for each of these conditions, the level of chemokines produced upon stimulation is not as high as with CD3/CD28 stimulation. 4-1BB (CD137) is a member of the TNF family of receptors that has been shown in mouse to costimulate T-cells and induce CD8$^+$ T-cell proliferation [130,131]. In human T-cells, 4-1BB has recently been shown to be a complement to CD28 in regulating IL-2 and IFN-γ production [131]. Thus, 4-1BB may serve to improve the efficiency of T-cell culture via CD3/CD28 stimulation, especially in a population of T-cells with a significant percentage of CD28$^-$ cells. T-cell stimulation by a variety of other surface receptors, including CD11a/CD18, CD27, and CD47 has been reported [132-134]. Future studies will aid in the characterization of these signals and may help in the development of improved polyclonal stimulation of T-cells.

Table 3. Cytokine and chemokine production from CD4$^+$ T-cells 24hrs after stimulation with anti-CD3 plus anti-CD28 immobilized on separate beads ('trans') vs the same bead ('cis'). Cells were stimulated at 1 x 10^6 cells/ml. The cytokine secretion is shown in pg/ml.

Cytokine/Chemokine	'trans' stimulation	'cis' stimulation
IL-2	3,907	189,600
IFN-γ	1,246	28,070
TNF-α	2,362	29,353
GM-CSF	1,355	8,557
RANTES	420	693
MIP-1α	830	6,858
MIP-1β	3,200	16,967
IL-4	63	306
IL-10	2,118	5,067

3.3.5 Superantigens

Human superantigens (SAg) are a group of microbial proteins that differ in several respects from conventional Ag. The large numbers of T-cells stimulated by SAg initially suggested that these molecules may function as T-cell mitogens by non-specifically stimulating T-cells. In fact, the recognition of SAgs is mediated by the TCR and specifically depends on the Vβ chain [135]. MHC class II molecules present SAgs, but they do not require processing and bind outside the conventional peptide groove [136]. Between 5% and 30% of the T-cell repertoire may be stimulated by a particular SAg, depending on the relative frequency of the Vβ targeted. This contrasts with conventional Ag, for which the precursor frequency is usually much less than 1 in 1000 [137]. Similar to conventional Ag, SAg stimulation of T-cells requires autologous or MHC class II-matched APC. B-lymphoblastoid cell lines have often been used for SAg presentation [138]. Two of the microbes that produce SAg stimulatory for human cells are *Staphylococcus aureus* and *Streptococcus pyogenes*. Staphylococcal enterotoxins A through E (SEA, etc.), toxic shock syndrome toxin-1 (TSST-1) and exfoliative toxins (ExFT) are produced by *Staphylococcus aureus*, while the streptococcal pyrogenic exotoxins (SPE-A-C) are produced by *Streptococcus pyogenes* [137]. SEB induces the specific expansion of human T-cells bearing TCR Vβ3, Vβ12, Vβ14, Vβ15, Vβ17 and Vβ20, whereas TSST-1 specifically stimulates T-cells with TCR Vβ2 [135,137]. *In vitro* culture with SAg permits specific analysis of expanding T-cell populations belonging to specific TCR Vβ lineages.

3.4 Development of Ag-Specific T-Cell Lines *In Vitro*

In vivo, T-lymphocytes recognize protein Ag as peptides presented by MHC on the surface of APC and receive co-stimulation by CD80 or CD86. APC can present exogenous Ag following phagocytosis and processing of whole Ag in endocytic compartments into antigenic peptides. These peptides then become associated with MHC Class II molecules and are transported to the cell surface [139]. The CD4 molecule aids in this interaction by binding MHC class II. In contrast, endogenous antigens are degraded by intracellular proteases in the APC cytosol, processed in proteasomes into peptides, and then transported to the ER lumen to be associated with MHC Class I molecules and β2 microglobulin before being transferred to the plasma membrane [140]. Binding of the CD8 molecule to MHC class I strengthens the recognition by the MHC-peptide/TCR interaction. As a result of the distinct processing mechanisms utilized for the two classes of MHC molecules, exogenous or vacuolar Ag are generally presented by MHC class II molecules, whereas endogenous or cytosolic Ag are presented primarily by MHC class I molecules. Efforts to culture Ag-specific T-cells *in vitro* attempt to mimic *in vivo* interactions. *In vitro*, the establishment of memory is obtained by successive enrichment in Ag-specific T-cells by cycles of restimulation with Ag in the presence of APC. With regard to immune memory, it is not clear *in vivo* whether this is maintained by continuous stimulation of T-cells by persisting Ag or by a non-specific stimulus [141]. This section will first describe the different types of APC used in the culture of Ag-specific T-cells, followed by a discussion of the culture of Ag-specific CD4$^+$ and CD8$^+$ T-cell lines and clones.

3.4.1 Choice of Antigen Presenting Cells (APC)

To obtain Ag-specific T-cell lines, the first issue to consider is the appropriate APC, which depends on the type of Ag, the quantity of APC available, and the quality of the APC needed. The quality of the APC depends on a number of factors, including levels of MHC molecules and co-stimulation molecules expressed and the cytokines secreted. The potency of a particular type of APC determines the APC:T-cell ratio needed for T-cell activation and growth. Macrophages are especially appropriate for antigen-processing studies using bacterial or particulate Ag because of their ability to phagocytose Ag. Large numbers of B-cells that can serve as APC can be obtained by activation with LPS, CD40L, or transformation by Epstein-Barr virus (EBV). Dendritic cells (DC) are the most potent APC as a result of the efficiency of Ag processing and high levels of co-stimulatory molecules expressed. Furthermore, DC are unique because they are the only APC that can initiate *in vivo* or *in vitro* an Ag-specific response from naive T-cell precursors [142]. To present a protein Ag to a T-cell, APC should be autologous or MHC-matched. In most protocols, APC are irradiated (3000 to 5000 Rad) or treated by mitomycin C to prevent their growth and competition with T-cells in culture. A large variety of other cell types such as fibroblasts may be utilized, but usually are only suitable to restimulate Ag-specific T-cells. The main obstacles to the use of these 'non-professional' APC are the general lack of MHC class II and co-stimulatory molecule expression and their relatively poor Ag-processing capacity. Soluble MHC class I and II molecules are being developed, that when associated with a peptide, can be used as 'artificial' APC.

Macrophages can be easily obtained by cultivating human peripheral blood-derived monocytes with G-CSF or GM-CSF. After ficoll, monocytes are isolated from PB MNC by plastic adherence (although such preparation may also contain rare DC). Adherence presents a technical advantage, as washing out Ag is easily performed without centrifugation, although it is difficult to recover the cells after culture to use in further experiments. EDTA from 0.2 to 1 mM in PBS can be used to detach the cells, whereas trypsin may cleave surface molecules important in adhesion or co-stimulation. Macrophages can be used at an APC:T-cell ratio of 1:5 to 1:1, depending on the specificity of the T-cell line. They do not express high levels of MHC class II unless activated by IFN-γ [143]. Macrophages are also highly efficient in internalizing and processing soluble proteins, and can be infected by various bacteria or viruses. Monocytes are not very efficient in taking up Ag and express low levels of MHC and co-stimulatory molecules. Monocytes obtained after centrifugation over percoll can be utilized at an APC:T-cell ratio of 1:1 fresh, or after cryopreservation. PB MNC include monocytes, macrophages, B-cells, and DC and thus perform better than monocytes alone. The advantage of PB MNC is that no additional separation is needed after ficoll separation. However, because of the low percentage of DC, they are only efficient at high APC:T-cell ratios of 1:1 to 5:1.

Human EBV-transformed B-lymphoblastoid cell lines (LCL), are readily available by culturing autologous PB MNC with infectious EBV supernatant obtained from the B95-8 marmoset cell line (ATCC # CRL-1612, ATCC, Manassas, VA). Culture is done in the presence of cyclosporin A at 1 μg/ml, which suppresses T-cell growth, for three weeks [144]. LCL express high levels of MHC class II and can either be irradiated or fixed with paraformaldehyde, and used at an APC:T-cell ratio of 1:5 to 1:1 for T-cell stimulation. Classically, LCL are utilized to present peptides to T-cells as they are relatively inefficient at processing soluble Ag. However, for Ag-specific B-cells, enhanced processing is observed because specific Ag uptake can be mediated by surface immunoglobulin. With the discovery that CD40 ligand (CD40L) on activated T-cells can activate B-cells and DC [145], a new method of B-cell expansion has been developed [146]. PB MNC are cocultured with murine fibroblasts transfected with the human CD40L, in the presence of IL-4 and cyclosporin A, and centrifuged over ficoll before use in functional assays.

DC are highly potent, specialized APC that are relatively rare in peripheral human blood (<0.2%). They can either be obtained from PB MNC in a complicated isolation procedure which yields relatively low numbers of cells [147-149], or differentiated from blood monocytes by culture in the presence of IL-4 and GM-CSF, followed by a maturation step with monocyte-conditioned medium [150,151]. Alternatively, TNF-α [152,153] or soluble CD40L [154,155] may be used in this maturation step. DC can also be derived from their precursors, $CD34^+$ hematopoietic progenitors, by culturing BM samples with c-kit ligand, IL-4, GM-CSF, IL-6 and TNF-α (see Chapter 7) [156]. DC can be pulsed with a soluble or particulate Ag when they are immature and remain highly potent for endocytosis and processing Ag. After the second culture period, the mature DC are less competent to uptake and process Ag, but are now highly efficient in activating T-cells because of high levels of MHC and co-stimulatory molecules expressed on their surface. In addition, mature DC are able to secrete cytokines like IL-12, IL-1β, or IL-7 that promote T-cell expansion [157]. DC are extremely potent in stimulating an allogeneic or a recall T-cell response, and can be used at APC to T-cell ratios as low as 1:50 to 1:1000. In addition, DC can induce primary proliferation to Ag from naive precursors. The disadvantage of using DC as APC is the lengthy time necessary for culture prior to mixture with T-cells, and the low yield in the differentiation/maturation steps. DC may also be more difficult to obtain and purify in sufficient quantities for the large-scale expansion of Ag-specific T-cells.

Several teams have developed soluble complexes with specific peptides linked to their restricted soluble MHC class I or class II molecules [158,159]. These complexes can provide signal 1, and if associated with anti-CD28 mAb, are capable of efficiently activating peptide-specific cell lines [160]. In addition, these complexes can be used to enrich Ag-specific T-cells [161]. Finally, if covalently linked to biotinylated sites for fluorochrome binding, they can be used to sort Ag-specific T-cells by FACS (see section 3.4.4).

3.4.2 Expansion of Ag-specific $CD4^+$ T-cell lines

Techniques for developing Ag-specific human $CD4^+$ T-cells have used initial stimulation with Ag, APC and IL-2, followed by restimulation to maintain growth and Ag specificity. Ag can be loaded on APC in the form of a soluble protein, which after uptake will be processed into peptides and associated with MHC class II molecules, or directly as a peptide which will bind spontaneously to its restricted MHC class II molecules. The traditional method of culture is based upon multiple sequential cycles of discontinuous stimulation followed by rest in the absence of antigen. PB MNC or purified $CD4^+$ T-lymphocytes are incubated with irradiated autologous APC in the presence of the soluble antigen for 4-7 days. After ficoll, T-cells are rested for seven days in the presence of feeder cells. Thereafter, three successive cycles of stimulation in the presence of antigen and APC are necessary to obtain Ag-specific $CD4^+$ T-cells (see Fig. 3B).

However, Ag-specific $CD4^+$ T-cells are now widely obtained using continuous stimulation, as this reduces the culture time *in vivo* to develop Ag specificity. In this protocol, PB MNC are stimulated with autologous APC and Ag for 4-7 days. After ficoll, T-cells are restimulated with APC and Ag for 4 days. Culture medium supplemented with IL-2 (20 to 100 IU/ml) is added at day 4 and 6. After a third round of stimulation, Ag-specific $CD4^+$ T-cell lines are evaluated for specificity in a T-cell proliferation assay or in an IL-2 production assay using APC pulsed with Ag. Cell lines can be further amplified by restimulation every two weeks with APC and Ag, and addition of fresh medium with IL-2 at day 4 and every three days thereafter. A rapid expansion of Ag-specific cell lines has been achieved by using a mix of anti-CD3 mAb and autologous EBV transformed cell lines together with PB MNC as APC, and Ag [162].

3.4.3 Expansion of Ag-specific CD8$^+$ T-cell lines

Immunotherapy with *ex vivo* expanded CD8$^+$ T-cells is being developed to treat a wide range of diseases including malignancies and viral infections. The identification of tumor Ags and viral Ags has permitted the expansion of Ag-specific CTL which can specifically lyse transformed or virally infected target cells. For treatment, polyclonal T-cell lines are preferred to CTL clones because they are likely to minimize the emergence of escape mutants to one epitope and are less costly to develop. However, CTL clones are a powerful tool to identify the precise sequence of tumor or viral epitopes, and by virtue of their specificity, are the least likely cell preparations to trigger adverse effects such as autoimmunity.

Strategies to develop Ag-specific CD8$^+$ T-cells involve cyclical activation of responder T-cells with MHC class I stimulator cells or APC [163,164]. To obtain MHC class I presentation, the antigen must be introduced into the cytosol in order to generate endogenously processed epitopes. The autologous APC can be either (i) infected with recombinant viral vectors like vaccinia virus or adenovirus modified to encode the antigen [165,166], (ii) transduced with a retroviral vector, (iii) transfected with RNA or DNA encoding the antigen using lipid complexes [167], (iv) injected with DNA using a gene gun [168], or (v) fused with a tumor cell or infected cell [169]. More classically, APC are pulsed with immunogenic peptides (the processing is then by-passed), but the sequence of the peptide has to be identified, as well as its class I restricted molecule [170]. Autologous LCL are generally used when the antigen is introduced by infection, transduction or transfection, whereas autologous DC are the APC of choice to be pulsed with peptides. For allogeneic specific CD8$^+$ T-cell lines, the APC should be obtained from a MHC class I-mismatched donor (the foreign histocompatibility antigens are then part of the APC), and the culture is called a mixed lymphocyte reaction (MLR).

Classically, purified CD8$^+$ T-cells, or PBL, or PB MNC (it is not necessary to remove the CD4$^+$ T-cells in most of the systems), are stimulated with autologous irradiated APC preloaded with peptide in the presence of IL-2 [171]. T-lymphocytes are restimulated on a weekly schedule with autologous pulsed APC. CTL activity or cytokine secretion is evaluated after three cycles of stimulation. Alternatively, T-cell lines derived from PB MNC stimulated by mitogens and IL-2 for 3-4 weeks are then exposed to autologous irradiated peptide-pulsed APC, and assessed for CTL activity after ten days [172].

Anti-CD3, with or without anti-CD28, added to Ag and APC can optimize the cloning efficiency and propagation of primed Ag-specific CD4$^+$ or CD8$^+$ T-cells [162]. Anti-CD3 mAb alone has been useful to expand T-cell lines when Ag quantity is limited or when virus-infected APC may have immunosuppressive effects [162]. However, specific killing was reduced in an allogeneic system if anti-CD3 mAb was not washed out of the culture after the initial 48 hours of stimulation [162].

3.4.4 T-cell clones

A T-cell clone is a population of T-cells with identical specificity because they have been derived from the same original cell. T-cell clones can be derived from naive (unprimed) lymphoid cell populations, but their precursor frequency is generally very low. For better efficiency, the frequency of cells reacting with Ag can be increased prior to cloning by sensitization *in vivo*, or by culturing cells with the relevant Ag in short-term *in vitro* culture. PB MNC can be utilized as a starting population because they already contain APC at a natural ratio. Ag is added for 5 days to medium supplemented with human serum, permitting Ag-specific precursor enrichment. Ficoll is then used to enrich T-cell blast populations [173]. These cells can then be plated at limiting dilution at concentrations of 0.3 cells/well, 1 cell/well and 3 cells/well and the likelihood of a clone growing in an individual well can be calculated based on the Poisson distribution [174]. Ag and APC are added to the wells at a high APC:T-cell ratio. Every 3-4 days for two weeks, half of the medium is replaced with IL-2 (30 to 100 IU/ml) and medium supplemented with human serum. If around 37% of the wells are growing at the lower initial cell density, one can assume that the positive wells are probably clones. This can be assured by multiple rounds of limiting dilution. If the frequency is greater than 37%, the positive wells are likely derived from multiple precursors, and are therefore not clonal. The growing clones are then transferred into a 96-well flat-bottom plate with fresh APC and Ag and the medium is renewed as above. When confluent, the clones are split in several wells of a 96-well plate and then transferred to a 24-well plate. Fresh APC and Ag should be provided every two weeks. Micromanipulation is another method that can ensure clonality [175]. Under the microscope, a single cell is selected and aspirated with a fine-bore Pasteur pipette before being transferred into one well of a 96-well flat-bottom plate containing Ag and APC at high APC:T-cell ratio. The culture can then be maintained as described in the limiting dilution method. A new method has been recently developed using soluble peptide-MHC complexes in tetramers [176]. Tetramers of MHC class I molecules were complexed to various peptides and used to bind peptide-specific cytotoxic T-cells *in vitro*, and to fresh T-cells isolated from peripheral blood [176,177]. In addition, tetramer binding correlated well with Ag-specific cytotoxicity [176]. Clones should be tested for specificity and cryopreserved for future use as soon as enough cells are available. Loss of valuable clones by contamination is the most common problem encountered in long-term culture.

3.4.5 γδ T-cells and CD4⁻CD8⁻ T-cells

γδ T-cells are a specialized subset of T-cells that react with non-peptide microbial Ag early during the immune response [178]. Human γδ T-cells can recognize a class of phosphorylated molecules known as isoprenoids through an extracellular presentation pathway that does not require Ag uptake, Ag processing, or Ag-presenting elements [179]. They can be expanded from PB MNC by culture in the presence of non-peptide Ag and IL-2, with or without mixtures of cytokines [180]. A new subset of αβ T-cells that are $CD4^-CD8^-$ have been shown to recognize lipid and lipoglycan antigen in the context of CD1 molecules, a non-polymorphic presentation pathway [181,182]. To date, all the CD1-restricted T-cells that have been derived recognize non-peptide Ag from *M. tuberculosis* or *M. leprae* [183,184]. They are found in healthy donors and patients with mycobacterial infections. These cells can produce high levels of IFN-γ and IL-12 [183], and can have a high level of cytolytic activities against Ag-pulsed and infected macrophages [184]. This double negative ($CD4^-CD8^-$) T-cell population can be derived and cloned as described for Ag-specific $CD4^+$ T-cells.

4. ASSAY TECHNIQUES

Characterization of T-lymphocytes in culture involves measurement of cell function such as proliferation, cytokine secretion and cytotoxicity. With the discovery of T-cells as a heterogeneous collection of subsets with differing roles in the immune response, it has become more important not only to know what a population of cells is doing, but also which subsets of the cells are responsible for the observed effect. In this section, a number of standard techniques are discussed along with newer techniques that allow analysis of the function of T-cell subsets contained within a larger population.

4.1 Proliferation

The number of cells in culture at any given time and the rate at which these cells are dividing are important measures of the status of the culture. T-cells can be enumerated on a hemacytometer and cell viability determined by trypan blue dye exclusion. When there are a large number of individual cultures to be counted, a Coulter Counter (Coulter, Miami, FL) may be used. The Coulter principle is based on changes in electrical resistance of cells or particles suspended in an electrolyte solution, which passes through two electrodes [185]. In addition, the size of cells can be determined by the magnitude of the change in resistance. Counting cells over a period of time allows calculation of the rate of cell propagation. Cell sizing may provide useful independent information during the initial purification of lymphocytes, from whole blood, as platelets, erythrocytes, lymphocytes and larger cells such as macrophages can be discerned as discrete peaks. In addition, while resting T-lymphocytes have a volume of approximately 170 femtoliters (fl), a population of activated T-cells can possess a mean volume of up to 800-900 fl. Our laboratory has found measurement of T-cell volume a useful tool in timing the restimulation of T-cells with anti-CD3 plus anti-CD28 mAb [29]. Shown in Fig. 4 are size histograms of post-ficoll PB MNC, CD4-enriched resting lymphocytes and activated T-cells. Electronic cell counters are relatively expensive, but provide greater accuracy, precision and speed than hemacytometers.

Alternatively, if proliferation in a large number of cultures or stimulation conditions needs to be measured, the rate of T-cell proliferation can be indirectly assayed by a number of methods. ^{3}H-thymidine added to stimulated cells becomes incorporated into newly synthesized DNA. Cells are generally pulsed for a few hours or overnight and harvested into filter paper that can be read in a scintillation counter [70]. Bromodeoxyuridine measures proliferation using the same principle, but it is detected by flow cytometry [186]. Hoechst 33342, a DNA-binding dye and Pyronin Y, a RNA-binding dye, can also be detected by flow cytometry and measure cell cycle phase [187]. Tetrazolium salts are converted by metabolically activated cells to formazan, which can be quantified by measuring absorbance at 490 nm [70]. Proliferation measured using these agents must take into account ongoing cell death in the population as a whole. An advantage of counting cells on a hemacytometer with trypan blue or sizing on a Coulter Multisizer is that the percentage of dead cells in a culture can be directly measured or estimated. Dead cells can be visualized on a flow cytometer by staining with propidium iodide [188] or ethidium monoazide [189]. For cell cultures that contain potential biohazards, use of ethidium monoazide is advantageous because flow cytometric analysis can be performed on fixed cells. Cells undergoing apoptosis can also be measured by propidium iodide or by staining DNA fragments using the TUNEL method [190].

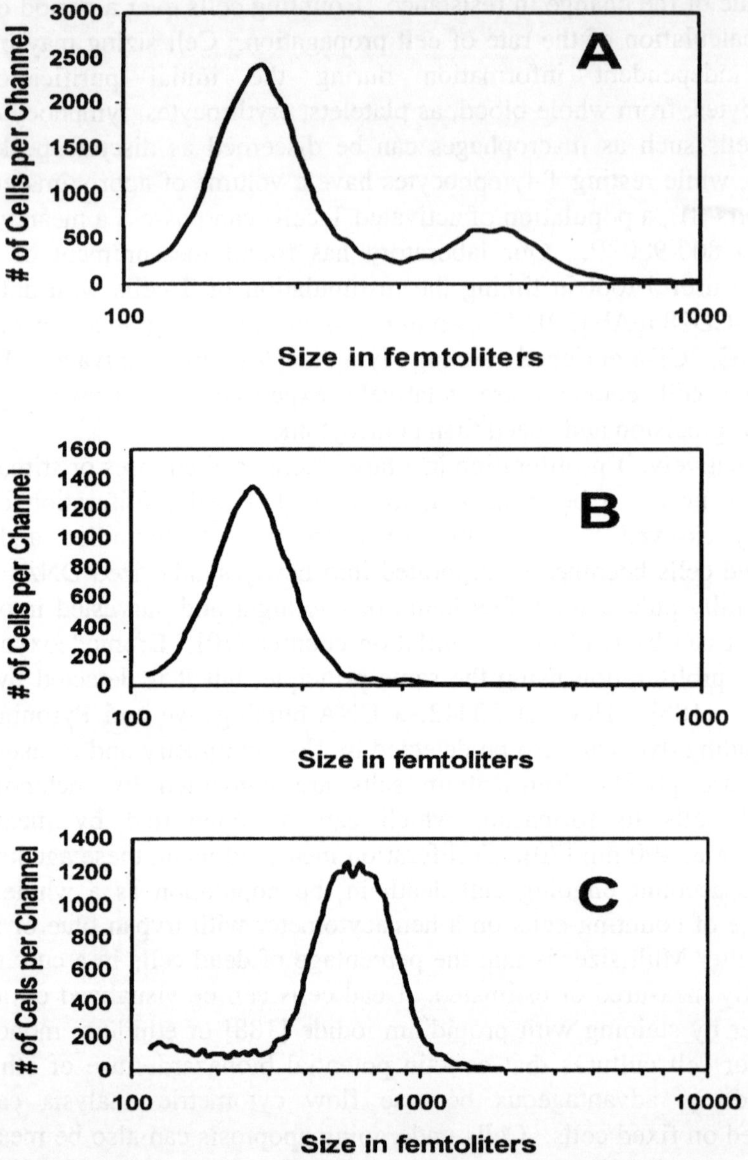

Figure 4. Cell population size histograms generated on a Coulter Multisizer IIe. The x-axis shows size distribution by femtoliter (10^{-15} liter) and the y-axis shows the number of cells per channel. Resting PB MNC isolated after ficoll gradient centrifugation of an apheresis leukopack are shown in A. The lymphocyte peak is to the left and the peak containing large monocytes and macrophages is to the right. B) By negative selection, resting CD4-enriched cells (86% $CD3^+CD4^+$) were obtained from the PB MNC's shown in panel A. C) The cells shown in B were stimulated with immobilized anti-CD3 plus anti-CD28 for 5 days. Note the difference in the scale of the x-axis.

To measure Ag-specificity of T-cell lines or clones, APC with or without Ag are added to T-cells for 2-4 days in 96-well plates. The cells are pulsed with ^{3}H-thymidine for the last 8-16 hours of the culture. Specificity is measured by calculation of a stimulation index (SI). The SI is obtained by first subtracting the counts per minute (cpm) obtained with T-cells alone from the cpm of T-cells plus APC pulsed with Ag (Ag-specific cpm). Second, the cpm obtained with T-cells alone is subtracted from the cpm obtained with APC alone (background cpm). Finally, the Ag-specific cpm is divided by the background cpm to give the SI. T-cell lines or clones are considered Ag-specific if the SI is greater than 3. Non-specific proliferation may result from T-cell autoreactivity to self-Ags.

4.2 Tracking Cell Division in T-Cell Subsets

The methods for assaying proliferation described above are useful to characterize populations of cells. However, when the proliferation of specific subpopulations of cells is to be measured, cell labeling techniques have been developed that allow two color staining and analysis by flow cytometry. PKH-26 is a membrane permeant dye that is excited at 488 nm and emits in the red wavelength similar to phycoerythrin [191]. The dye is retained in the cell membrane in a stable manner and upon cell division, it is assumed that each daughter cell is of a fluorescence intensity one-half of its parent cell. In this way, the number of cell divisions a particular subset has undergone can be tracked by counter-staining with fluorescent-labeled mAb [192]. The software program ModFit (Verity, Topsham, ME) can discern PKH-26 fluorescence peaks and calculate the percentage of cells that have undergone 1, 2 or more cell divisions. Another dye, carboxy-fluorescein diacetate succinimidyl ester (CFDASE) can also be used to track cell divisions. Upon entering a cell, the lipophilic blocking groups of CFDASE are cleaved by non-specific esterases, and the resulting charged molecule (CFSE) is trapped in the cytoplasm by covalent linkage to lysine residues. The advantage of CFSE is that the difference in intensity between fluorescence peaks is clearly distinguished [193]. Wells et al. have shown that the dye loading is homogeneous and bright for T-cells, is stable for several days *in vitro* and could track up to eight cell divisions [194]. Two color staining with CFSE allows the calculation of cell division in cell subsets. A recent report has demonstrated the feasibility of CFSE staining and cell division analysis in concert with intracellular cytokine staining (discussed in section 4.5) [195]. Each of the above studies using CFSE was performed using murine cells. CFSE can be used to label and perform functional analysis on primary human T-cells as well. Shown in Fig. 5 is a fluorescence histogram of human CD4$^+$ T-cells loaded with CFSE and stimulated for four days with DC pulsed with SAg. Multiparameter flow cytometry using cell tracking dyes such as PKH-26 and CFSE has the potential in the next few years to vastly improve knowledge of how subsets of T-cells propagate and function in relation to whole cell populations.

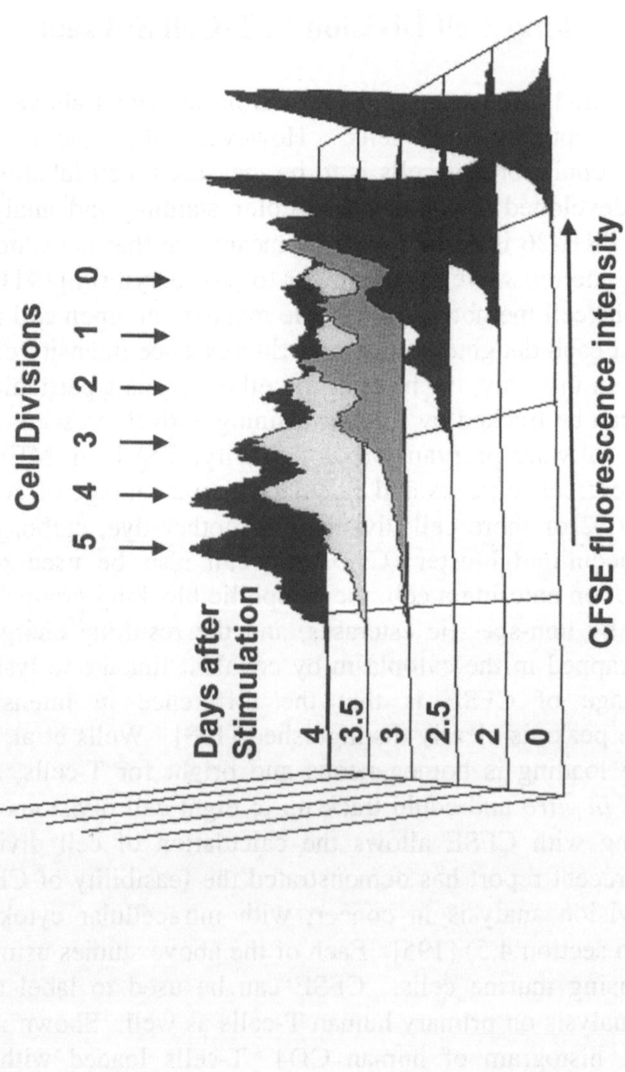

Figure 4. T-cell division as tracked by the fluorescent dye CFSE. $CD4^+$ T-cells were loaded with the fluorescent dye CFSE and stimulated with 0.5 µg/ml TSST-1 on day 0. At various times, cells were counterstained with APC-labeled anti-TCR Vβ2. Flow cytometry analysis was performed by gating on TCR $V\beta2^+$ cells and plotting CFSE fluorescence.

4.3 TCR Repertoire

A polyclonal population of T-cells isolated from a healthy donor should contain a representative distribution of the T-cell repertoire. *In vivo*, infection, autoimmunity or malignancy may skew the repertoire. *In vitro*, oligoclonal or monoclonal expansion of certain families may occur during selection of Ag-specific cells. The ability to assess the relative clonality of an *in vitro* population of cells can aid in tailoring culture techniques to either maintain a polyclonal population or select for an Ag-specific or clonal population of T-cells. The characterization of T-cell receptor gene segments in humans has facilitated the development of a polymerase chain reaction (PCR)-based technique for analyzing the T-cell repertoire as represented by the complementarity determining region (CDR) variability of T-cell receptor Vβ chains [196]. The CDR3 of the TCR Vβ chain is the only non-germline encoded hypervariable region and is one of the primary sites of antigen specificity. Because nucleotide transferases can add or delete bases, the length of the CDR3 may vary by up to eight amino acids. The distribution of the various CDR3 lengths in a polyclonal population of cells is gaussian-like, in an oligoclonal population non-gaussian and, in a clonal population, consists of one CDR3 length. For analysis, TCR transcripts undergo RT-PCR and are amplified using 24 different TCR Vβ and one fluorescent labeled Cβ primer. The products then undergo electrophoresis on an automated sequencer where the different size CDR3 bands can be separated and quantified based upon peak fluorescent intensity. Area under the curve can then be obtained to perform semi-quantitative analysis [196-198].

This technique has been used to type the T-lymphocyte infiltrate in diseased organs or tissues and also to track immune reconstitution following BM transplant [199]. Stimulation and prolonged culture of $CD4^+$ T-cells with anti-CD3 and anti-CD28 is able to maintain most of the input diversity, as determined by TCR Vβ PCR analysis [29]. TCR Vβ analysis can also be used to validate methods of T-cell stimulation and culture as suitable for the study of oligoclonal populations isolated from donors with infection, malignancy or autoimmune disease. Arenz et al. isolated liver-infiltrating T-cells from patients with autoimmune hepatitis and cultured these cells *in vitro* with PHA or anti-CD3 plus irradiated feeder cells [200]. They were able to show that antigen-independent activation did not result in preferential expansion of single TCR Vβ families, and thus *in vitro* culture with PHA or anti-CD3 provides a method to analyze small populations of organ-infiltrating T-cells. Ag-specific stimulation *in vitro* of TIL has been shown by TCR Vβ analysis to either enrich or eliminate T-cell subpopulations present in freshly isolated TIL [201,202]. Refining the techniques for enrichment of clones specific for tumor will have important implications for immunotherapy. It is interesting to note that certain T-cell populations were specifically deleted by *in vitro* culture with tumor cells. TCR Vβ analysis may thus prove useful in revealing the mechanisms of tumor-specific immune suppression. mAb are available for many TCR Vβ families. However, because of the low frequency of any given Vβ family in a polyclonal population of cells, analysis of TCR Vβ changes during *in vitro* culture by flow cytometry are more suitable for oligoclonal or clonal populations of T-cells.

4.4 Flow Cytometry and T-Cell Culture

Flow cytometry is an increasingly essential tool in the phenotypic and functional analysis of T-lymphocytes. As with other assay techniques, the power of flow cytometry is related to the specificity with which it is applied and the purity or characterization of the population of T-cells being studied. Recently the ability to perform eight color, 10-parameter flow cytometric analysis was reported [203]. However, the equipment, software and some of the dye conjugates necessary for this type of analysis are quite complex and not yet commercially available. As more is learned about T-cell subsets and the differences in their function, it is important to be able to distinguish which cells within a larger population are responsible for the measurements observed. It is beyond the scope of this chapter to cover all aspects and techniques of flow cytometry relevant to T-cells. For excellent detailed examinations of flow cytometry in general and applications to T-lymphocytes in particular, see references [204-207]. In this section, the use of flow method to detect Ag-specific T-cells will be discussed. The measurement of intracellular cytokine levels by flow cytometry is discussed in the next section.

Detection of Ag-specific T-cells in a bulk population is a tedious proposition by the standard method of limiting dilution analysis [174], particularly if the desired population is a small fraction of the total cell population. Recently, a new technique for the detection of Ag-specific T-cells by flow cytometry was described [176]. This technique utilizes the affinity of the peptide-MHC complex for the TCR and the ability of tetrameric complexes to bind several TCR, and thus increase the binding efficiency of a fluorescent reagent to detect Ag-specific T-cells at frequencies less than 1%. The disadvantage of this method is that it is technically demanding since reagents must be custom designed for individual haplotypes and antigenic specificities. These reagents can be used to sort Ag-specific T-cells from a bulk population and the sorted cells can then be expanded in culture (Fig. 3C). Cells sorted by this method are able to lyse Ag-specific targets [208]. The enrichment of Ag-specific cells during *in vitro* culture can be monitored as well [115]. Ag-specific flow cytometry can conceivably be combined with intracellular cytokine staining to provide functional analysis of a specific immune response. In the future, the wider availabilty of dimeric and tetrameric MHC-peptide complexes will aid greatly in the ability to isolate, grow and characterize Ag-specific T-cells.

4.5 Cytokine Secretion and Intracellular Cytokine Flow Cytometry

A recurring motif in most current assay techniques for T-cells is the importance of mAb. This is true as well for the detection of cytokine production by T-cells. Prior to the wide availability of commercial ELISA kits to measure cytokine secretion, the standard assay was a bioassay based on the cytokine requirement of certain cell lines for growth [37,209]. The bioassay relies on the requirement of defined cells lines for exogenous growth factors in order to proliferate. By titrating an unknown supernatant to a defined standard, a measure of the cytokine present can be calculated. While this method is an indirect measure of the absolute amount of cytokine present in supernatants, it is a direct measure of the amount of biologically active cytokine present and thus holds an advantage over ELISA or intracellular cytokine flow cytometry in that respect. However, bioassays may also be subject to the effects of other cytokines in supernatant that may non-specifically activate or inhibit the growth of the indicator cell line. ELISA is based on the ability of a mAb to recognize a specific cytokine and the amplification of this recognition by a secondary antibody linked to an enzyme, which changes color following the addition of a substrate. The advantage of ELISA is that it is quick and reproducible. The disadvantage of ELISA is that it is relatively expensive and measures an antigenic site on a cytokine that may or may not directly correlate to biologic activity.

ELISA and bioassays can only measure the total amount of cytokine produced by a whole population of cells. Intracellular cytokine flow cytometry is a recently developed technique that allows individual cells to be analyzed for cytokine production [210]. Using two or more colors, cells can be phenotyped and analyzed for single or multiple cytokine production. Furthermore, many samples can be analyzed in a short period of time. In the standard technique, cells are stimulated or restimulated *in vitro* by PMA and ionomycin for 4-6 hours in order to activate the cells. Resting cells or cells in steady-state produce little or no measurable cytokines. This stimulation is performed in the presence of monensin or brefeldin A, which disrupts intracellular trafficking and prevents secretion of cytokines. After incubation, cells are fixed in paraformadehyde and then permeabilized with saponin to allow the anti-cytokine mAb into the cell. The choice and titration of anti-cytokine mAb is critical to detect specific binding as cells fixed with paraformadehyde can non-specifically bind Ab. Specificity of mAb binding can be demonstrated by the blocking of staining with recombinant cytokine or unconjugated anti-cytokine mAb. One criticism of this technique is that pharmacologic stimulation does not mimic physiologic or antigenic stimulation. In fact, cells can be stimulated with cytomegalovirus (CMV) plus anti-CD28 to measure the Ag-specific cytokine response [211,212]. The characterization of T-cell subsets and Ag-specific cells should benefit greatly from the wide application of this technique.

4.6 CTL Activity

Measurement of *in vitro* cytotoxic function of human T-cells can be accomplished using anti-CD3 antibody which stimulates all cytolytic effector T-cells, or with a specific stimulating Ag. Anti-CD3 mediated cytotoxicity allows measurement of lytic capacity of T-cell populations, lines and clones in a polyclonal fashion as mediated via the TCR complex [213]. Classically, target cells are radiolabelled with ^{51}Cr overnight [214]. The T-cell effector population can consist of PB MNC, T-cells or T-cell subpopulations. However, PB MNC contain NK cells, which may cause non-MHC restricted lysis of target cells, so additional controls should be included. Effector cells, ^{51}Cr labeled target cells and anti-CD3 antibodies are added together for four hours at varying effector-to-target (E/T) ratios. Specific lysis is calculated by comparing ^{51}Cr release from target cells in presence of the effectors to spontaneous release or maximum release by target cells alone treated with detergent.

In Ag-mediated T-cell cytotoxicity assays, Ag is used to stimulate cytotoxic T-cells instead of anti-CD3 antibody. This permits the measurement of cytolytic capacity in a more physiologic manner. It also allows the examination of specific antigenic epitopes recognized by the TCR [215]. Best results are obtained with an expanded population of specific cytolytic T-cells obtained from humans recently immunized or sensitized *in vitro* to a specific Ag. Autologous or MHC class I-matched target cells, generally B-lymphoblastoid cell lines, can be sensitized with tumor or viral Ag during an overnight incubation with ^{51}Cr or infected with recombinant virus encoding for specific proteins [216]. Ag-mediated cytotoxicity may also be quantitated using target cells labeled with a green fluorescent marker as a non-radioactive alternative to ^{51}Cr [217,218]. In this method, after four hours incubation of targets with effector cells, counter-staining with propidium iodide permits discrimination between live and dead cells within both cell populations. Results are analyzed by flow cytometry and show an excellent correlation with data obtained from a standard ^{51}Cr release assay. In addition to being a method free of radioactivity, this technique also allows multiparameter analysis of T-cell subsets.

5. UTILITY OF THE SYSTEM

Effective T-cell culture can be both a means and an end in itself. In this section, the use of T-cells cultured *in vitro* as a model for replicative senescence and immunosenescence will be discussed as an example of a research oriented application. The use and potential of cultured T-cells in adoptive immunotherapy and gene therapy of infection and malignancy will also be reviewed.

The discovery by Hayflick and Moorehead of the limited *in vitro* lifespan of human fetal lung fibroblasts led to a hypothesis of programmed senescence for somatic cells [219]. This hypothesis that cells have a limited number of divisions before growth arrest has had great implications for aging research. The application of this hypothesis to the immune system, and lymphocytes in particular, has recently been tested with the advent of effective methods of T-cell culture. However, one concern is whether effective methods of *in vitro* culture can ever approximate the ideal conditions present *in vivo*. In a 1984 review, Effros and Walford [220] noted that reports of long-term cultures of T-cells and T-cell clones seemed to contradict the notion of the Hayflick limit of 60-70 population doublings. However, in earlier studies, the function and clonality of these long-term cultures was not sufficiently addressed, leaving open the possibility of clones with chromosomal abnormalities responsible for these very long-term cultures. Subsequently, using mitogens and irradiated feeder cells a mean T-cell population doubling of 23 ± 7 hours was observed [95]. Other studies have noted replicative capacities of 50-70 population doublings following lectin and IL-2 stimulation of adult T-cell clones [221]. However, this does not appear to reflect the capacity of the vast majority of T-cell clones, and more likely represents the selective survival of cells having extensive replication potential. CD3/CD28 stimulation in cis results in a polyclonal expansion of adult $CD4^+$ T-cells of 30-40 population doublings [29]. Whether these culture conditions can be improved upon for polyclonal T-cells using newer techniques has yet to be determined. Nevertheless, *in vitro* culture of lymphocytes, and T-cells in particular, has provided a model system for the study of immunosenescence.

In vitro, culture of T-cells has also played a role in the examination of the telomere hypothesis in which the lengths of telomeres are thought to correlate with replicative capacity. Telomeres are structures at the end of chromosomes that appear to be involved in chromosomal integrity and replicative capacity [222]. With each round of cell division, telomeres are shortened 50-100 bases. Telomere length has been observed to decrease *in vitro* with cell division and *in vivo* with increased age [223,224]. Extension of *in vitro* lifespan has been demonstrated by the transfection of normal epithelial cells or fibroblasts with telomerase, an enzyme previously shown to elongate telomeres in germ cells [225]. In T-cells stimulated with PHA and IL-2, the decrease in telomere length was first demonstrated by Vaziri et al. [224]. Subsequently, in purified naive and memory T-cells, a decrease in telomere length was observed following CD3/CD28 stimulation [226]. Interestingly, there was a relatively constant difference in telomere length between naive and memory cells observed throughout the culture period and *in vivo* with increased age, pointing to a more extensive replicative history of memory cells compared to naive cells. The activity of telomerase can also be upregulated in $CD4^+$ T-cells following activation with anti-CD3, anti-CD3/anti-CD28 or PMA/ionomycin [227], and CD3/CD28 stimulation may also contribute to stabilization of telomere length during long-term culture [228]. The contribution of telomerase activation to the stabilization of telomere length of T-cells *in vivo* has yet to be determined.

Adoptive immunotherapy involves the *ex vivo* manipulation of immune cells and reinfusion in order to augment or reconstitute immune system function. Technical advancements in cell purification and culture have facilitated the development of T-cell therapy. These technical hurdles were first addressed in the growth of LAK cells by culture with high doses of IL-2 for the treatment of advanced cancer [229], followed by the culture and infusion of TIL [230]. Activation of T-cells by anti-CD3 mAb and IL-2 *in vitro* has been used in pre-clinical studies and demonstrated the feasibility of this method in the adoptive immunotherapy of cancer patients [85,87]. A phase I trial involving short-term culture with anti-CD3 mAb and IL-2 of $CD4^+$ T-cells and reinfusion to cancer patients was performed at the National Cancer Institute [231]. These investigators noted a significant increase in $CD4^+$, $CD4^+HLA\text{-}DR^+$, and $CD4^+CD45RO^+$ T-cells *in vivo* following infusion. The identification of tumor specific antigens has enabled the development of culture methods for CTLs specific for these antigens (reviewed in [232,233]). T-cells may also be redirected by *in vitro* culture with bispecific Ab against CD3 and ovarian carcinoma cells to provide a local immunomodulation by reinfused cells [234].

Walter et al. have demonstrated the feasibility of adoptive immunotherapy using *in vitro* cultured T-cells [235]. In this study, $CD8^+$ T-cell clones specific for CMV were isolated from allogeneic BM donors and reinfused into recipients in order to reconstitute CMV immunity following BM transplantation. In the 14 patients who received cells, neither CMV viremia nor CMV disease was observed. Autologous polyclonal $CD8^+$ T-cells cultured *in vitro* with PHA and IL-2 have been reinfused to AIDS patients [236] demonstrating safety and feasibility. Subsequently, Lieberman et al. have cultured T-cells with PHA plus IL-2 followed by incubation with autologous PB MNC pulsed with HIV peptides [172]. This approach has generated $CD8^+$ CTLs enriched for HIV-specific cytotoxicity and results in a small increase in HIV-specific CTL activity up to six months after infusion. The first disease to be treated by T-lymphocyte directed gene therapy was adenosine deaminase deficiency in 1990. The *in vitro* culture protocol involved stimulation with anti-CD3 mAb and IL-2 followed by transduction with a vector containing an intact adenosine deaminase gene and additional culture prior to reinfusion [122]. A similar study cultured T-cells with PHA and IL-2 prior to transduction [237]. Each of these studies demonstrated the long-term survival of gene-modified T-cells. For EBV, gene-marked virus-specific CTLs were prepared by repetitive stimulation with EBV-transformed cells and anti-CD3 mAb plus IL-2 [238]. The infusion of these CTLs restored EBV-specific immunity. Anti-CD3 stimulation plus IL-2 has also been used to culture and transduce T-cells between HIV-discordant identical twins [88] or in an autologous setting [239]. In each study, gene marked cells could be detected in the lymphoid tissue or peripheral blood up to six months after infusion. These gene marking and gene therapy studies have proven the feasibility of *ex vivo* T-cell culture and manipulation for the treatment of disease. Although space limits a more detailed discussion of T-cell immunotherapy and gene therapy, it is clear that *ex vivo* culture methods for the polyclonal and antigen-specific expansion of T-cells hold great promise for therapy of many diseases.

6. SUMMARY

Methods for T-cell culture, manipulation and functional evaluation continue to evolve at an astonishing rate. The discovery of IL-2 and other cytokines as well as the surface receptors and coreceptors important in T-cell activation has led to the design of reagents that have improved T-cell culture. Because lymphoid cells are composed of a heterogeneous mixture of cell lineages, it is most important to work with a well-defined and phenotyped population of T-cells. The development of techniques such as membrane cell tracking dyes and intracellular cytokine staining now allow the characterization of individual T-cells and T-cell subsets. Previously, cells had to be separated and cultured as individual subsets or clones to acquire the information now available through these techniques. Because of the interdependence of T-cell subsets such as helper and cytotoxic cells, separate culture does not permit physiologic interaction among cells. *In vitro* culture itself is only a rough approximation of the interactions *in vivo*. As more is learned about the cytokines and surface receptors important to T-cell interactions with other cells in the body, *in vitro* models will likely evolve to emulate the structure of a lymph node, an evolutionarily optimized T-cell culture vessel.

ACKNOWLEDGEMENTS

The authors wish to thank Ms. Nancy Craighead for helpful comments and suggestions. The views expressed in this article are those of the authors and do not reflect the official policy or position of the Department of the Army, Navy, Department of Defense, nor the United States Government. Supported in part by Army contract #DAMD17-93-V-3004.

REFERENCES

1. Bretscher, PA and Cohn, M (1970) A theory of self discrimination. *Science* 169:1042-1049.
2. Lafferty, KJ and Cunningham, AJ (1975) A new analysis of allogeneic interactions. *Australian Journal of Experimental Biology and Medical Science* 53:27-42.
3. Carter, LL and Dutton, RW (1996) Type 1 and type 2: a fundamental dichotomy for all T-cell subsets. *Current Opinion in Immunology* 8:336-342.
4. Abbas, AK, Murphy, KM, Sher, A (1996) Functional diversity of helper T-lymphocytes. *Nature* 383:787-793.
5. Halverson, DC, Schwartz, GN, Carter, C, Gress, RE, Fowler, DH (1997) *In vitro* generation of allospecific human CD8+ T-cells of Tc1 and Tc2 phenotype. *Blood* 90:2089-2096.
6. Poppema, S, Lai, R, Visser, L, Yan, XJ (1996) CD45 (leucocyte common antigen) expression in T and B lymphocyte subsets. *Leukemia and Lymphoma.* 20:217-222.
7. Picker, LJ, Treer, JR, Ferguson-Darnell, B, Collins, PA, Buck, D, Terstappen, LW (1993) Control of lymphocyte recirculation in man. I. Differential regulation of the peripheral lymph node homing receptor L-selection on T-cells during the virgin to memory cell transition. *Journal of Immunology* 150:1105-1121.
8. Baars, PA, Maurice, MM, Rep, M, Hooibrink, B, van, LR (1995) Heterogeneity of the circulating human CD4+ T-cell population. Further evidence that the CD4+CD45RA-CD27- T-cell subset contains specialized primed T-cells. *Journal of Immunology* 154:17-25.
9. Xu, X, Beckman, I, Ahern, M, Bradley, J (1993) A comprehensive analysis of peripheral blood lymphocytes in healthy aged humans by flow cytometry. *Immunology and Cell Biology* 71:549-557.
10. Fauci, AS (1996) Host factors and the pathogenesis of HIV-induced diseases. *Nature* 384:529-534.

11. Roederer, M (1995) T-cell dynamics of immunodeficiency. *Nature Medicine* 1:621-627.
12. Cantrell, D (1996) T-cell antigen receptor signal transduction pathways. *Annual Review of Immunology* 14:259-74:259-274.
13. Robey, E and Allison, JP (1995) T-cell activation: integration of signals from the antigen receptor and co-stimulatory molecules. *Immunology Today* 16:306-310.
14. June, CH, Bluestone, JA, Nadler, LM, Thompson, CB (1994) The B7 and CD28 receptor families. *Immunology Today* 15:321-331.
15. Altman, A, Coggeshall, KM, Mustelin, T (1990) Molecular events mediating T-cell activation. *Advances in Immunology* 48:227-360:227-360.
16. Siegel, JN and June, CH (1995) Signal transduction and T-lymphocyte activation, in RR Rich (ed.), *Principles of Clinical Immunology, Mosby Year Book, Inc.*, St. Louis, pp. 192-216
17. Nishizuka, Y (1995) Protein kinase C and lipid signaling for sustained cellular responses. *FASEB Journal* 9:484-496.
18. Weiss, A (1993) T-cell antigen receptor signal transduction: a tale of tails and cytoplasmic protein-tyrosine kinases. *Cell* 73:209-212.
19. June, CH, Ledbetter, JA, Linsley, PS, Thompson, CB (1990) Role of the CD28 receptor in T-cell activation. *Immunology Today* 11:211-216.
20. Thompson, CB, Lindsten, T, Ledbetter, JA, Kunkel, SL, Young, HA, Emerson, SG, Leiden, JM, June, CH (1989) CD28 activation pathway regulates the production of multiple T- cell-derived lymphokines/cytokines. *Proceedings of the National Academy of Sciences USA* 86:1333-1337.
21. Lindsten, T, Lee, KP, Harris, ES, Petryniak, B, Craighead, N, Reynolds, PJ, Lombard, DB, Freeman, GJ, Nadler, LM, Gray, GS, Thompson, CB, June, CH (1993) Characterization of CTLA-4 structure and expression on human T- cells. *Journal of Immunology* 151:3489-3499.
22. Thompson, CB and Allison, JP (1997) The emerging role of CTLA-4 as an immune attenuator. *Immunity* 7:445-450.
23. Larsen, CP and Pearson, TC (1997) The CD40 pathway in allograft rejection, acceptance, and tolerance. *Current Opinion in Immunology* 9:641-647.
24. DeBenedette, MA, Chu, NR, Pollok, KE, Hurtado, J, Wade, WF, Kwon, BS, Watts, TH (1995) Role of 4-1BB ligand in co-stimulation of T-lymphocyte growth and its upregulation on M12 B lymphomas by cAMP. *Journal of Experimental Medicine* 181:985-992.
25. Shuford, WW, Klussman, K, Tritchler, DD, Loo, DT, Chalupny, J, Siadak, AW, Brown, TJ, Emswiler, J, Raecho, H, Larsen, CP, Pearson, TC, Ledbetter, JA, Aruffo, A, Mittler, RS (1997) 4-1BB co-stimulatory signals preferentially induce CD8+ T-cell proliferation and lead to the amplification *in vivo* of cytotoxic T-cell responses. *Journal of Experimental Medicine* 186:47-55.
26. Paul, WE, ed. (1993) *Fundamental Immunology,* Raven Press, New York
27. Levine, BL, Ueda, Y, Craighead, N, Huang, ML, June, CH (1995) CD28 ligands CD80 (B7-1) and CD86 (B7-2) induce long-term autocrine growth of CD4+ T-cells and induce similar patterns of cytokine secretion *in vitro*. *International Immunology* 7:891-904.
28. Riley, JL, Carroll, RG, Levine, BL, Bernstein, W, St.Louis, DC, Weislow, OS, June, CH (1997) Intrinsic resistance to T-cell infection with HIV type 1 induced by CD28 co-stimulation. *Journal of Immunology* 158:5545-5553.
29. Levine, BL, Bernstein, WB, Connors, M, Craighead, N, Lindsten, T, Thompson, CB, June, CH (1997) Effects of CD28 co-stimulation on long-term proliferation of CD4+ T-cells in the absence of exogenous feeder cells. *Journal of Immunology* 159:5921-5930.

30. Nowell, PC (1960) Phytohemagglutinin: An Initiator of Mitosis in Cultures of Normal Human Leukocytes. *Cancer Research* 20:462-466.
31. Stobo, JD and Paul, WE (1973) Functional heterogeneity of murine lymphoid cells. 3. Differential responsiveness of T-cells to phytohemagglutinin and concanavalin A as a probe for T-cell subsets. *Journal of Immunology* 110:362-375.
32. Weksler, ME and Kuntz, MM (1976) Synergy between human T and B lymphocytes in their response to phythaemagglutinin and pokeweed mitogen. *Immunology* 31:273-281.
33. Svedmyr, E (1975) Long-term maintenance *in vitro* of human T-cells by repeated exposure to the same stimulator cells. Differences when using repeated stimulation in allogeneic mixed leukocyte culture and when using stimulation with autologous lymphoblastoid cells. *Scandinavian Journal of Immunology* 4:421-427.
34. Morgan, DA, Ruscetti, FW, Gallo, R (1976) Selective *in vitro* growth of T-lymphocytes from normal human BMs. *Science* 193:1007-1008.
35. Ruscetti, FW, Morgan, DA, Gallo, RC (1977) Functional and morphologic characterization of human T-cells continuously grown *in vitro*. *Journal of Immunology* 119:131-138.
36. Gillis, S, Baker, PE, Ruscetti, FW, Smith, KA (1978) Long-term culture of human antigen-specific cytotoxic T-cell lines. *Journal of Experimental Medicine* 148:1093-1098.
37. Gillis, S, Ferm, MM, Ou, W, Smith, KA (1978) T-cell growth factor: parameters of production and a quantitative microassay for activity. *Journal of Immunology* 120:2027-2032.
38. Kohler, G and Milstein, C (1975) Continuous cultures of fused cells secreting antibody of predefined specificity. *Nature* 256:495-497.
39. Bauer, JD (1982) Collection and preparation of the specimen, in D Ladig (ed.), *Clinical Laboratory Methods, C.V. Mosby Company,* St. Louis, pp. 27-32
40. Hannet, I, Erkeller-Yuksel, F, Lydyard, P, Deneys, V, DeBruyere, M (1992) Developmental and maturational changes in human blood lymphocyte subpopulations. *Immunology Today* 13:215, 218.
41. Rubinstein, P, Dobrila, L, Rosenfield, RE, Adamson, JW, Migliaccio, G, Migliaccio, AR, Taylor, PE, Stevens, CE (1995) Processing and cryopreservation of placental/umbilical CB for unrelated BM reconstitution. *Proceedings of the National Academy of Sciences USA* 92:10119-10122.
42. Areman, EM, Cullis, H, Sacher, RA, Cottler-Fox, M, Deeg, HJ (1990) Automated isolation of MNC using the Fenwal CS3000 blood cell separator. *Progress in Clinical and Biological Research* 333:379-85:379-385.
43. Keilholz, U, Klein, HG, Korbling, M, Brado, B, Carter, CS, Cullis, H, Galm, F, Hunstein, W (1991) Peripheral blood mononuclear cell collection from patients undergoing adoptive immunotherapy or peripheral blood-derived stem cell transplantation and from healthy donors. *Journal of Clinical Apheresis* 6:131-136.
44. Bambi, F, Faulkner, LB, Azzari, C, Gelli, AM, Tamburini, A, Tintori, V, Lippi, AA, Tucci, F, Bernini, G, Genovese, F (1998) Pediatric peripheral blood progenitor cell collection: haemonetics MCS 3P versus COBE Spectra versus Fresenius AS104. *Transfusion* 38:70-74.
45. Almici, C, Carlo-Stella, C, Donnenberg, AD, Rizzoli, V (1993) Counterflow centrifugal elutriation: present and future. *BM Transplant.* 12:105-108.
46. Abrahamsen, TG, Carter, CS, Read, EJ, Rubin, M, Goetzman, HG, Lizzio, EF, Lee, YL, Hanson, M, Pizzo, PA, Hoffman, T (1991) Stimulatory effect of counterflow centrifugal elutriation in large- scale separation of peripheral blood monocytes can be reversed by storing the cells at 37 degrees C. *Journal of Clinical Apheresis* 6:48-53.

47. Quinones, RR, Gutierrez, RH, Dinndorf, PA, Gress, RE, Ney, AB, Taylor, B, Karandish, S, Carter, CS, Luban, NL, Reaman, GH (1993) Extended-cycle elutriation to adjust T-cell content in HLA-disparate BM transplantation. *Blood* 82:307-317.
48. Kay, HD, Petrie, HT, Burge, JJ, Klassen, LW (1986) Rapid recovery of non-hemolyzed serum and untraumatized cells by using a new method of blood defibrination *in vitro*. *Journal of Immunological Methods* 92:251-260.
49. Boyum, A (1968) Isolation of MNC and granulocytes from human blood. Isolation of monuclear cells by one centrifugation, and of granulocytes by combining centrifugation and sedimentation at 1g. *Scandinavian Journal of Clinical Laboratory Investigation* Supplement 97:77-89.
50. Thiele, DL and Lipsky, PE (1985) Regulation of cellular function by products of lysosomal enzyme activity: elimination of human natural killer cells by a dipeptide methyl ester generated from L-leucine methyl ester by monocytes or polymorphonuclear leukocytes. *Proceedings of the National Academy of Sciences USA* 82:2468-2472.
51. Thiele, DL and Lipsky, PE (1986) The immunosuppressive activity of L-leucyl-L-leucine methyl ester: selective ablation of cytotoxic lymphocytes and monocytes. *Journal of Immunology* 136:1038-1048.
52. Townsend, RM and Simon, P (1992) The preferential expansion of functional CD4+ lymphocyte populations *in vitro*. *Journal of Immunotherapy.* 12:256-264.
53. Strelkauskas, AJ, Teodorescu, M, Dray, S (1975) Enumeration and isolation of human T and B lymphocytes by rosette formation with antibody-coated erythrocytes. *Clinical and Experimental Immunology* 22:62-71.
54. Linscott, WD (1998) *Linscott's directory of immunological and biological reagents,* William D. Linscott, Santa Rosa, CA
55. Stanciu, LA, Shute, J, Holgate, ST, Djukanovic, R (1996) Production of IL-8 and IL-4 by positively and negatively selected CD4+ and CD8+ human T-cells following a four-step cell separation method including magnetic cell sorting (MACS). *Journal of Immunological Methods* 189:107-115.
56. Payne, SM, Sharrow, SO, Shearer, GM, Biddison, WE (1981) Preparative separation of human T-cells reactive with the OKT4 monoclonal antibody. *International Journal of Immunopharmacology* 3:227-232.
57. Thomas, TE, Abraham, SJ, Otter, AJ, Blackmore, EW, Lansdorp, PM (1992) High gradient magnetic separation of cells on the basis of expression levels of cell surface antigens. *Journal of Immunological Methods* 154:245-252.
58. Miltenyi, S, Muller, W, Weichel, W, Radbruch, A (1990) High gradient magnetic cell separation with MACS. *Cytometry* 11:231-238.
59. Geretti, AM, Van Els, CA, Van Baalen, CA, Poelen, MC, Osterhaus, AD (1993) Preservation of phenotype and function of positively selected virus- specific CD8+ T-lymphocytes following anti-Fab detachment from immunomagnetic beads. *Journal of Immunological Methods* 161:129-133.
60. June, CH, Ledbetter, JA, Gillespie, MM, Lindsten, T, Thompson, CB (1987) T-cell proliferation involving the CD28 pathway is associated with cyclosporine-resistant interleukin 2 gene expression. *Molecular and Cellular Biology* 7:4472-4481.
61. Zwerner, RK, Schmittling, RJ, Russell, TR (1996) A simple and rapid method for removal of specific cell populations from whole blood. *Journal of Immunological Methods* 198:199-202.
62. Patel, D, Rubbi, CP, Rickwood, D (1995) Separation of T and B lymphocytes from human peripheral blood MNC using density perturbation methods. *Clinica Chimica Acta* 240:187-193.

63. Glassman, AB and Bennett, CE (1979) Cryopreservation of human lymphocytes: a brief review and evaluation of an automated liquid nitrogen freezer. *Transfusion* 19:178-181.
64. Rowley, SD (1992) Hematopoietic stem cell cryopreservation: a review of current techniques. *Journal of Hematotherapy* 1:233-250.
65. Glassman, AB and Christopher, JB (1984) Effect of cryopreservation on lymphocyte markers evaluated by monoclonal antibodies. *Transfusion* 24:538-539.
66. Preece, G, Murphy, G, Ager, A (1996) Metalloproteinase-mediated regulation of L-selectin levels on leucocytes. *Journal of Biological Chemistry* 271:11634-11640.
67. Qian, D and Weiss, A (1997) T-cell antigen receptor signal transduction. *Current Opinion in Cell Biology* 9:205-212.
68. Lanier, LL (1993) Distribution and function of lymphocyte surface antigens. Molecules costimulating T-lymphocyte activation and effector function. *Annals of the New York Academy of Sciences* 677:86-93:86-93.
69. Levine, BL, Mond, JJ,June, CH (1998) Lymphocyte Proliferation, in IM Roitt, PJ Delves (eds.), *Encyclopedia of Immunology,* Academic Press, London, pp. 2017-2023
70. Freshney, RI (1994) *Culture of Animal Cells,* Wiley-Liss, Inc., New York
71. Rutzky, LP and Pumper, RW (1974) Supplement to a survey of commercially available tissue culture media (1970). *In vitro* 9:408-409.
72. Morton, HJ (1970) A survey of commercially available tissue culture media. *In vitro* 6:89-108.
73. Moore, GE, Gerner, RE, Franklin, HA (1967) Culture of normal human leukocytes. *Journal Of The American Medical Association* 199:519-524.
74. Wang, RJ (1976) Effect of room fluorescent light on the deterioration of tissue culture medium. *In vitro* 12:19-22.
75. Herzberg, VL and Smith, KA (1987) T-cell growth without serum. *Journal of Immunology* 139:998-1004.
76. Tanaka, M, Suda, T, Haze, K, Nakamura, N, Sato, K, Kimura, F, Motoyoshi, K, Mizuki, M, Tagawa, S, Ohga, S, Hatake, K, Drummond, AH, Nagata, S (1996) Fas ligand in human serum. *Nature Medicine* 2:317-322.
77. Selvaggi, TA, Walker, RE, Fleisher, TA (1997) Development of antibodies to fetal calf serum with arthus-like reactions in human immunodeficiency virus-infected patients given syngeneic lymphocyte infusions. *Blood* 89:776-779.
78. Macy, E, Bulpitt, K, Champlin, RE, Saxon, A (1989) Anaphylaxis to infusion of autologous BM: an apparent reaction to self, mediated by IgE antibody to bovine serum albumin. *Journal of Allergy and Clinical Immunology* 83:871-875.
79. Trimble, LA, Perales, M-A, Knazek, RA, Lieberman, J (1996) Serum Enhances the *Ex vivo* Generation of HIV-Specific Cytotoxic T Cells. *Biotechnology and Bioengineering* 50:521-528.
80. Knazek, RA, Wu, YW, Aebersold, PM, Rosenberg, SA (1990) Culture of human tumor infiltrating lymphocytes in hollow fiber bioreactors. *Journal of Immunological Methods* 127:29-37.
81. Freedman, RS, Edwards, CL, Kavanagh, JJ, Kudelka, AP, Katz, RL, Carrasco, CH, Atkinson, EN, Scott, W, Tomasovic, B, Templin, S (1994) Intraperitoneal adoptive immunotherapy of ovarian carcinoma with tumor-infiltrating lymphocytes and low-dose recombinant interleukin-2: a pilot trial. *Journal of Immunotherapy Emphasis on Tumor Immunology* 16:198-210.
82. Hatton, JP, Lewis, ML, Roquefeuil, SB, Chaput, D, Cazenave, JP, Schmitt, DA (1998) Use of an adaptable cell culture kit for performing lymphocyte and monocyte cell cultures in microgravity. *Journal of Cellular Biochemistry* 70:252-267.

83. Muul, LM, Nason-Burchenal, K, Carter, CS, Cullis, H, Slavin, D, Hyatt, C, Director, EP, Leitman, SF, Klein, HG, Rosenberg, SA (1987) Development of an automated closed system for generation of human lymphokine-activated killer (LAK) cells for use in adoptive immunotherapy. *Journal of Immunological Methods* 101:171-181.
84. Levine, BL, Cotte, J, Small, CC, Carroll, RG, Riley, JL, Bernstein, WB, Van Epps, D, Hardwick, RA, June, CH (1998) Large scale production of CD4+ T-cells from HIV-1 infected donors following CD3/CD28 co-stimulation. *Journal of Hematotherapy* (in press).
85. Nakamura, Y, Tokuda, Y, Iwasawa, M, Tsukamoto, H, Kidokoro, M, Kobayashi, N, Kato, S, Mitomi, T, Habu, S, Nishimura, T (1992) Large-scale culture system of human CD4+ helper/killer T-cells for the application to adoptive tumour immunotherapy. *British Journal of Cancer* 66:20-26.
86. Carter, CS, Leitman, SF, Cullis, H, Muul, LM, Nason-Burchenal, K, Rosenberg, SA, Klein, HG (1988) Technical aspects of lymphokine-activated killer cell production. *Journal of Clinical Apheresis.* 4:113-117.
87. Sekine, T, Shiraiwa, H, Yamazaki, T, Tobisu, K, Kakizoe, T (1993) A feasible method for expansion of peripheral blood lymphocytes by culture with immobilized anti-CD3 monoclonal antibody and interleukin-2 for use in adoptive immunotherapy of cancer patients. *Biomedical Pharmacotherapeutics* 47:73-78.
88. Walker, RE, Carter, CS, Muul, L, Natarajan, V, Herpin, BR, Leitman, SF, Klein, HG, Mullen, CA, Metcalf, JA, Baseler, M, Falloon, J, Davey, RTJ, Kovacs, JA, Polis, MA, Masur, H, Blaese, RM, Lane, HC (1998) Peripheral expansion of pre-existing mature T-cells is an important means of CD4+ T-cell regeneration HIV-infected adults. *Nature Medicine* 4:852-856.
89. Robinet, E, Certoux, JM, Ferrand, C, Maples, P, Hardwick, A, Cahn, JY, Reynolds, CW, Jacob, W, Herve, P, Tiberghien, P (1998) A closed culture system for the *ex vivo* transduction and expansion of human T-lymphocytes. *Journal of Hematotherapy* 7:205-215.
90. de Waal, M, Yssel, H, de Vries, JE (1993) Direct effects of IL-10 on subsets of human CD4+ T-cell clones and resting T-cells. Specific inhibition of IL-2 production and proliferation. *Journal of Immunology* 150:4754-4765.
91. Groux, H, Bigler, M, de Vries, JE, Roncarolo, MG (1998) Inhibitory and stimulatory effects of IL-10 on human CD8+ T-cells. *Journal of Immunology* 160:3188-3193.
92. Licastro, F, Davis, LJ, Morini, MC (1993) Lectins and superantigens: membrane interactions of these compounds with T-lymphocytes affect immune responses. *International Journal of Biochemistry* 25:845-852.
93. Hadden, JW (1988) Transmembrane signals in the activation of T-lymphocytes by lectin mitogens. *Molecular Immunology* 25:1105-1112.
94. Ahmann, GB, Sachs, DH, Hodes, RJ (1978) Requirement for an Ia-bearing accessory cell in Con A-induced T-cell proliferation. *Journal of Immunology* 121:1981-1989.
95. Perillo, NL, Walford, RL, Newman, MA, Effros, RB (1989) Human T-lymphocytes possess a limited *in vitro* life span. *Experimental Gerontology* 24:177-187.
96. Truneh, A, Albert, F, Golstein, P, Schmitt-Verhulst, AM (1985) Early steps of lymphocyte activation bypassed by synergy between calcium ionophores and phorbol ester. *Nature* 313:318-320.
97. Niedel, JE, Kuhn, LJ, Vandenbark, GR (1983) Phorbol diester receptor copurifies with protein kinase C. *Proceedings of the National Academy of Sciences USA* 80:36-40.
98. Blumberg, PM (1988) Protein kinase C as the receptor for the phorbol ester tumor promoters: sixth Rhoads memorial award lecture. *Cancer Research* 48:1-8.

99. Berkow, RL and Kraft, AS (1985) Bryostatin, a non-phorbol macrocyclic lactone, activates intact human polymorphonuclear leukocytes and binds to the phorbol ester receptor. *Biochemical and Biophysical Research Communications* 131:1109-1116.
100. Weiss, A and Littman, DR (1994) Signal transduction by lymphocyte antigen receptors. *Cell* 76:263-274.
101. Abb, J, Bayliss, GJ, Deinhardt, F (1979) Lymphocyte activation by the tumor-promoting agent 12-O- tetradecanoylphorbol-13-acetate (TPA). *Journal of Immunology* 122:1639-1642.
102. Hess, AD, Silanskis, MK, Esa, AH, Pettit, GR, May, WS (1988) Activation of human T-lymphocytes by bryostatin. *Journal of Immunology* 141:3263-3269.
103. Schuchter, LM, Esa, AH, May, S, Laulis, MK, Pettit, GR, Hess, AD (1991) Successful treatment of murine melanoma with bryostatin 1. *Cancer Research* 51:682-687.
104. van der Hem, KG, Schuurhuis, GJ, Drager, AM, Odding, JH, Huijgens, PC (1996) Heterogenous effects of bryostatin on human myeloid leukemia clonogenicity: dose and time scheduling dependency. *Leukemia Research* 20:743-750.
105. Chatila, T, Silverman, L, Miller, R, Geha, R (1989) Mechanisms of T-cell activation by the calcium ionophore ionomycin. *Journal of Immunology* 143:1283-1289.
106. Weiss, A and Imboden, JB (1987) Cell surface molecules and early events involved in human T-lymphocyte activation. *Advances in Immunology* 41:1-38.
107. Topalian, SL, Muul, LM, Solomon, D, Rosenberg, SA (1987) Expansion of human tumor infiltrating lymphocytes for use in immunotherapy trials. *Journal of Immunological Methods* 102:127-141.
108. Bich-Thuy, LT, Lane, HC, Fauci, AS (1986) Recombinant interleukin-2-induced polyclonal proliferation of *in vitro* unstimulated human peripheral blood lymphocytes. *Cellular Immunology* 98:396-410.
109. Wu, CY, Demeure, CE, Gately, M, Podlaski, F, Yssel, H, Kiniwa, M, Delespesse, G (1994) *In vitro* maturation of human neonatal CD4 T-lymphocytes. I. Induction of IL-4-producing cells after long-term culture in the presence of IL-4 plus either IL-2 or IL-12. *Journal of Immunology* 152:1141-1153.
110. Liu, SQ, Shiba, R, Kim, BS, Saijo, K, Ohno, T (1994) Long-term serum/plasma-free culture of human cytotoxic T-lymphocytes induced from peripheral blood MNC. *Cancer Immunology and Immunotherapy* 39:279-285.
111. Tsujitani, S, Nakashima, M, Watanabe, T, Kaibara, N, Koprowski, H, Steplewski, Z (1995) Cytokine combinations for induction of antigen-specific cytolytic T-lymphocytes from peripheral blood lymphocytes. *Anticancer Research* 15:655-660.
112. Unutmaz, D, Baldoni, F, Abrignani, S (1995) Human naive T-cells activated by cytokines differentiate into a split phenotype with functional features intermediate between naive and memory T-cells. *International Immunology* 7:1417-1424.
113. Fukui, T, Katamura, K, Abe, N, Kiyomasu, T, Iio, J, Ueno, H, Mayumi, M, Furusho, K (1997) IL-7 induces proliferation, variable cytokine-producing ability and IL- 2 responsiveness in naive CD4+ T-cells from human CB. *Immunology Letters* 59:21-28.
114. Webb, LM, Foxwell, BM, Feldmann, M (1997) Interleukin-7 activates human naive CD4+ cells and primes for interleukin-4 production. *European Journal of Immunology* 27:633-640.
115. Lalvani, A, Dong, T, Ogg, G, Patham, AA, Newell, H, Hill, AV, McMichael, AJ, Rowland-Jones, S (1997) Optimization of a peptide-based protocol employing IL-7 for *in vitro* restimulation of human cytotoxic T-lymphocyte precursors. *Journal of Immunological Methods* 210:65-77.
116. Trinchieri, G (1998) Immunobiology of interleukin-12. *Immunologic Research* 17:269-278.

117. Mehrotra, PT, Grant, AJ, Siegel, JP (1995) Synergistic effects of IL-7 and IL-12 on human T-cell activation. *Journal of Immunology* 154:5093-5102.

118. Kubin, M, Kamoun, M, Trinchieri, G (1994) Interleukin 12 synergizes with B7/CD28 interaction in inducing efficient proliferation and cytokine production of human T-cells. *Journal of Experimental Medicine* 180:211-222.

119. Parada, NA, Center, DM, Kornfeld, H, Rodriguez, WL, Cook, J, Vallen, M, Cruikshank, WW (1998) Synergistic activation of CD4+ T-cells by IL-16 and IL-2. *Journal of Immunology* 160:2115-2120.

120. Gullberg, M and Smith, KA (1986) Regulation of T-cell autocrine growth. T4+ cells become refractory to interleukin 2. *Journal of Experimental Medicine* 163:270-284.

121. Uberti, JP, Joshi, I, Ueda, M, Martilotti, F, Sensenbrenner, LL, Lum, LG (1994) Preclinical studies using immobilized OKT3 to activate human T-cells for adoptive immunotherapy: optimal conditions for the proliferation and induction of non-MHC-restricted cytotoxicity. *Clinical Immunology and Immunopathology* 70:234-240.

122. Blaese, RM, Culver, KW, Miller, AD, Carter, CS, Fleisher, T, Clerici, M, Shearer, G, Chang, L, Chiang, Y, Tolstoshev, P (1995) T-lymphocyte-directed gene therapy for ADA-SCID: initial trial results after 4 years. *Science* 270:475-480.

123. Liu, Y and Janeway, CA, Jr. (1992) Cells that present both specific ligand and co-stimulatory activity are the most efficient inducers of clonal expansion of normal CD4 T-cells. *Proceedings of the National Academy of Sciences USA* 89:3845-3849.

124. Levine, BL, Mosca, JD, Riley, JL, Carroll, RG, Vahey, MT, Jagodzinski, LL, Wagner, KF, Mayers, DL, Burke, DS, Weislow, OS, St.Louis, DC, June, CH (1996) Antiviral effect and *ex vivo* CD4+ T-cell proliferation in HIV- positive patients as a result of CD28 co-stimulation. *Science* 272:1939-1943.

125. Carroll, RG, Riley, JL, Levine, BL, Feng, Y, Kaushal, S, Ritchey, DW, Bernstein, W, Weislow, OS, Brown, CR, Berger, EA, June, CH, St.Louis, DC (1997) Differential regulation of HIV-1 fusion cofactor expression by CD28 co-stimulation of CD4+ T-cells. *Science* 276:273-276.

126. Rowland-Jones, S and Tan, R (1997) Control of HIV co-receptor expression: implications for pathogenesis and treatment. *Trends in Microbiology* 5:300-2; discussion 302-3.

127. Pohl, C, Denfeld, R, Renner, C, Jung, W, Bohlen, H, Sahin, U, Hombach, A, van Lier, R, Schwonzen, M, Diehl, V (1993) CD30-antigen-specific targeting and activation of T-cells via murine bispecific monoclonal antibodies against CD3 and CD28: potential use for the treatment of Hodgkin's lymphoma. *International Journal of Cancer* 54:820-827.

128. Lamers, CH, van de Griend RJ, Braakman, E, Ronteltap, CP, Benard, J, Stoter, G, Gratama, JW, Bolhuis, RL (1992) Optimization of culture conditions for activation and large-scale expansion of human T-lymphocytes for bispecific antibody-directed cellular immunotherapy. *International Journal of Cancer* 51:973-979.

129. Costello, R, Cerdan, C, Pavon, C, Brailly, H, Hurpin, C, Mawas, C, Olive, D (1993) The CD2 and CD28 adhesion molecules induce long-term autocrine proliferation of $CD4^+$ T-cells. *European Journal of Immunology* 23:608-613.

130. Hurtado, JC, Kim, YJ, Kwon, BS (1997) Signals through 4-1BB are co-stimulatory to previously activated splenic T-cells and inhibit activation-induced cell death. *Journal of Immunology* 158:2600-2609.

131. Kim, YJ, Kim, SH, Mantel, P, Kwon, BS (1998) Human 4-1BB regulates CD28 co-stimulation to promote Th1 cell responses. *European Journal of Immunology* 28:881-890.

132. Carrera, AC, Rincön, M, S:anchez-Madrid, F, Löpez-Botet, M, de Landazuri, MO (1988) Triggering of co-mitogenic signals in T-cell proliferation by anti-LFA-1 (CD18,

CD11a), LFA-3, and CD7 monoclonal antibodies. *Journal of Immunology* 141:1919-1924.

133. Waclavicek, M, Majdic, O, Stulnig, T, Berger, M, Baumruker, T, Knapp, W, Pickl, WF (1997) T-cell stimulation via CD47: agonistic and antagonistic effects of CD47 monoclonal antibody 1/1A4. *Journal of Immunology* 159:5345-5354.
134. Hintzen, RQ, Lens, SM, Lammers, K, Kuiper, H, Beckmann, MP, van, LR (1995) Engagement of CD27 with its ligand CD70 provides a second signal for T-cell activation. *Journal of Immunology* 154:2612-2623.
135. Kappler, J, Kotzin, B, Herron, L, Gelfand, EW, Bigler, RD, Boylston, A, Carrel, S, Posnett, DN, Choi, Y, Marrack, P (1989) V beta-specific stimulation of human T-cells by staphylococcal toxins. *Science* 244:811-813.
136. Chatila, T and Geha, RS (1993) Signal transduction by microbial superantigens via MHC class II molecules. *Immunological Reviews* 131:43-59:43-59.
137. Kotzin, BL, Leung, DY, Kappler, J, Marrack, P (1993) Superantigens and their potential role in human disease. *Advances in Immunology* 54:99-166:99-166.
138. Heeg, K and Wagner, H (1995) Induction of responsiveness in superantigen-induced anergic T-cells. Role of ligand density and co-stimulatory signals. *Journal of Immunology* 155:83-92.
139. Harding, CV (1996) Class II antigen processing: analysis of compartments and functions. *Critical Reviews in Immunology* 16:13-29.
140. Koopmann, JO, Hammerling, GJ, Momburg, F (1997) Generation, intracellular transport and loading of peptides associated with MHC class I molecules. *Current Opinion in Immunology* 9:80-88.
141. Ahmed, R and Gray, D (1996) Immunological memory and protective immunity: understanding their relation. *Science* 272:54-60.
142. Steinman, RM (1991) The dendritic cell system and its role in immunogenicity. *Annual Review of Immunology* 9:271-96:271-296.
143. Hockett, RD, Cook, JR, Findlay, K, Harding, CV (1996) Interferon-gamma differentially regulates antigen-processing functions in distinct endocytic compartments of macrophages with constitutive expression of class II major histocompatibility complex molecules. *Immunology* 88:68-75.
144. Sugden, B and Mark, W (1977) Clonal transformation of adult human leukocytes by Epstein-Barr virus. *Journal of Virology* 23:503-508.
145. Grewal, IS and Flavell, RA (1996) The role of CD40 ligand in co-stimulation and T-cell activation. *Immunological Reviews* 153:85-106:85-106.
146. Schultze, JL, Michalak, S, Seamon, MJ, Dranoff, G, Jung, K, Daley, J, Delgado, JC, Gribben, JG, Nadler, LM (1997) CD40-activated human B-cells: an alternative source of highly efficient APC to generate autologous antigen-specific T-cells for adoptive immunotherapy. *Journal of Clinical Investigation* 100:2757-2765.
147. Zhou, LJ and Tedder, TF (1995) Human blood dendritic cells selectively express CD83, a member of the immunoglobulin superfamily. *Journal of Immunology* 154:3821-3835.
148. Mehta-Damani, A, Markowicz, S, Engleman, EG (1995) Generation of antigen-specific CD4+ T-cell lines from naive precursors. *European Journal of Immunology* 25:1206-1211.
149. Thomas, R, Davis, LS, Lipsky, PE (1993) Isolation and characterization of human peripheral blood dendritic cells. *Journal of Immunology* 150:821-834.
150. O'Doherty, U, Steinman, RM, Peng, M, Cameron, PU, Gezelter, S, Kopeloff, I, Swiggard, WJ, Pope, M, Bhardwaj, N (1993) Dendritic cells freshly isolated from human blood express CD4 and mature into typical immunostimulatory dendritic cells after culture in monocyte-conditioned medium. *Journal of Experimental Medicine* 178:1067-1076.

151. Romani, N, Reider, D, Heuer, M, Ebner, S, Kampgen, E, Eibl, B, Niederwieser, D, Schuler, G (1996) Generation of mature dendritic cells from human blood. An improved method with special regard to clinical applicability. *Journal of Immunological Methods* 196:137-151.
152. Sallusto, F and Lanzavecchia, A (1994) Efficient presentation of soluble antigen by cultured human dendritic cells is maintained by granulocyte/macrophage colony-stimulating factor plus interleukin 4 and downregulated by tumor necrosis factor alpha. *Journal of Experimental Medcine* 179:1109-1118.
153. Zhou, LJ and Tedder, TF (1996) CD14+ blood monocytes can differentiate into functionally mature CD83+ dendritic cells. *Proceedings of the National Academy of Sciences USA* 93:2588-2592.
154. Caux, C, Massacrier, C, Vanbervliet, B, Dubois, B, van Kooten, C, Durand, I, Banchereau, J (1994) Activation of human dendritic cells through CD40 cross-linking. *Journal of Experimental Medcine* 180:1263-1272.
155. Cella, M, Scheidegger, D, Palmer-Lehmann, K, Lane, P, Lanzavecchia, A, Alber, G (1996) Ligation of CD40 on dendritic cells triggers production of high levels of interleukin-12 and enhances T-cell stimulatory capacity: T-T help via APC activation. *Journal of Experimental Medcine* 184:747-752.
156. Caux, C, Dezutter-Dambuyant, C, Schmitt, D, Banchereau, J (1992) GM-CSF and TNF-alpha cooperate in the generation of dendritic Langerhans cells. *Nature* 360:258-261.
157. de Saint-Vis, B, Fugier-Vivier, I, Massacrier, C, Gaillard, C, Vanbervliet, B, Ait-Yahia, S, Banchereau, J, Liu, YJ, Lebecque, S, Caux, C (1998) The cytokine profile expressed by human dendritic cells is dependent on cell subtype and mode of activation. *Journal of Immunology* 160:1666-1676.
158. Greten, TF, Slansky, JE, Kubota, R, Soldan, SS, Jaffee, EM, Leist, TP, Pardoll, DM, Jacobson, S, Schneck, JP (1998) Direct visualization of antigen-specific T-cells: HTLV-1 Tax11-19- specific CD8(+) T-cells are activated in peripheral blood and accumulate in cerebrospinal fluid from HAM/TSP patients. *Proceedings of the National Academy of Sciences USA* 95:7568-7573.
159. Kozono, H, White, J, Clements, J, Marrack, P, Kappler, J (1994) Production of soluble MHC class II proteins with covalently bound single peptides. *Nature* 369:151-154.
160. Zhu, X, Bavari, S, Ulrich, R, Sadegh-Nasseri, S, Ferrone, S, McHugh, L, Mage, M (1997) A recombinant single-chain human class II MHC molecule (HLA-DR1) as a covalently linked heterotrimer of alpha chain, beta chain, and antigenic peptide, with immunogenicity *in vitro* and reduced affinity for bacterial superantigens. *European Journal of Immunology* 27:1933-1941.
161. Bousso, P, Michel, F, Pardigon, N, Bercovici, N, Liblau, R, Kourilsky, P, Abastado, JP (1997) Enrichment of antigen-specific T-lymphocytes by panning on immobilized MHC-peptide complexes. *Immunology Letters* 59:85-91.
162. Riddell, SR and Greenberg, PD (1990) The use of anti-CD3 and anti-CD28 monoclonal antibodies to clone and expand human antigen-specific T-cells. *Journal of Immunological Methods* 128:189-201.
163. MacDonald, HR, Cerottini, JC, Ryser, JE, Maryanski, JL, Taswell, C, Widmer, MB, Brunner, KT (1980) Quantitation and cloning of cytolytic T-lymphocytes and their precursors. *Immunological Reviews* 51:93-123:93-123.
164. Modlin, RL, Kato, H, Mehra, V, Nelson, EE, Fan, XD, Rea, TH, Pattengale, PK, Bloom, BR (1986) Genetically restricted suppressor T-cell clones derived from lepromatous leprosy lesions. *Nature* 322:459-461.
165. Yee, C, Gilbert, MJ, Riddell, SR, Brichard, VG, Fefer, A, Thompson, JA, Boon, T, Greenberg, PD (1996) Isolation of tyrosinase-specific CD8+ and CD4+ T-cell clones from

the peripheral blood of melanoma patients following *in vitro* stimulation with recombinant vaccinia virus. *Journal of Immunology* 157:4079-4086.

166. Gong, J, Chen, L, Chen, D, Kashiwaba, M, Manome, Y, Tanaka, T, Kufe, D (1997) Induction of antigen-specific antitumor immunity with adenovirus- transduced dendritic cells. *Gene Therapy* 4:1023-1028.

167. Boczkowski, D, Nair, SK, Snyder, D, Gilboa, E (1996) Dendritic cells pulsed with RNA are potent antigen-presenting cells *in vitro* and *in vivo*. *Journal of Experimental Medicine* 184:465-472.

168. Yang, NS and Sun, WH (1995) Gene gun and other non-viral approaches for cancer gene therapy. *Nature Medicine* 1:481-483.

169. Gong, J, Chen, D, Kashiwaba, M, Kufe, D (1997) Induction of antitumor activity by immunization with fusions of dendritic and carcinoma cells. *Nature Medicine* 3:558-561.

170. Wentworth, PA, Celis, E, Crimi, C, Stitely, S, Hale, L, Tsai, V, Serra, HM, Del Guercio, MF, Livingston, B, Alazard, D (1995) *In vitro* induction of primary, antigen-specific CTL from human peripheral blood MNC stimulated with synthetic peptides. *Molecular Immunology* 32:603-612.

171. Peshwa, MV, Page L.A., Qian, L, Yang, D, van Schooten W.C.A. (1996) Generation and *Ex vivo* Expansion of HTLV-1 Specific CD8+ T-Lymphocytes for Adoptive Immunotherapy. *Biotechnology and Bioengineering* 50:529-540.

172. Lieberman, J, Skolnik, PR, Parkerson, GR3, Fabry, JA, Landry, B, Bethel, J, Kagan, J (1997) Safety of autologous, *ex vivo*-expanded human immunodeficiency virus (HIV)-specific cytotoxic T-lymphocyte infusion in HIV- infected patients. *Blood* 90:2196-2206.

173. Bach, FH, Inouye, H, Hank, JA, Alter, BJ (1979) Human T-lymphocyte clones reactive in primed lymphocyte typing and cytotoxicity. *Nature* 281:307-309.

174. Lefkovits, I and Waldman, H (1979) *Limiting dilution analysis of cells in the immune system,* Cambridge University Press, Cambridge

175. Glasebrook, AL and Fitch, FW (1984) Derivation of T-cell clones, in RH Kennet, KB Bechtol, TJ McKearn (eds.), *Monoclonal antibodies and functional cell lines,Plenum,* New York, pp. 413-417

176. Altman, JD, Moss, PH, Goulder, PR, Barouch, DH, McHeyzer-Williams, MG, Bell, JI, McMichael, AJ, Davis, MM (1996) Phenotypic analysis of antigen-specific T-lymphocytes. *Science* 274:94-96.

177. Dunbar, PR, Ogg, GS, Chen, J, Rust, N, van der Bruggen, P, Cerundolo, V (1998) Direct isolation, phenotyping and cloning of low-frequency antigen- specific cytotoxic T-lymphocytes from peripheral blood. *Current Biology* 8:413-416.

178. De Libero, G (1997) Sentinel function of broadly reactive human gamma delta T-cells. *Immunology Today* 18:22-26.

179. Morita, CT, Beckman, EM, Bukowski, JF, Tanaka, Y, Band, H, Bloom, BR, Golan, DE, Brenner, MB (1995) Direct presentation of nonpeptide prenyl pyrophosphate antigens to human gamma delta T-cells. *Immunity* 3:495-507.

180. Garcia, VE, Sieling, PA, Gong, J, Barnes, PF, Uyemura, K, Tanaka, Y, Bloom, BR, Morita, CT, Modlin, RL (1997) Single-cell cytokine analysis of gamma delta T-cell responses to nonpeptide mycobacterial antigens. *Journal of Immunology* 159:1328-1335.

181. Porcelli, S, Morita, CT, Brenner, MB (1992) CD1b restricts the response of human CD4-8- T-lymphocytes to a microbial antigen. *Nature* 360:593-597.

182. Beckman, EM, Porcelli, SA, Morita, CT, Behar, SM, Furlong, ST, Brenner, MB (1994) Recognition of a lipid antigen by CD1-restricted alpha beta+ T-cells. *Nature* 372:691-694.

183. Sieling, PA, Chatterjee, D, Porcelli, SA, Prigozy, TI, Mazzaccaro, RJ, Soriano, T, Bloom, BR, Brenner, MB, Kronenberg, M, Brennan, PJ (1995) CD1-restricted T-cell recognition of microbial lipoglycan antigens. *Science* 269:227-230.

184. Stenger, S, Mazzaccaro, RJ, Uyemura, K, Cho, S, Barnes, PF, Rosat, JP, Sette, A, Brenner, MB, Porcelli, SA, Bloom, BR, Modlin, RL (1997) Differential effects of cytolytic T-cell subsets on intracellular infection. *Science* 276:1684-1687.

185. Kachel, V(1990) Electrical Resistance Pulse Sizing:Coulter Sizing, in MR Melamed, T Lindmo, ML Mendelsohn (eds.), *Flow Cytometry and Sorting, Wiley-Liss,* New York, pp. 45-80

186. Kubbies, M, Schindler, D, Hoehn, H, Rabinovitch, PS (1985) Cell cycle kinetics by BrdU-Hoechst flow cytometry: an alternative to the differential metaphase labelling technique. *Cell and Tissue Kinetics* 18:551-562.

187. Shapiro, HM (1981) Flow cytometric estimation of DNA and RNA content in intact cells stained with Hoechst 33342 and pyronin Y. *Cytometry* 2:143-150.

188. Darzynkiewicz, Z, Li, X,Gong, J (1994) Assays of cell viability. Discrimination of cells dying by apoptosis, in Z Darzynkiewicz, HA Crissman, JP Robinson (eds.), *Methods in Cell Biology: Flow Cytometry, Academic Press*

189. Riedy, MC, Muirhead, KA, Jensen, CP, Stewart, CC (1991) Use of a photolabeling technique to identify nonviable cells in fixed homologous or heterologous cell populations. *Cytometry* 12:133-139.

190. Gorczyca, W, Gong, J, Darzynkiewicz, Z (1993) Detection of DNA strand breaks in individual apoptotic cells by the in situ terminal deoxynucleotidyl transferase and nick translation assays. *Cancer Research* 53:1945-1951.

191. Horan, PK, Melnicoff, MJ, Jensen, BD, Slezak, SE (1990) Fluorescent cell labeling for *in vivo* and *in vitro* cell tracking. *Methods in Cell Biology* 33:469-90:469-490.

192. Yamamura, Y, Rodriguez, N, Schwartz, A, Eylar, E, Bagwell, B, Yano, N (1995) A new flow cytometric method for quantitative assessment of lymphocyte mitogenic potentials. *Cellular and Molecular Biology (Noisy-le-grand)* 41 Suppl 1:S121-32.

193. Lyons, AB and Parish, CR (1994) Determination of lymphocyte division by flow cytometry. *Journal of Immunological Methods* 171:131-137.

194. Wells, AD, Gudmundsdottir, H, Turka, LA (1997) Following the fate of individual T-cells throughout activation and clonal expansion. Signals from T-cell receptor and CD28 differentially regulate the induction and duration of a proliferative response. *Journal of Clinical Investigation* 100:3173-3183.

195. Gett, AV and Hodgkin, PD (1998) Cell division regulates the T-cell cytokine repertoire, revealing a mechanism underlying immune class regulation. *Proceedings of the National Academy of Sciences USA* 95:9488-9493.

196. Pannetier, C, Even, J, Kourilsky, P (1995) T-cell repertoire diversity and clonal expansions in normal and clinical samples. *Immunology Today* 16:176-181.

197. Cochet, M, Pannetier, C, Regnault, A, Darche, S, Leclerc, C, Kourilsky, P (1992) Molecular detection and *in vivo* analysis of the specific T-cell response to a protein antigen. *European Journal of Immunology* 22:2639-2647.

198. Pannetier, C, Cochet, M, Darche, S, Casrouge, A, Zoller, M, Kourilsky, P (1993) The sizes of the CDR3 hypervariable regions of the murine T-cell receptor beta chains vary as a function of the recombined germ- line segments. *Proceedings of the National Academy of Sciences USA* 90:4319-4323.

199. Gorski, J, Yassai, M, Zhu, X, Kissella, B, Keever, C, Flomenberg, N (1994) Circulating T-cell repertoire complexity in normal individuals and BM recipients analyzed by CDR3 size spectratyping. Correlation with immune status. *Journal of Immunology* 152:5109-5119.

200. Arenz, M, Herzog-Hauff, S, Meyer, zum Buschenfelde KH, Lohr, HF (1997) Antigen-independent *in vitro* expansion of T-cells does not affect the T-cell receptor V beta repertoire. *Journal of Molecular Medicine* 75:678-686.

201. Caignard, A, Guillard, M, Gaudin, C, Escudier, B, Triebel, F, Dietrich, PY (1996) In situ demonstration of renal-cell-carcinoma-specific T-cell clones. *International Journal of Cancer* 66:564-570.
202. Dietrich, PY, Walker, PR, Schnuriger, V, Saas, P, Perrin, G, Guillard, M, Gaudin, C, Caignard, A (1997) TCR analysis reveals significant repertoire selection during *in vitro* lymphocyte culture. *International Immunology* 9:1073-1083.
203. Roederer, M, De Rosa, S, Gerstein, R, Anderson, M, Bigos, M, Stovel, R, Nozaki, T, Parks, D, Herzenberg, L (1997) 8 color, 10-parameter flow cytometry to elucidate complex leukocyte heterogeneity. *Cytometry* 29:328-339.
204. Shevach, EV, Sharrow, SO, Holmes, KL, Fowlkes, BJ, Schmid, I, Giorgi, JV, Otten, G, Yokoyama, WM, June, CH, Abe, R, Rabinaovitch, PS, Segal, DM, Noguchi, PD (1995) Immunofluorescence and Cell Sorting, in JE Coligan, AM Kruisbeek, DH Margulies, EM Shevach, W Strober (eds.), *Current Protocols in Immunology,John Wiley & Sons, Inc.*, pp. 5.0.1-5.8.8
205. Ault, KA (1990) Applications in Immunology and Lymphocyte Analysis, in MR Melamed, T Lindmo, ML Mendelsohn (eds.), *Flow Cytometry and Sorting,John Wiley & Sons, Inc.*, New York, pp. 685-696
206. Darzynkiewicz, Z, Robinson, JP, Crissman, HA (1994) Flow Cytometry, Second Edition, Part A. *Methods in Cell Biology 41.*
207. Darzynkiewicz, Z, Robinson, JP, Crissman, HA (1994) Flow Cytometry, Second Edition, Part B. *Methods in Cell Biology 42.*
208. Kuroda, MJ, Schmitz, JE, Barouch, DH, Craiu, A, Allen, TM, Sette, A, Watkins, DI, Forman, MA, Letvin, NL (1998) Analysis of Gag-specific cytotoxic T-lymphocytes in simian immunodeficiency virus-infected rhesus monkeys by cell staining with a tetrameric major histocompatibility complex class I-peptide complex. *Journal of Experimental Medcine* 187:1373-1381.
209. Mire-Sluis, AR, Page, L, Thorpe, R (1995) Quantitative cell line based bioassays for human cytokines. *Journal of Immunological Methods* 187:191-199.
210. Prussin, C (1997) Cytokine flow cytometry: understanding cytokine biology at the single-cell level. *Journal of Clinical Immunology* 17:195-204.
211. Waldrop, SL, Pitcher, CJ, Peterson, DM, Maino, VC, Picker, LJ (1997) Determination of antigen-specific memory/effector CD4+ T-cell frequencies by flow cytometry: evidence for a novel, antigen- specific homeostatic mechanism in HIV-associated immunodeficiency. *Journal of Clinical Investigation* 99:1739-1750.
212. Suni, MA, Picker, LJ, Maino, VC (1998) Detection of antigen-specific T-cell cytokine expression in whole blood by flow cytometry. *Journal of Immunological Methods* 212:89-98.
213. Mentzer, SJ, Barbosa, JA, Burakoff, SJ (1985) T3 monoclonal antibody activation of nonspecific cytolysis: a mechanism of CTL inhibition. *Journal of Immunology* 135:34-38.
214. Brunner, KT, Mauel, J, Cerottini, JC, Chapuis, B (1968) Quantitative assay of the lytic action of immune lymphoid cells on 51- Cr-labelled allogeneic target cells *in vitro*; inhibition by isoantibody and by drugs. *Immunology* 14:181-196.
215. Shearer, GM, Salahuddin, SZ, Markham, PD, Joseph, LJ, Payne, SM, Kriebel, P, Bernstein, DC, Biddison, WE, Sarngadharan, MG, Gallo, RC (1985) Prospective study of cytotoxic T-lymphocyte responses to influenza and antibodies to human T-lymphotropic virus-III in homosexual men. Selective loss of an influenza-specific, human leukocyte antigen- restricted cytotoxic T-lymphocyte response in human T-lymphotropic virus-III positive individuals with symptoms of acquired immunodeficiency syndrome and in a patient with acquired immunodeficiency syndrome. *Journal of Clinical Investigation* 76:1699-1704.

216. Riviere, Y, Tanneau-Salvadori, F, Regnault, A, Lopez, O, Sansonetti, P, Guy, B, Kieny, MP, Fournel, JJ, Montagnier, L (1989) Human immunodeficiency virus-specific cytotoxic responses of seropositive individuals: distinct types of effector cells mediate killing of targets expressing gag and env proteins. *Journal of Virology* 63:2270-2277.
217. Lichtenfels, R, Biddison, WE, Schulz, H, Vogt, AB, Martin, R (1994) CARE-LASS (calcein-release-assay), an improved fluorescence-based test system to measure cytotoxic T-lymphocyte activity. *Journal of Immunological Methods* 172:227-239.
218. Mattis, AE, Bernhardt, G, Lipp, M, Forster, R (1997) Analyzing cytotoxic T-lymphocyte activity: a simple and reliable flow cytometry-based assay. *Journal of Immunological Methods* 204:135-142.
219. Hayflick L. and Moorhead P.S. (1961) The serial cultivation of human diploid cell strains. *Experimental Cell Research* 25:585.
220. Effros, RB and Walford, RL (1984) T-cell cultures and the Hayflick limit. *Human Immunology* 9:49-65.
221. Adibzadeh, M, Pohla, H, Rehbein, A, Pawelec, G (1995) Long-term culture of monoclonal human T-lymphocytes: models for immunosenescence? *Mechanisms in Ageing and Development* 83:171-183.
222. Harley, CB and Villeponteau, B (1995) Telomeres and telomerase in aging and cancer. *Current Opinion in Genetics and Development* 5:249-255.
223. Harley, CB, Futcher, AB, Greider, CW (1990) Telomeres shorten during ageing of human fibroblasts. *Nature* 345:458-460.
224. Vaziri, H, Schachter, F, Uchida, I, Wei, L, Zhu, X, Effros, R, Cohen, D, Harley, CB (1993) Loss of telomeric DNA during aging of normal and trisomy 21 human lymphocytes. *American Journal of Human Genetics* 52:661-667.
225. Bodnar, AG, Ouellette, M, Frolkis, M, Holt, SE, Chiu, CP, Morin, GB, Harley, CB, Shay, JW, Lichtsteiner, S, Wright, WE (1998) Extension of life-span by introduction of telomerase into normal human cells. *Science* 279:349-352.
226. Weng, NP, Levine, BL, June, CH, Hodes, RJ (1995) Human naive and memory T-lymphocytes differ in telomeric length and replicative potential. *Proceedings of the National Academy of Sciences USA* 92:11091-11094.
227. Weng, NP, Levine, BL, June, CH, Hodes, RJ (1996) Regulated expression of telomerase activity in human T-lymphocyte development and activation. *Journal of Experimental Medicine* 183:2471-2479.
228. Weng, NP, Palmer, LD, Levine, BL, Lane, HC, June, CH, Hodes, RJ (1997) Tales of tails: regulation of telomere length and telomerase activity during lymphocyte development, differentiation, activation, and aging. *Immunological Reviews* 160:43-54.
229. Rosenberg, SA, Lotze, MT, Muul, LM, Leitman, S, Chang, AE, Ettinghausen, SE, Matory, YL, Skibber, JM, Shiloni, E, Vetto, JT (1985) Observations on the systemic administration of autologous lymphokine-activated killer cells and recombinant interleukin-2 to patients with metastatic cancer. *New England Journal Medicine* 313:1485-1492.
230. Rosenberg, SA, Yannelli, JR, Yang, JC, Topalian, SL, Schwartzentruber, DJ, Weber, JS, Parkinson, DR, Seipp, CA, Einhorn, JH, White, DE (1994) Treatment of patients with metastatic melanoma with autologous tumor-infiltrating lymphocytes and interleukin 2. *Journal of the National Cancer Institute* 86:1159-1166.
232. Yee, C, Riddell, SR, Greenberg, PD (1997) Prospects for adoptive T-cell therapy. *Current Opinion in Immunology* 9:702-708.
233. Hall, SS (1997) *A commotion in the blood: Life, death, and the immune system,* Henry Holt and Company, New York.

234. Lamers, CH, Bolhuis, RL, Warnaar, SO, Stoter, G, Gratama, JW (1997) Local but no systemic immunomodulation by intraperitoneal treatment of advanced ovarian cancer with autologous T-lymphocytes re-targeted by a bi-specific monoclonal antibody. *International Journal of Cancer* 73:211-219.
235. Walter, EA, Greenberg, PD, Gilbert, MJ, Finch, RJ, Watanabe, KS, Thomas, ED, Riddell, SR (1995) Reconstitution of cellular immunity against cytomegalovirus in recipients of allogeneic BM by transfer of T-cell clones from the donor. *New England Journal of Medicine* 333:1038-1044.
236. Klimas, N, Patarca, R, Walling, J, Garcia, R, Mayer, V, Moody, D, Okarma, T, Fletcher, MA (1994) Clinical and immunological changes in AIDS patients following adoptive therapy with activated autologous CD8 T-cells and interleukin-2 infusion. *AIDS* 8:1073-1081.
237. Bordignon, C, Notarangelo, LD, Nobili, N, Ferrari, G, Casorati, G, Panina, P, Mazzolari, E, Maggioni, D, Rossi, C, Servida, P (1995) Gene therapy in peripheral blood lymphocytes and BM for ADA- immunodeficient patients. *Science* 270:470-475.
238. Heslop, HE, Ng, CY, Li, C, Smith, CA, Loftin, SK, Krance, RA, Brenner, MK, Rooney, CM (1996) Long-term restoration of immunity against Epstein-Barr virus infection by adoptive transfer of gene-modified virus-specific T-lymphocytes. *Nature Medicine* 2:551-555.
239. Ranga, U, Woffendin, C, Verma, S, Xu, L, June, CH, Bishop, DK, Nabel, GJ (1998) Enhanced T-cell engraftment after retroviral delivery of an antiviral gene in HIV-infected individuals. *Proceedings of the National Academy of Sciences USA* 95:1201-1206.

Chapter 4

The Culture, Characterization and Triggering of B-Lymphocytes

[1]Gerrard Teoh and [2]Kenneth C Anderson
[1]Division of Adult Oncology, Dana-Farber Cancer Institute and [2]Department of Medicine, Harvard Medical School, 44 Binney Street, Boston, MA 02115. Tel: 001-617-632-2144; Fax: 001-617-632-2569; E-mail: kenneth_anderson@dfci.harvard.edu

1. INTRODUCTION

B-lymphocytes are important effectors of the immune system. Activation of B-lymphocytes induces both proliferation and terminal differentiation into Ig secreting plasma cells. The intracellular processes that occur during B-lymphocyte activation often include structural changes to the Ig genes and involve double-stranded DNA strand break repair. Since malignant B-lymphocytes (i.e. chronic lymphocytic leukemia (CLL) and multiple myeloma (MM)) progressively acquire genetic aberrations during their evolution [1,2], it is tempting to speculate that signals dictating the growth and differentiation of B-lymphocytes might also trigger abnormal genetic events, including mutagenesis. Signals such as IL-6 [3,4] and CD40L [4-7], which have been shown to directly and indirectly influence the growth and survival of malignant MM B-cells, could potentially mediate genetic instability and clonal evolution in tumor cells. It is therefore essential when culturing and studying B-lymphocytes not only to define clearly the stages of differentiation but also to determine the responses of these cells to growth factors and other molecules.

1.1 Origin of Normal B-Lymphocytes

Greater than 25% of cells in the normal adult bone marrow (BM) are immature hematopoietic precursors. Normal B-lymphocytes are derivatives of lymphoid hematopoietic precursors from the BM, fetal liver and fetal omentum [8-10]. Immunocompetent B-lymphocytes survive for approximately 3 days unless rescued by antigen-specific and non-specific stimuli [11,12], and throughout life this normal B-lymphocyte repertoire is continually renewed by the growth and differentiation of pluripotent hematopoietic stem cells [13,14]. After migrating to germinal centers of lymphoid organs, B-lymphocytes undergo Ig isotype class switching (from IgM to IgG, IgA, IgD and IgE) and somatic hypermutation of the variable region of the Ig genes. Post-switched plasmablasts then re-enter the BM and under the influence of a host of factors, including contact with BM stromal cells via adhesion molecules, differentiate into long-lived (about 30 days) Ig secreting plasma cells [15,16].

1.2 Historical Aspects of the Culture of B-Lymphocytes

Since normal B-lymphocytes survive approximately three days when explanted, unless rescued by appropriate antigenic stimulation [11,12], culturing B-cells for longer periods *ex vivo* necessitates the delivery of mitogenic signals throughout the culture period. Short-term (2-4 weeks) culture of normal B-lymphocytes can be achieved by T-cell-dependent and independent methods, with or without additional B-cell mitogens [17-24]. These stimuli induce B-lymphocyte activation, proliferation and differentiation into Ig-secreting plasma cells [25]. Support of normal B-lymphocytes in T-cell-dependent culture systems can be accomplished by culturing B-cells with lethally γ-irradiated (CsCl source, 1.5-9.5 Gy) T-cells, with mitomycin C (25 μg/ml, 20 minutes, Sigma Diagnostics, St Louis, MO) treated T-cells, or with untreated T-cells cultured in a diffusion tissue culture chamber separated from B-cells by a filter (0.4-0.45 μm, Transwell #3413, Costar, Cambridge, MA) [19,26,27]. Additional T-cell mitogenic stimulation using phytohemagglutinin (PHA) (5-10 μg/ml, Sigma Diagnostics) or pokeweed mitogen (PWM) (1:100-3200, Sigma Diagnostics) may be included if necessary. Although PWM is principally a B-cell mitogen and causes capping of surface Ig on B-cells in the absence of T-cells [28], it requires T-cell help for full mitogenic activity [29].

In T-cell-independent B-lymphocyte culture systems, B-cell mitogens including *Staphylococcus aureus* cells (SAC) (1 μg/ml, Pansorbin, Calbiochem, La Jolla, CA), *Staphylococcus* Protein A (SPA) (10 μg/ml, Pharmacia Biotech AB, Uppsala, Sweden), and lipopolysaccharide (LPS)

(50 µg/ml, Sigma Diagnostics) may be used to support normal B-cells [22,30-32]. Cross-linking (i.e. by antigen (Ag), anti-IgM antibody (Ab), or dextran sulphate) cell surface Ig induces proliferation and differentiation of B-lymphocytes [33,34], and is associated with up-regulation of Src-family protein kinases (i.e. p53lyn and p56lyn) and tyrosine phosphorylation of several downstream peptides [35,36]. The generation of T-cell-independent long-term (>12 months) B-lymphocyte cultures was first achieved by transformation of normal B-lymphocytes into B-cell lines by Epstein-Barr virus (EBV) [37,38]. Epstein-Barr virus-infected B-cells are clonally transformed and immortalized, permitting very important studies of clonal variations within normal B-cells.

At about the same time, long-term (3-4 months) culture of stem cells, granulocytes and monocytes was made possible by the prior establishment of an adherent layer of feeder BM stromal cells [39]. Indeed, differentiation of B-lymphocytes is dependent on direct contact with BM stromal cells [40] mediated by adhesion molecules, which may be induced by CD40L activation [7,41,42]. B-lymphocyte longevity arises from antigenic stimulation of the B-cell receptor by BM stromal cells, which produces a weak but sufficient signal for the cell to survive [12]. Culturing mononuclear cells (MNC) at a high cell density initially facilitates the development of an adherent layer consisting of macrophages, adipocytes and dendritic cells (DCs) which ultimately support the long-term (>12 months) culture of B-lymphocytes [43,44]. Moreover, using this method, B-lymphocyte enrichment is also simultaneously achieved through the selective depletion of erythroid and myeloid lineage cells. More recently, cultures of B-lymphocytes on specialized feeder cells (S17 stromal cell line) [45] have accelerated (3-4 weeks vs. >5 weeks) the establishment of long-term B-lymphocyte cultures [46], providing an important tool for B-lymphocyte studies.

BM stromal cells are essential for providing a microenvironment for normal hematopoiesis by mediating intercellular signals through cell-to-cell contact or cytokines [47,48]. We have shown that adhesion of MM cells to BM stromal cells induces the secretion of IL-6 from BM stromal cells [49]. By combining exogenous growth factors (i.e. CD40 + IL-4), non-EBV transformed normal B- as well as malignant B-lymphocytes (i.e. CLL and non-Hodgkin's lymphoma) can be maintained in long-term culture without the need for feeder cells [50-57].

Finally, cultured B-lymphocytes can now be targets for gene therapy. However, stably transfected B-cells have been difficult to clone. Recently, MRC-5 fibroblasts transduced with the hygromycin B-resistance gene were used as feeder cells to produce stable B-lymphocyte transfectants [58]. The efficiency of cloning human B-lymphocytes was greatly increased (81-100%

vs 20-35%) when compared to cultures of transfected B-cells on semi-solid media (i.e agarose), with or without untransduced feeder cells [34,59,60].

2. TISSUE PROCUREMENT AND PROCESSING

2.1 Sources of Normal B-Lymphocytes

In normal humans, pools of B-lymphocytes can be found in peripheral blood, BM, tonsils, spleen, lymph nodes and Peyer's patches. Significant numbers (20-30 million/ml) of MNC can be obtained by leukapheresis of peripheral blood (i.e. leukopacks) on an apheresis cell sorter (e.g. Cobe Spectra, Cobe BCT, Inc., Lakewood, CO) without substantial distress to the donor. Conversely, obtaining B-lymphocytes from BM requires a sterile minor surgical field and local or regional anesthesia. The yield from BM harvesting (8-10 million/ml) is also less than from leukopacks. Unlike resting peripheral blood and BM B-lymphocytes, tonsillar B-lymphocytes obtained from patients undergoing tonsillectomies for chronic tonsillitis have often been activated *in vivo*. Finally, normal B splenocytes are an uncommon source of B-lymphocytes, usually obtained from cadaveric organ donors.

2.2 Separation of Peripheral Blood and Bone Marrow B-Lymphocytes

2.2.1 Ficoll-Hypaque density gradient sedimentation

Heparinized peripheral blood and BM samples are first diluted with 2-3 volumes of HBSS or RPMI 1640 medium (both Gibco BRL, Gaithersburg, MD) and thoroughly mixed to ensure that cells are well-dispersed in the suspension. Twenty-five milliliters of Ficoll-Hypaque (Pharmacia Biotech AB) are dispensed into each 50 ml conical centrifuge tube (Falcon #2070, Becton-Dickinson, Lincoln Park, NJ) and the cell suspension, equally distributed over an appropriate number of conical tubes, is carefully overlaid on the ficoll layer. Density gradient sedimentation is achieved by centrifugation at 200 *g* for 10 minutes at 16 ^{o}C followed by 400 *g* for 15 minutes at 16 ^{o}C in a bench centrifuge. Next, 80% of the upper medium phase is discarded and the MNC layer at the medium-ficoll interface is carefully isolated by aspiration. MNC are resuspended in 2-3 volumes of fresh HBSS or RPMI 1640 medium in a new conical tube and washed thrice with douncing and centrifugation at 400 *g* for 5 minutes at room temperature to remove the excess ficoll.

2.2.2 Erythrocyte lysis

The final pellet is resuspended in 1 ml of erythrocyte lysing solution and incubated for 5 minutes at room temperature. Next, pellet the cells at 400 *g* for 3 minutes at room temperature and discard the red supernatant containing hemoglobin from lysed erythrocytes. This procedure should be repeated until the supernatant is clear.

2.2.2.1 Erythrocyte lysing solution

0.83 g NH_4Cl (Sigma Diagnostics)
0.1 g $KHCO_3$ (Sigma Diagnostics)
3.72 mg EDTA (Sigma Diagnostics)
100 ml distilled water (DW)
Adjust solution to pH 7.2-7.4 and 280-290 mOsm, and filter through 0.22 μm cellulose acetate filter (Corning, Inc., Corning, NY).

2.3 Separation of Splenic and Tonsillar B-Lymphocytes

Solid tissue specimens such as the spleen and tonsils are first teased apart either manually using tissue-dissecting forceps or dispersed mechanically using a tissue homogenizer (e.g. Tissue Tearor Homogenizer, Fisher Scientific, Pittsburg, PA). This is done in calcium magnesium-free (CMF) phosphate-buffered saline (PBS) with or without 1 mM EDTA, which is used additionally if clumping is extensive.

2.3.1 CMF PBS

0.2 g KCl (Sigma Diagnostics)
0.2 g KH_2PO_4 (Sigma Diagnostics)
8.0 g NaCl (Sigma Diagnostics)
2.16 g $NaHPO_4.7H_20$ (Sigma Diagnostics)
1000 ml DW
Adjust to pH 7.4 and filter through 0.22 μm cellulose acetate filter.

2.3.2 Manual method

The tissue is diced with a sterile surgical blade in a large tissue culture dish (Nunc #168381, Naperville, IL). Ten milliliters of CMF PBS is added and each piece of tissue gently teased apart into a suspension. Excess tissue (e.g. splenic capsule) is discarded.

2.3.3 Mechanical method

The tissue is placed in a 50 ml conical tube (Falcon #2070) and 20 ml of CMF PBS added. Using the mechanical homogenizer, begin with low power and gradually increase power until the tissue is minced into a suspension. This should not take more than 5 minutes. There is no excess tissue as all tissue is homogenized.

2.3.4 Isolation of MNC

Cells in suspension are first filtered through a 70 µm nylon tissue strainer (Falcon #2350) to exclude excess tissue and clumps. MNC are isolated by Ficoll-Hypaque density gradient sedimentation as described above. Remember to use CMF PBS instead of HBSS or RPMI 1640 medium for washing. It is also usually necessary to lyse erythrocytes.

2.4 Purification of B-Lymphocytes

2.4.1 E-rosetting

Sheep erythrocytes (SRBC) treated with 2-amino-ethylisothiouronium bromide (AET) spontaneously form rosettes with human T-lymphocytes [61] and can be used to remove T-lymphocytes from suspensions of purified MNC.

2.4.1.1 0.2 M AET formula

400 mg AET (Sigma Diagnostics)
10 ml DW
Adjust to pH 8.0, filter through 0.22 µm cellulose acetate filter, and use immediately as it cannot be stored.

2.4.1.2 Preparing 25% AET-SRBC stock suspension

Wash 5 ml of SRBC (Biowhittaker, Watersville, MD) thrice with PBS to remove the buffer. Pellet SRBC by centrifuging at 300-400 *g* for 10 minutes at room temperature in a bench centrifuge and carefully remove and discard the supernatant by aspiration. To the pellet (approximately 2 ml of packed SRBC), add 4 volumes of 0.2 M AET and incubate with continuous mixing in a water bath for 20 minutes at 37 °C. Wash the pellet 3-4 times with PBS until the supernatant is clear (i.e. no further lysis is observed) and resuspend in 3 volumes of fetal bovine serum (FBS), producing 25% AET-SRBC stock suspension.

2.4.1.3 Method

Resuspend MNC in 80% HBSS and 20% FBS (Sigma Diagnostics) and enumerate viable cells by trypan blue (Gibco BRL) exclusion. Dilute the 25% AET-SRBC stock suspension with 4 volumes of HBSS (producing 5% AET-SRBC suspension) and use 1 ml of 5% AET-SRBC suspension for every 10 million MNC. Mix MNC and 5% AET-SRBC thoroughly and incubate for 10 minutes at room temperature. Centrifuge at 200 *g* for 10 minutes at room temperature in a bench centrifuge and incubate the pellet for a further 60 minutes at room temperature, or 20 minutes on ice. Carefully remove and discard the supernatant by aspiration and very gently underlay 1 volume of Ficoll-Hypaque. Centrifuge first at 200 *g* for 10 minutes at room temperature followed by 600 *g* for 15 minutes at room temperature in a bench centrifuge. Remove and wash the T-lymphocyte-depleted MNC at the medium-ficoll interface thrice with PBS and resuspend the cells in 80% RPMI 1640 medium and 20% human AB serum (N.A.B.I., Miami, FL). Incubate in a medium-sized tissue culture dish (Falcon #3003) for 15 minutes in a humidified 5% CO_2 atmosphere at 37 °C to allow monocytes to adhere. Pellet nonadherent B-lymphocytes and culture cells at a high density (1-5 million cells/ml).

2.4.2 Immunoaffinity columns and beads

In recent years, a variety of immunoaffinity columns and beads have become available commercially leading to the relative simplification of the purification process. These may vary in the composition of the capture mAb, the capacity of the system, as well as the method of separation (e.g. maxi/mini columns, magnetic beads). Examples of the use of these systems can be found throughout this volume (e.g. see Chapter 9).

2.4.3 Immunofluorescence cell sorting

Normal B-lymphocytes can be selected by marking for universal B-cell Ags (e.g. CD19 or CD20, Coulter, Corp., Miami, FL) with fluorescence-tagged monoclonal (m)Ab and indirect immunofluorescence cell sorting (Coulter Epics 753 cell sorter, Coulter, Corp.).

2.4.3.1 Setting gates

Pilot tubes containing negatively- (isotype control) and positively- (e.g. CD19 or CD20) stained aliquots of cells are first prepared. It is useful to include a known true positive control (e.g. a specific cell line) to accurately set the immunofluorescence scatter gates. Using 0.5 million cells/tube (Falcon #2054), wash the cells thrice with PBS (remember to use CMF PBS for splenocytes and tonsillar B-lymphocytes). Next, block non-specific Fc binding by incubating cells with 20% human AB serum in PBS for 30 minutes at room temperature. After washing thrice in buffer, the appropriate fluorescence-labeled mAb is added and incubated for 30 minutes on ice. Excess mAb is removed by washing thrice in buffer, and cells are resuspended in 1% Formalin (Sigma Diagnostics) in buffer. Indirect immunofluorescence is analyzed by flow cytometry.

2.4.3.2 Cell sorting

In addition to the positive and negative controls, the rest of the cells should be similarly stained with the amount of mAb used adjusted to the number of cells in the sample. After the final wash, cells are suspended at a density of 1 million cells/ml in buffer with 25 IU/ml penicillin and 25 μg/ml streptomycin (both Gibco BRL) without Formalin. Cells should be placed on ice and sorted immediately.

3. ASSAY TECHNIQUES

3.1 B-Lymphocyte Antigens

The expression of various cell surface Ags on normal B-lymphocytes characterizes the stage of differentiation [14,62]. In addition, normal B-lymphocytes also express non-lineage dependent Ags, including CD40, which may affect growth, survival and differentiation of these cells. Below is an outline of the B-lymphocyte immunophenotype:

Stem cells: $CD34^+$, $CD19^-$, $CD20^-$, surface Ig^-, $CD40^-$
Pro-B-lymphocyte: $CD34^-$, $CD19^{dim}$, $CD20^-$, surface Ig^-, $CD40^+$
Pre-B-lymphocyte: $CD34^-$, $CD19^+$, $CD20^{dim/+}$, surface Ig^{dim}, $CD40^+$
B-lymphocyte: $CD34^-$, $CD19^+$, $CD20^+$, surface Ig^+, $CD40^+$

3.1.1 Immunophenotype of Chronic Lymphocytic Leukemia cells

CLL cells are $CD19^+$, $CD20^+$, $CD45RA/RO^+$, $CD5^+$, $CD38^+$, $CD40^+$, surface Ig^+, and $CD138/syndecan^-$.

3.1.2 Immunophenotype of Multiple Myeloma cells

MM cells are $CD38^+$, $CD138/syndecan^+$, $CD40^+$, $CD19^{dim}$, $CD20^{dim}$, and $CD45RA^-$.

3.2 Assays of B-Lineage Active Cytokines

Cytokines active on B-lymphocytes (i.e. IL-4, IL-6 and IL-10) can be measured using either a specific ELISA or bioassays.

3.2.1 B9 bioassay

An example of such a bioassay is the B9 bioassay [63] which specifically measures the proliferative response of IL-6-dependent B9 murine hybridoma cells to exogenous IL-6 in control as well as in test culture supernatants from normal or malignant B-lymphocytes. DNA synthesis is assayed using ^{3}H-thymidine (Dupont NEN, Boston, MA) incorporation, and the assay is sensitive to 0.5 pg/ml of IL-6 [64]. Although specificity is largely confined to IL-6, some activity may be observed in cultures containing murine IL-4 [64] or lipopolysaccharide (LPS) [65].

3.2.1.1 B9 murine hybridoma cell culture medium

B9 cells are easily cultured in complete medium at 37 °C in a humidified 5% CO_2 atmosphere, and can be transferred at low cell densities. It is not uncommon for cells to grow at a high viable cell density (1-5 million cells/ml) as a monolayer as well as in suspension. Cells are fed every 2 to 3 days with the culture medium below:

95% RPMI 1640 with L-glutamine medium, or 95% IMDM (Sigma Diagnostics)
5% FBS
25 IU/ml penicillin
25 µg/ml streptomycin
5 mM L-Glutamine (Gibco BRL)
50 µM 2-mercaptoethanol (Sigma Diagnostics)
10 mM HEPES buffer (Intergen Co., Purchase, NY)
1 ng/ml rhIL-6 (Kirin Brewery, Tokyo, Japan)

3.2.1.2 B9 bioassay method

B9 cells (5000 cells/well) are cultured in triplicate in 200 µl of complete medium with titrations of rhIL-6 (0.0 pg/ml, 0.1 pg/ml, 1 pg/ml, 10 pg/ml, 100 pg/ml, 1 ng/ml, 10 ng/ml, and 100 ng/ml) in a 96-well flat-bottom tissue culture plate (Falcon #3072) for 72 hours to establish a standard curve. Titrations (e.g. 1:5, 1:10, 1:50, 1:100, 1:500, and 1:1000) of test sample supernatants are added to B9 murine hybridoma cells similarly cultured, without additional rhIL-6. One µCi of ^{3}H-thymidine is added to each well 16 hours prior to harvesting onto glass filters (HARVESTAR® 96 MACH II harvester, Tomtec Inc., Orange, CT), and analysis in a beta-scintillation counter (1205 BETAPLATE® counter, Wallac, Gaithersburg,

MD). The responsiveness (stimulation index = test cpm / baseline cpm) of B9 cells is calculated, and IL-6 concentrations are derived by comparing the stimulation index of test samples with the standard curve.

3.2.2 HepG2 bioassay

More recently, an alternative IL-6 bioassay was reported using HepG2 human hepatoma cells transfected with the rat IL-6R gene (HepG2-rIL-6-R) [66]. An ELISA reporting the production of haptoglobin, which is secreted by HepG2 cells following stimulation by IL-6, is used to measure the concentration of IL-6 in the test sample.

4. CULTURE TECHNIQUES

4.1 Media and Conditions

4.1.1 B-lymphocyte culture medium

The basic medium for short- and long-term B-lymphocyte culture consists of:

90% RPMI 1640 with L-glutamine medium
10% FBS
25 IU/ml penicillin
25 μg/ml streptomycin
5 mM L-glutamine
10 mM HEPES buffer
50 μM 2-mercaptoethanol

In well-established cultures with log-phase cell growth, antibiotics, HEPES buffer and 2-mercaptoethanol may be omitted.

4.1.2 B-lymphocyte culture conditions

Optimal conditions for B-lymphocyte culture are 37 °C in a humidified 5% CO_2 atmosphere of a sterile tissue culture incubator, using standard tissue culture flasks or dishes. To minimize contamination and drying, tissue culture flasks are preferred because most B-lymphocytes and derivatives grow in suspension. Some B-cells also concomitantly develop an extensive adherent monolayer when stably established.

It is common for the culture medium to turn yellow when cells are in log-phase growth. Tissue culture medium should be changed when yellowing occurs by removing 75-90% of spent medium. Cells in suspension are pelleted by centrifugation at 300-400 *g* for 5 minutes at room temperature in a bench centrifuge. The supernatant is discarded and the cells resuspended in an equal volume of fresh medium.

4.2 Influence and Dependence on Growth Factors

Normal B-cells have been shown to secrete several growth factors, including IL-1, IL-2, IL-6, IL-10, TNF and TGFβ [67-72]. In addition, many groups have reported the effects of cytokines on the growth of B-lymphocytes [73-77]. Interleukin-6 is a major growth factor regulating the differentiation and Ig secretion of normal B-lymphocytes [78,79]. Cross-linking cell surface CD40 or triggering with CD40L promotes B-cell growth (including malignant B-cells, CLL and non-Hodgkin's lymphoma) [42-49], prevents apoptosis [51,80,81], induces adhesion molecules [7,41,42,51] and promotes Ig isotype class switching [55,82,83]. The Ig isotype secreted after CD40 activation is linked to specific cytokines (i.e. IL-4, IL-10, TGFβ) [80,82-85].

4.3 Long-Term Culture Using Adherent Feeder Cell Layers

Normal B-lymphocytes, unlike cell lines, often do not survive long in culture when explanted. Since growth depends to a large extent on contact with BM stromal cells, co-culturing B-lymphocytes on feeder BM stromal cells *in vitro* will greatly enhance the recovery of viable explanted B-lymphocytes. In addition, pre-coating tissue culture flasks and dishes with concanavalin A (Con A) (Sigma Diagnostics, 7.5 mg per 35 mm diameter tissue culture dish, Falcon #3001) may also assist in long term culture of B-lymphocytes, by selectively promoting the adherence of viable cells [86].

4.3.1 Whitlock-Witte method

4.3.1.1 Cell culture inoculum density

The method developed by Whitlock and Witte [44] uses different inoculum cell densities to select for the growth of either adherent feeder cells or suspension cells. Initiating long-term cultures with a viable MNC density of 1 million MNC/ml (up to 5-10 million MNC/ml) induces the development of both adherent feeder cells as well as nonadherent B-lymphocytes, whereas only adherent feeder cells proliferate when a density of 0.3 million cells/ml is used. Importantly, the culture medium used includes HEPES buffer and 2-mercaptoethanol, but no cytokines.

4.3.1.2 Method

Resuspend BM MNC in tissue culture flasks in complete medium (90% RPMI 1640 with L-glutamine, 10% FBS, 25 IU/ml penicillin, 25 μg/ml streptomycin, 5 mM L-Glutamine, 10 mM HEPES buffer, and 50 μM 2-mercaptoethanol) without cytokines at initial cell densities depending upon the desired cell type. Exchange 80% of the medium weekly. Carefully aspirate spent medium with a Pasteur pipet; do not disturb the adherent cell layer. Check for contamination.

4.3.1.3 Establishment

In the first 1-2 weeks of culture, death of nonadherent cells occurs, whereas adherent cells continue to proliferate as a monolayer. The adherent feeder cell layer becomes established in 3-4 weeks, at which time there are very few cells in suspension. Foci of B-lymphocytes attached to the top of adherent feeder cells are noted at 4-5 weeks, which continue to proliferate and eventually become detached from the adherent layer. The establishment of nonadherent B-lymphocytes (up to 1 million cells/ml) occurs only after five weeks of culture.

4.3.1.4 Immunophenotype

The immunophenotype of these long-term cultures is typically pre-B (i.e. $CD3^-$, $CD10^+$, $CD19^+$, $CD20^+$, $CD40^+$, $DF3/MUC1^-$ and dual $CD38^+/CD45RA^+$) lymphocytes (Fig. 1), although we have also successfully cultured cells with mature B-lymphocyte phenotype (i.e. $CD3^-$, $CD10^-$, $CD19^+$, $CD20^+$, $CD40^+$, $DF3/MUC1^-$ and dual $CD38^+/CD45RA^+$) (Fig. 2).

4.3.1.5 Advantages

This method of generating long-term (>12 months) pre-B and B-lymphocyte cultures is relatively simple, and has the advantage over Dexter's method [39] of not requiring a separate incubator for culturing adherent cells at 33 °C, or the need to recharge with fresh MNC.

4.3.2 Lymphoid progenitor cell (LPC) method

4.3.2.1 Cell culture inoculum density

The LPC method [46] of generating long-term B-lymphoid progenitors uses a pre-cultured monolayer of normal murine BM feeder cells (i.e. S17 cells) prepared using mycophenolic acid selection. In this way, cultures of murine BM MNC are more rapidly (3-4 weeks vs >5 weeks) established giving rise to long-term B-lymphocytes growing in suspension above adherent S17 cells. The adherent feeder cell layer of murine S17 cells [45] is first pre-cultured as a monolayer. Murine BM MNC (1 million cells/ml) are next cultured on the monolayer of S17 cells.

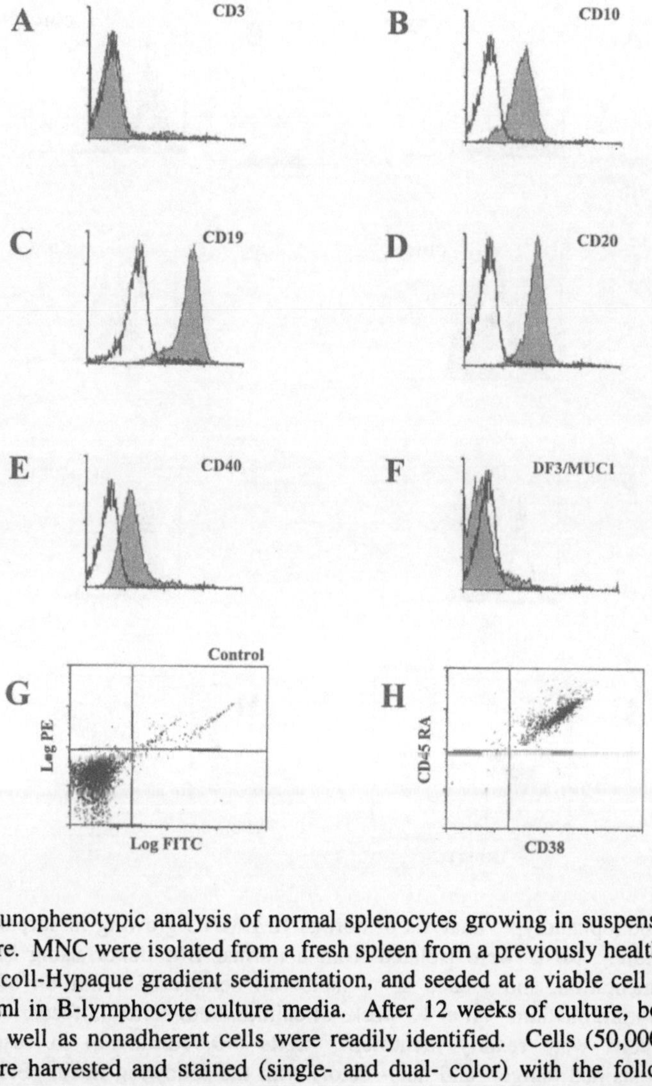

Figure 1. Immunophenotypic analysis of normal splenocytes growing in suspension after 12 weeks of culture. MNC were isolated from a fresh spleen from a previously healthy cadaveric donor using Ficoll-Hypaque gradient sedimentation, and seeded at a viable cell density of 5 million MNC/ml in B-lymphocyte culture media. After 12 weeks of culture, both adherent feeder cells as well as nonadherent cells were readily identified. Cells (50,000/sample) in suspension were harvested and stained (single- and dual- color) with the following mAb: FITC-conjugated anti-CD3 (A), FITC-conjugated anti-CD10 (B), PE-conjugated B4 anti-CD19 (C), FITC-conjugated B1 anti-CD20 (D), PE-conjugated anti-CD45RA (H), FITC-conjugated mouse IgG_1 (isotype control; A, B, D, E, F and G), and PE-conjugated mouse IgG_1 (isotype control; C and G) (all from Coulter, Corp,); FITC-conjugated anti-CD40 (PharMingen, San Diego, CA; E); DF3 anti-MUC1 (F); and HB7 anti-CD38 (H). Purified DF3 anti-MUC1 and HB7 anti-CD38 mAb were secondarily stained with FITC-conjugated anti-mouse mAb (Coulter Corp.). Indirect immunofluorescence flow cytometric analysis was performed on the Coulter Epics XL flow cytometer. Purity of these cells was greater than 95%. As can be seen in Fig. 1, these normal splenocytes growing in long-term suspension cultures demonstrated a pre-B-cell immunophenotype (i.e. $CD3^-$, $CD10^+$, $CD19^+$, $CD20^+$, $CD40^+$, $DF3/MUC1^-$ and dual $CD38^+/CD45RA^+$).

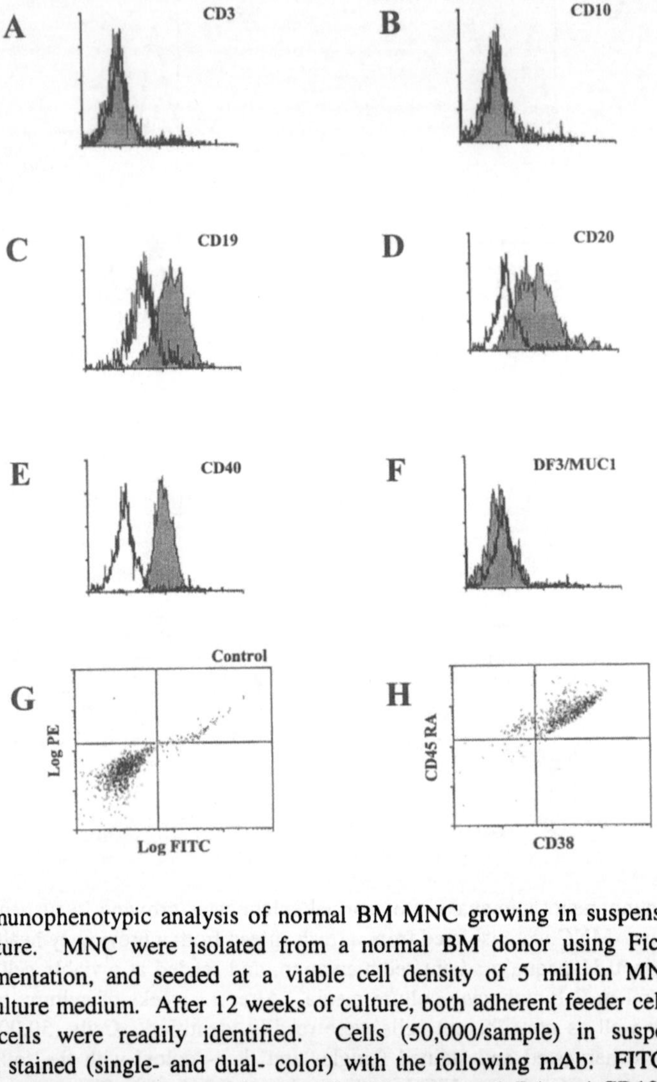

Figure 2. Immunophenotypic analysis of normal BM MNC growing in suspension after 12 weeks of culture. MNC were isolated from a normal BM donor using Ficoll-Hypaque gradient sedimentation, and seeded at a viable cell density of 5 million MNC/ml in B-lymphocyte culture medium. After 12 weeks of culture, both adherent feeder cells as well as nonadherent cells were readily identified. Cells (50,000/sample) in suspension were harvested and stained (single- and dual- color) with the following mAb: FITC-conjugated anti-CD3 (A), FITC-conjugated anti-CD10 (B), PE-conjugated B4 anti-CD19 (C), FITC-conjugated B1 anti-CD20 (D), PE-conjugated anti-CD45RA (H), FITC-conjugated mouse IgG_1 (isotype control; A, B, D, E, F and G), and PE-conjugated mouse IgG_1 (isotype control; C and G) (all from Coulter); FITC-conjugated anti-CD40 (PharMingen)(E); DF3 anti-MUC1 (F); and HB7 anti-CD38 (H). Purified DF3 anti-MUC1 and HB7 anti-CD38 mAb were secondarily stained with FITC-conjugated anti-mouse mAb. Indirect immunofluorescence flow cytometric analysis was performed on the Coulter Epics XL flow cytometer. Purity of these cells was greater than 95%. As can be seen in Fig. 2, these normal BM MNC growing in long-term suspension cultures demonstrated a B-cell immunophenotype (i.e. $CD3^-$, $CD10^-$, $CD19^+$, $CD20^+$, $CD40^+$, DF3/$MUC1^-$ and dual $CD38^+$/$CD45RA^+$).

4.3.2.2 Method

Resuspend BM MNC in tissue culture flasks pre-cultured with a monolayer of S17 feeder cells in complete medium without cytokines. Feed cells after three days and exchange 75% of the medium after another four days. Carefully aspirate spent medium with a Pasteur pipet; do not disturb the adherent cell layer. Check for contamination. Repeat the cycle weekly.

4.3.2.3 Establishment

Similar to the previous method, death of nonadherent cells occurs after 1-2 weeks of culture. This is followed rapidly by proliferation of cells that initially attach to the adherent cell layer, but later grow mostly in suspension. By 3-4 weeks, growth of nonadherent cells becomes established.

4.3.2.4 Immunophenotype

Microscopically, small to medium lymphoid cells can be identified. Immunophenotyping confirms that these cells are pre-B (cytoplasmic μ^+, surface Ig^-, $CD45R/B220^+$) lymphocytes.

4.3.2.5 Advantages

This method has the advantage of rapid (3-4 weeks vs >5 weeks) establishment of long-term B-lymphocyte cultures. Although it was originally performed in mice, it is equally possible that the LPC method could be used in humans for rapidly generating long-term normal B-lymphocyte cultures.

4.3.3 Hygromycin B-resistant feeder cells for cloning

4.3.3.1 Cell culture inoculum density

The development of long-term B-lymphocyte cultures has facilitated the introduction of genes into these cells. In this setting, both adherent feeder cells as well as B-lymphocytes are transfected with the hygromycin B-resistance gene [60] to allow selection of the transfected B-cell without loss of the adherent feeder layer of cells arising from the treatment with hygromycin B. The MRC-5 fibroblast has previously been successfully transfected with the hygromycin B-resistance gene [58] and can be pre-cultured as an adherent monolayer of cells. Using adherent feeder cell layers in this way has allowed for greater cloning efficiency of B-lymphocytes compared to B-lymphocytes cultured alone. As few as 25-50 hygromycin B-resistant gene transfected B-lymphocytes/ml, mixed with 10 million untransfected B-lymphocytes/ml, can be introduced at initiation of culture. Approximately 20,000 hygromycin B-resistant gene-transfected MRC-5 fibroblast feeder cells/ml, previously cultured as a monolayer, constitute the adherent feeder cell layer that efficiently supports the initiating B-cells.

4.3.3.2 Method

Resuspend B-lymphocytes in tissue culture flasks pre-cultured with a monolayer of MRC-5 fibroblast feeder cells in complete medium without cytokines. Select for transfectants using 200 μg/ml of hygromycin B. Exchange 50% of the medium every 3-5 days. Carefully aspirate spent medium with a Pasteur pipet; do not disturb the adherent cell layer. Check for contamination. Transfer nonadherent cloned B-lymphocytes to plain tissue culture flasks (i.e. without MRC-5 fibroblast feeder cells) after 3-4 weeks.

4.3.3.3 Establishment

Similar to the above methods, death of nonadherent B-lymphocytes occurs after 1-2 weeks of culture. Cloned B-lymphocytes are established at 3-4 weeks of culture.

4.3.3.4 Immunophenotype

The immunophenotypic analysis of these cells confirms that they are B-lymphocytes. Standard gene reporter assays can be used to demonstrate the presence of the transgene within cloned B-lymphocytes.

4.3.3.5 Advantages

This method of introducing genes into B-lymphocytes using a layer of hygromycin-resistant gene-transfected adherent feeder cells has greatly increased the number of cells obtained, as compared to cultures of B-lymphocytes without adherent feeder cells. Moreover, the establishment of cloned B-lymphocytes is rapid (i.e. within 3-4 weeks) and can be maintained in long-term culture.

4.4 NIH3T3 CD40-Ligand Transfectants

NIH3T3/wt, NIH3T3/CD40LT and NIH3T3/vt cells, generously provided by Dr. Gordon J. Freeman (Dana-Farber Cancer Institute, Boston, MA) are cultured in F12/DMEM medium [7,42].

4.4.1 F12/DMEM medium

45% F12 (Ham) nutrient mixture with L-glutamine medium (Gibco BRL)
45% DMEM with 4500 mg/l D-Glucose and L-glutamine medium (Gibco BRL)
10% FBS
2 mM L-glutamine
10 mM HEPES buffer
15 μg/ml gentamicin (Elkins-Sinn, Inc., Cherry Hill, NJ)

4.4.2 Culture technique

Selection for expression of the transgene (NIH3T3/CD40LT and NIH3T3/vt cells, but not NIH3T3/wt cells) is done with 400 (100-800) μg/ml of GENETICIN® (G418, Gibco BRL). Cells are grown as a confluent monolayer in medium-sized tissue culture dishes (Falcon #3003) at 37 °C in a humidified 5% CO_2 atmosphere. Tissue culture medium should be replaced every 3-5 days by aspirating spent medium and replacing with 8 ml of fresh medium. Since all three lines should be simultaneously cultured for most experiments, it is important to make sure that G418 is used only for NIH3T3/CD40LT and NIH3T3/vt cells, but not NIH3T3/wt cells.

Cells can be harvested at 70% confluency by scraping (Costar #3010, Costar, Cambridge, MA). Trypsinization should be avoided as cell surface membrane CD40L expression may be diminished by the procedure. If the cells are to be used for subsequent flow cytometric analysis, it is important to disperse the cells mechanically, e.g. by repeatedly pipetting cells suspended in PBS with or without 1 mM EDTA. Residual clumps of cells can be removed by filtration through a 70 μm nylon tissue strainer (Falcon #2350).

4.4.3 CD40L activation

CD40L activation is performed by co-culturing B-cells with γ-irradiated (CsCl source, 10 Gy) or formalin-fixed NIH3T3/CD40LT cells (100 B-lymphocytes:1 NIH3T3 cell) to trigger B-cells *in vitro* [42]. It is important to remove G418 from NIH3T3/CD40LT and NIH3T3/vt cells before co-culturing with B-lymphocytes and tumor cells, since G418 is toxic to B-cells that do not express the G418-resistant gene product. This is easily done by washing cells six times with 50 ml of PBS and centrifugation at 400 *g* for 5 minutes at room temperature in a bench centrifuge.

4.4.4 Formalin fixation of NIH3T3/CD40LT cells

NIH3T3/CD40LT cells are fixed in 1 ml of 1% Formalin (Sigma Diagnostics) for 10 minutes, washed a further six times with PBS to remove Formalin, counted, and resuspended in RPMI 1640 medium with 25 IU/ml penicillin and 25 μg/ml streptomycin. Formalin-fixed NIH3T3/CD40LT cells can conveniently be stored "ready-to-use" in a sterile incubator at 37°C for up to three days.

5. UTILITY OF SYSTEM

These systems for culture of normal and malignant B-lymphocytes allow for the characterization of the cellular phenotype and molecular events which accompany B-cell growth and differentiation, as well as those factors affecting B-cell survival. For example, cross-linking surface Ig or triggering with CD40L leads to B-cell activation [50-57], whereas PWM stimulates B-cell differentiation [28]. Culture systems *in vitro* demonstrate that IL-6 mediates differentiation of normal B-lymphocytes [77-79], whereas it mediates growth of malignant B-cells (i.e. MM cells) [3,7].

5.1 B-Cell Differentiation

Standard methods of immunophenotyping are used for characterizing differentiation antigens on normal and malignant B-lymphocytes. Half a million cells/tube (Falcon #2054) are washed thrice with PBS, incubated with 20% human AB serum in PBS for 30 minutes at room temperature, washed a further three times with PBS, and incubated with the appropriate fluorescence-labeled mAb for 30 minutes on ice. Cells are again washed thrice with PBS, resuspended in 1% Formalin in PBS, and analyzed by indirect immunofluorescence flow cytometry (Coulter Epics XL, Coulter, Corp.).

5.2 B-Cell Growth

The IL-6 proliferation assay is an example of a B-lymphocyte growth assay. Normal B-cells, either resting or activated by cross-linking cell surface Ig, or malignant B-lymphocytes (50,000 cells/ml; 200 μl/well) are first cultured in serum-free medium alone or with rhIL-6 (100 ng/ml) for 42 hours in a flat-bottom 96-well tissue culture plate (Falcon #3072). Next, ^{3}H-thymidine (1 μCi/well) is added and cells are incubated for six hours followed by harvesting onto glass filters (HARVESTAR® 96 MACH II harvester, Tomtec Inc.), and analysis in a beta-scintillation counter (1205 BETAPLATE® counter, Wallac). The difference in ^{3}H-thymidine incorporation between cells cultured in medium alone or with rhIL-6 reflects cellular response to IL-6.

5.3 B-Cell Signaling

Culture systems are useful for studying molecular events associated with activation, proliferation and survival of B-lineage cells. For example, IL-6 mediates differentiation of normal B-lymphocytes and growth of malignant B-lymphocytes. Signaling via the IL-6R transmembrane complex, which is made up of gp80 and gp130 subunits [87], occurs by tyrosine phosphorylation of a series of substrates initiated by activated gp130. Changes in tyrosine phosphorylation of signaling molecules are analyzed by immunoprecipitation and immunoblotting with anti-phosphotyrosine mAb (Transduction Laboratories, Inc., Lexington, KY) [88,89]. Normal and malignant B-cells (10 million) are first cultured in serum-free medium for two hours, followed by culturing in rhIL-6 (100 ng/ml) without serum for 15 minutes. The reaction is stopped by washing the cells in ice-cold Tris buffered saline (20 mM Tris pH 7.6, 150 mM NaCl, both from Fisher Scientific) containing 1 mM sodium orthovanadate (Sigma Diagnostics).

5.4 B-Cell Survival and Apoptosis

Glucocorticoids cause apoptosis of normal and malignant B-lymphocytes, and have been used to treat B-cell tumors (i.e. MM). At least one component of their anti-tumor activity arises from inhibition of IL-6 transcription in MM cells [90,91]. We have previously shown that MM cells undergo dexamethasone-induced apoptosis in MAP kinase-dependent, but JNK/SAP kinase- and cytochrome-*c*-independent fashion [92,93]. Culturing MM cells in rhIL-6 (100 ng/ml) also rescues tumor cells from apoptosis induced by dexamethasone (10 μM).

6. CONCLUSION

Normal B-lymphocytes depend greatly on contact with other cells such as T-cells or BM stromal cells, as well as growth factors such as IL-4 and IL-6, for their growth, differentiation, and survival. Although short-term culture of B-lymphocytes is possible using T-cell co-cultures and/or mitogens, long-term culture requires the establishment of an adherent layer of feeder BM stromal cells, use of combinations of growth factors (i.e.

CD40L + IL-4), or immortalization (i.e. EBV transformation). Currently, the efficiency of cloning stably transfected normal B-lymphocytes is optimally achieved by using feeder cell layers.

LITERATURE CITED

1. Dono M, Hashimoto S, Fais F, Trejo V, Allen SL, Lichtman SM, Schulman P, Vinciguerra VP, Sellars B, Gregersen PK, Ferrarini M, and Chiorazzi N (1996) Evidence for progenitors of chronic lymphocytic leukemia B-cells that undergo intraclonal differentiation and diversification. *Blood* 87:1586-1594.
2. Hallek M, Bergsagel PL, and Anderson KC (1998) Multiple myeloma: Increasing evidence for a multistep transformation process. *Blood* 91:3-21.
3 Kawano M, Hirano T, Matsuda T, Taga T, Horii Y, Iwato K, Asaoku H, Tang B, Tanabe O, Tanaka H, Kuramota A, and Kishimoto T (1988) Autocrine generation and requirement of BSF2/IL6 for human multiple myeloma. *Nature* 332:83-85.
4 Urashima M, Chauhan D, Hatziyanni M, Ogata A, Hollenbaugh D, Aruffo A, and Anderson KC (1996) CD40 ligand triggers interleukin-6 mediated B-cell differentiation. *Leuk. Res.* 20:507-515.
5 Westendorf J, Ahmann G, Armitage R, Spriggs M, Lust J, Greipp P, Katzmann J, and Jelinek D (1994) CD40 expression in malignant plasma cells: Role in stimulation of autocrine IL-6 secretion by a human myeloma cell line. *J. Immunol.* 152:117-128.
6 Tong AW, Zhang B-q, Mues G, Solano M, Hanson T, and Stone MJ (1994) Anti-CD40 antibody binding modulates human multiple myeloma clonogenicity in vitro. *Blood* 84:3026-3033.
7 Urashima M, Chauhan D, Uchiyama H, Freeman GJ, and Anderson KC (1996) CD40 ligand triggered interleukin-6 secretion in multiple myeloma. *Blood* 85:1903-1912.
8 Owen JJT, Cooper MD, and Raff MC (1974) In vitro generation of B-lymphocytes in mouse foetal liver, a mammalian 'bursa equivalent'. *Nature* 249:361-363.
9 Coffman RL, and Weissman IL (1981) A monoclonal antibody that recognizes B-cells and B-cell precursors in mice. *J. Exp. Med.* 153:269-279.
10 Solvason N, and Kearney JF (1992) The human fetal omentum: A site of B-cell generation. *J. Exp. Med.* 175:397-403.
11 Freitas AA, and Countinho A (1981) Very rapid decay of mature lymphocytes in the periphery. *J. Exp. Med.* 154:994-999.
12 Neuberger MS (1997) Antigen receptor signaling gives lymphocytes a long life. *Cell* 92:971-973.
13 Freitas AA, Rocho B, Forni L, and Counthino A (1982) Population dynamics of B-lymphocytes: Demonstration of high turnover rates in the central and peripheral lymphoid organs. *J. Immunol.* 128:54-60.
14 Billips LG, Lassoued K, Nunez C, Wang J, Kubagawa H, Gartleand GL, Burrows PD, and Cooper MD (1997) Human B-cell development. *Ann. New York Acad. Sci.* 764:1-8.
15 Nishikawa SI, Era T, Ogawa M, Nishikawa S, Ohno N, Hayashi SI, and Kunisada T (1992) Control of intramarrow B-cell genesis by stromal cell-derived molecules. *Cur. Topics Microbiol. Immunol.* 182:27-35.
16 Kincade PW (1994) B-lymphopoiesis: Global factors, local control. *Proc. Natl. Acad. Sci. U.S.A.* 91:2888-2889.
17 Shen J, Wilson F, Shifrine M, and Gershwin ME (1977) Select growth of human T-lymphocytes in single phase semisolid culture. *J. Immunol.* 119:1299-1305.

18 Radnay J, Goldman I, and Rosenszajn LA (1977) Growth of human B-lymphocyte colonies *in vitro*. *Nature* 278:351-353.
19 Izaquirre CA, Minden DA, Howatson AF, and McCulloch EA (1980) Colony formation by normal and malignant human B-lymphocytes. *Br. J. Cancer* 42:430-437.
20 Whisler RL, Bobak DA, and Newhouse YG (1983) Phenotypic characteristics of peripheral blood T-cells regulating colony growth by nontransformed human B-lymphocytes. *J. Immunol.* 127:1758-1762.
21 Zola H (1985) Differentiation and maturation of human B-lymphocytes: A review. *Pathology* 17:365-381.
22 Kishimoto T (1985) Factors affecting B-cell growth and differentiation. *Ann. Rev. Immunol.* 3:133-157.
23 Fernandez LA, MacSween JM, and Langley GR (1986) Immunoregulation of B-lymphocyte colony formation by T-cell subsets in patients with chronic lymphocytic leukemia. *Blood* 67:294-298.
24 Lederman S, Yellin MJ, Inghirami G, Lee JJ, Knowles DK, and Chess L (1992) Molecular interactions mediating T-B-lymphocyte collaboration in human lymphoid follicles. *J. Immunol.* 149:3817-3826.
25 Howard M, and Paul WE (1983) Regulation of B-cell growth and differentiation by soluble factors. *Ann. Rev. Immunol.* 1:307-333.
26 Fernandez LA, MacSween JM, and Robson DA (1989) Growth of B-cell colonies independent of T-cell contact. *Immunol. Lett.* 22:167-182.
27 Fernandez LA, MacSween JM, and Robson DA (1995) Growth of B-cell colonies independent of T-cells: A new perspective. *Immunol. Invest.* 24:669-678.
28 Gausset P, Delespesse G, Huber C, Keenes B, and Govaerts AJ (1976) *In vitro* response of subpopulations of human lymphocytes. II. DNA synthesis induced by anti-immunoglobulin antibodies. *J. Immunol.* 116:446-453.
29 Keightley RG, Cooper MD, and Lawton AR (1976) The T-cell dependence of B-cell differentiation induced by pokeweed mitogen. *J. Immunol.* 117:1538-1544.
30 Fauci AS, and Pratt KR (1976) Polyclonal activation of bone-marrow-derived lymphocytes from human peripheral blood measured by a direct plaque-forming cell assay. *Proc. Natl. Acad. Sci. U.S.A.* 73:3676-3679.
31 Smith CI, Hammarström L, Bird AG, Kunori T, Gustafsson B, and Holme T (1979) Lipopolysaccharide and lipid A-induced human blood lymphocyte activation as detected by a protein A plaque assay. *Eur. J. Immunol.* 9:619-625.
32 Shankey TV, Daniele RP, and Nowell PC (1981) Kinetics of mitogen-induced proliferation and differentiation of human peripheral blood B-lymphocytes. *Clin. Exp. Immunol.* 44:156-166.
33 DeFranco AL, Kung JT, and Paul WE (1982) Regulation of growth and proliferation in B-cell subpopulations. *Immunol. Rev.* 64:161-182.
34 Diamantstein T, Rühl H, Vogt W, and Bochert G (1973) Stimulation of B-cells by dextran sulphate in vitro. *Immunology* 25:743-747.
35 Yamanashi Y, Kakiuchi T, Mizuguchi J, Yamamoto T, and Toyoshima K (1991) Association of B-cell antigen receptor with protein tyrosine kinase Lyn. *Science* 251:192-194.
36 Yamamoto T, Yamanashi Y, and Toyoshima K (1993) Association of Src-family kinase Lyn with B-cell antigen receptor. *Immunol. Rev.* 132:187-206.
37 Yamamoto N, and Hinuma Y (1975) Clonal transformation of human leukocytes by Epstein-Barr virus on soft agar. *Int. J. Cancer* 17:191-196.
38 Sugden B, and Mark W (1977) Clonal transformation of adult human leukocytes by Epstein-Barr virus. *J. Virol.* 23:503-508.

39 Dexter TM, Allen TD, and Lajtha LG (1977) Conditions controlling the in vitro proliferation of hemopoietic stem cells in vitro. *J. Cell. Physiol.* 91:335-340.

40 Kierney PC, and Dorshkind K (1987) B-lymphocyte precursors and myeloid progenitors survive in diffusion chamber cultures but B-cell differentiation requires close association with stromal cells. *Blood* 70:1418-1424.

41 Teoh G, and Anderson KC (1997) Interaction of tumor and hosT-cells with adhesion and extracellular matrix molecules in the development of multiple myeloma. *Hematol. Oncol. Clin. North Am.* 11:27-42.

42 Teoh G, Urashima M, Greenfield EA, Nguyen KA, Lee JF, Chauhan D, Ogata A, Treon SP, and Anderson KC (1998) The 86 kD subunit of Ku autoantigen mediates homotypic and heterotypic adhesion of multiple myeloma cells. *J. Clin. Invest.* 101:1379-1414.

43 Whitlock CA, and Witte ON (1982) Long term culture of B-lymphocytes and their precursors from murine bone marrow. *Proc. Natl. Acad. Sci. U.S.A.* 79:3608-3612.

44 Whitlock CA, Robertson D, and Witte ON (1984) Murine B-cell lymphopoiesis in long term culture. *J. Immunol. Methods.* 67:353-369.

45 Collins L, and Dorshkind K (1987) A stromal cell line from myeloid long-term bone marrow cultures can support myelopoiesis. *J. Immunol.* 138:1082-1087.

46 Saffran DC, Faust EA, and Witte ON (1992) Establishment of a reproducible culture technique for the selective growth of B-cell progenitors. *Curr. Topics Microbiol. Immunol.* 182:37-43.

47 King AG, Wierda D, and Landreth KS (1988) Bone marrow stromal cell regulation of B-lymphopoiesis. I. The role of macrophages, IL-1, and IL-4 in pre-B-cell maturation. *J. Immunol.* 141:2016-2026.

48 Kittler EL, McGrath H, Temeles D, Crittenden RB, Kister VK, and Queensberry PJ (1992) Biologic significance of constitutive and subliminal growth factor production by bone marrow stroma. *Blood* 79:3168-3178.

49 Uchiyama H, Barut BA, Mohrbacher AF, Chauhan D, and Anderson KC (1993) Adhesion of human myeloma-derived cell lines to bone marrow stromal cells stimulates interleukin-6 secretion. *Blood* 82:3712-3720.

50 Clark EA, and Ledbetter JA (1986) Activation of human B-cells mediated through two distincT-cell surface differentiation antigens, Bp35 and Bp50. *Proc. Natl. Acad. Sci. U.S.A.* 83:4494-4498.

51 Banchereau J, de Paoli P, Valle A. Garcia E, and Rousset F. (1991) Long-term human B-cell lines dependent on interleukin-4 and antibody to CD40. *Science* 251:70-72.

52 Banchereau J, and Rousset F (1991) Growing human B-lymphocytes in the CD40 system. *Nature* 353:678-679.

53 Armitage RJ, Fanslow WC, Strockbine L, Sato TA, Clifford KN, Macduff BM, Anderson DM, Gimpel SD, Davis-Smith T, Maleszewski CR, Clark EA, Smith CA, Grabstein KH, Cosman D, Spriggs MK (1992) Molecular and biological characterization of a murine ligand for CD40. *Nature* 357:80-82.

54 Hollenbaugh D, Grosmaire LS, Kullas CD, Chalupny NJ, Braesch-Andersen S, Noelle RJ, Stamenkovic I, Ledbetter JA, and Aruffo A (1992) The human T-cell antigen gp39, a member of the TNF gene family, is a ligand for the CD40 receptor: Expression of a soluble form of gp39 with B-cell co-stimulatory activity. *EMBO J.* 11:4313-4321.

55 Spriggs MK, Armitage RJ, Strockbine L, Clifford KN, Macduff BM, Sato TA, Maliszewski CR, and Fanslow WC (1992) Recombinant human CD40 ligand stimulates B-cell proliferation and immunoglobulin E secretion. *J. Exp. Med.* 176:1543-1550.

56 Crawford DH, and Catovsky D (1993) *In vitro* activation of leukemic B-cells by interleukin-4 and antibodies to CD40. *Immunology* 80:40-44.

57 Johnson PW, Watt SM, Betts DR, Davies D, Jordan S, Noroton AJ, and Lister TA (1993) Isolated follicular lymphoma cells are resistant to apoptosis and can be grown *in vitro* in the CD40/stromal cell system. *Blood* 82:1848-1857.
58 Huschtscha LI, and Holliday R (1983) Limited and unlimited growth of SV40-transformed cells from human diploid MRC-5 fibroblasts. *J. Cell Sci.* 63:77-99.
59 Ayers KN (1982) High cloning efficiency of human lymphoid cells in agarose without feeder layers. *J. Natl. Cancer Inst.* 68:919-923.
60 Fabb SA, and Ragoussis J (1997) High-efficiency human B-cell cloning using hygromycin B-resistant feeder cells. *Biotechniques* 22:814-822.
61 Fernandez LA, MacSween JM, and Langley GR (1977) T-cell function in untreated B-cell chronic lymphocytic leukemia. *Cancer* 39:1168-1174.
62 Pribyl JR, Dittel N, and LeBien TW (1997) *In vitro* studies of human B-lymphopoiesis. *Ann. New York Acad. Sci.* 764:9-18.
63 Aarden, LA, De Groot ER, Schaap OL, and Lansdorp PM (1987) Production of hybridoma growth factor by human monocytes. *Eur. J. Immunol.* 17:1411-1416.
64 Helle M, Boeije L, and Aarden LA (1988) Functional discrimination between interleukin 6 and interleukin 1. *Eur. J. Immunol.* 18:1535-1540.
65 Pedersen MR, Jensen S, Christensen JD, and Hansen EW (1995) Lipopolysaccharide in concentrations above 40 ng/ml stimulates proliferation of the IL-6-dependent B9 cell line. *J. Immunol. Methods* 180:159-163.
66 Oppmann B, Stoyan T, Fischer M, Voltz N, Marz P, and Rose-John S (1996) Alternative assay procedures for cytokines and soluble receptors of the IL-6 family. *J. Immunol. Methods* 195:153-159.
67 Gordon J, Ley SC, Melamed MD, English LS, and Hughes-Jones NC (1984) Immortalized B-lymphocytes produce B-cell growth factor. *Nature* 310:145-147.
68 Matsushima K, Procopio A, Abe H, Scala G, Ortaldo JR, and Oppenheim JJ (1985) Production of interleukin 1 activity by normal human peripheral blood B-lymphocytes. *J. Immunol.* 135:1132-1136.
69 Muraguchi A, Nishimoto H, Kawamura N, Hori A, and Kishimoto T (1986) B-cell-derived BCGF functions as autocrine growth factor(s) in normal and transformed B-lymphocytes. *J. Immunol.* 137:179-186.
70 Horii Y, Muraguchi A, Suematsu S, Matsuda T, Yoshizaki K, Hirano T, and Kishimoto T (1988) Regulation of BSF-2/IL-6 production by human mononuclear cells. Macrophage-dependent synthesis of BSF-2/IL-6 by T-cells. *J. Immunol.* 141:1529-1535.
71 Freeman GJ, Freedman AS, Rabinowe SN, Segil JM, Horowitz J, Rosen K, Whitman JF, and Nadler LM (1989) Interleukin 6 gene expression in normal and neoplastic B-cells. *J. Clin. Invest.* 83:1512-1518.
72 Matthes T, Werner-Favre C, Tang H, Zhang X, Kindler V, and Zubler RH (1993) Cytokine mRNA expression during an in vitro response of human B-lymphocytes: Kinetics of B-cell tumor necrosis factor α, interleukin (IL)6, IL-10, and transforming growth factor β1 mRNAs. *J. Exp. Med.* 178:521-528.
73 McKearn JP, McCubrey J, and Fagg B (1985) Enrichment of hematopoietic precursor cells and cloning of multipotential B-lymphocyte precursors. *Proc. Natl. Acad. Sci. U.S.A.* 82:7414-7418.
74 Palacios R, Karasuyama H, and Rolink A (1987) $Ly1^+$ PRO-B-lymphocyte clones. Phenotype, growth requirements and differentiation *in vitro* and *in vivo*. *EMBO J.* 6:3687-3693.
75 Kinashi T, Inaba K, Tsubata T, Tashiro K, Palacios R, and Honjo T (1988) Differentiation of an interleukin 3-dependent precursor B-cell clone into immunoglobulin-producing cells in vitro. *Proc. Natl. Acad. Sci. U.S.A.* 85:4473-4477.

76 Rolink A, Kudo A, Karasuyama H, Kikuchi Y, and Melchers F (1991) Long term proliferating early pre B-cell lines and clones with the potential to develop to surface Ig-positive, mitogen reactive B-cells *in vitro* and *in vivo*. *EMBO J.* 10:327-336.
77 Rousset F, Garcia E, and Banchereau J (1991) Cytokine-induced proliferation and immunoglobulin production of human B-lymphocytes triggered through their CD40 antigen. *J. Exp. Med.* 173:705-710.
78 Anderson KC, Morimoto C, Paul SR, Chauhan D, Williams D, Cochran M, and Barut BA (1992) Interleukin-11 promotes accessory cell-dependent B-cell differentiation in humans. *Blood* 80:2797-2804.
79 Tanner JE, and Tosato G (1992) Regulation of B-cell growth and immunoglobulin gene transcription by interleukin-6. *Blood* 79:452-459.
80 Liu YJ, Joshua DE, Williams GT, Smith CA, Gordon J, and MacLennon IC (1989) Mechanism of antigen-driven selection in germinal centres. *Nature* 364:929-931.
81 Tsubata T, Wu J, and Honjo T (1993) B-cell apoptosis induced by antigen receptor crosslinking is blocked by a T-cell signal through CD40. *Nature* 364:645-648.
82 Jabara HH, Fu SM, Geha RS, and Vercelli D (1990) CD40 and IgE: Synergism between anti-CD40 monoclonal antibody and interleukin 4 in the induction of IgE synthesis by highly purified human B-cells. *J. Exp. Med.* 172:1862-1864.
83 Armitage RJ, Macduff BM, Spriggs MK, and Fanslow WC (1993) Human B-cell proliferation and Ig secretion induced by recombinant CD40 ligand are modulated by soluble cytokines. *J. Immunol.* 150:3671-3680.
84 Defrance T, Vanbervliet B, Briere F, Durand I, Rousset F, and Banchereau J (1992) Interleukin 10.and transforming growth factor β cooperate to induce anti-CD40-activated naïve human B-cells to secrete immunoglobulin A. *J. Exp. Med.* 175:671-682.
85 Nonoyama S, Hollenbaugh D, Aruffo T, Ledbetter JA, and Ochs HD (1993) B-cell activation via CD40 is required for specific antibody production by antigen-stimulated human B-cells. *J. Exp. Med.* 178:1097-1102.
86 Andersen A-ST, and Junker S (1994) Simple and efficient recovery of rare living lymphoid cells from a vast majority of dead cells. *Nucleic Acids Res.* 22:5769-5770.
87 Murakami M, Hibi M, Nakagawa N, Nakagawa T, Yasukawa K, Yamanishi K, Taga T, and Kishimoto T (1993) IL-6-induced homodimerization of gp130 and associated activation of a tyrosine kinase. *Science* 260:1808-1810.
88 Ogata A, Chauhan D, Teoh G, Treon SP, Urashima M, Schlossman RL, and Anderson KC (1997) IL-6 triggers cell growth via the Ras-dependent mitogen-activated protein kinase cascade. *J. Immunol.* 1559:2212-2221.
89 Ogata A, Chauhan D, Urashima M, Teoh G, Treon SP, and Anderson KC (1997) Blockade of mitogen-activated protein kinase cascade signaling in interleukin 6-independent multiple myeloma cells. *Clin. Cancer Res.* 3:1017-1022.
90 Ishikawa H, Tanaka H, Iwato K, Tanabe O, Asaoku H, Nobuyoshi M, Yamamoto I, Kawano M, and Kuramoto A (1990) Effect of glucocorticoids on the biologic activities of plasma cells: Inhibition of interleukin-1β osteoclast activating factor-induced bone resorption. *Blood* 75:715-720.
91 Hardin J, MacLeod S, Grigorieva I, Chang R, Barlogie B, Xiao H, and Epstein J (1991) Interleukin-6 prevents dexamethasone-induced myeloma cell death. *Blood* 84:3063-3070.
92 Chauhan D, Pandey P, Ogata A, Teoh G, Treon S, Urashima M, Kharbanda S, and Anderson KC (1997) Dexamethasone induces apoptosis of multiple myeloma cells in a JNK/SAP kinase independent mechanism. *Oncogene* 15:837-843.
93 Chauhan D, Pandey P, Ogata A, Teoh G, Krett N, Halgren R, Rosen S, Kufe D, Kharbanda S, and Anderson K (1997) Cytochrome *c*-dependent and -independent induction of apoptosis in multiple myeloma cells. *J. Biol. Chem.* 272:29995-29997.

Chapter 5

Monocytes and Macrophages

Ivan N Rich
Center for Cancer Treatment and Research, Palmetto Richland Memorial Hospital, Seven Richland Medical Park, Columbia SC 29203. Tel: 001-803-434-4734; Fax: 001-803-434-4950; E-mail: irich@rmhntpdc.rmh.edu

1. INTRODUCTION

The macrophage is probably the most ubiquitous and versatile cell in the body. It is involved in a multitude of different functions depending on its location. By monitoring its external environment, it can modify its function in order to maintain homeostasis. Without the macrophage, the body would be open to infection and the immune system would not function. It is also safe to say that without this amoeba-like cell, the regulation of the hematopoietic system would probably spiral into chaos. This is because the macrophage, depending on its location and functional capacity, has the capability of producing one of the largest repertoires of growth regulatory factors of any cell type, thereby not only influencing cells in immediate contact with it, but also making its presence known indirectly to cells. The macrophage also has the enviable position of not only being a member of the lympho-hematopoietic system, but also plays a pivotal role in its regulation by being part of the hematopoietic microenvironment.

Developmentally, the macrophage is probably one of the first, if not the first, hematopoietic cell to be produced in the embryo. Although macrophages, as we know them in the fetus and adult, have not been detected in the visceral yolk sac blood islands, it is certainly possible that these may consist of a central primitive macrophage-like cell that is produced from the hematopoietic-initiating cell [1]. It has been hypothesized that the latter are actually primordial germ cells which are found at the base of the allantois in the extraembryonic mesoderm in late day 6 mouse embryos. These cells may not only be responsible for the initiation of extraembryonic hematopoiesis, because upon reentry into the embryo between days 8 and 8.5 of development, on their migratory pathway to the gonads via the para-aortic splanchnopleura (also called the AGM or aorta, gonads, mesonephros region), a subpopulation may give rise to a temporary, but definitive, hematopoiesis in this area before fetal hepatic hematopoiesis is initiated [2]. It is interesting to note that the macrophage is also probably the first morphologically identifiable hematopoietic cell in the fetal liver anlage [3], paving the way for the wave of hepatic erythropoiesis to occur.

Although the production of primitive embryonic macrophages has been postulated [4], the production lineage to mature macrophages, as with all other hematopoietic cell lineages, is extremely well-defined, both at the cellular and molecular levels. As with all hematopoietic cells, the macrophage is derived from the hematopoietic stem cell compartment and may be considered to be produced from a bipotential progenitor cell called the granulocyte-macrophage colony-forming unit, or CFU-GM (these cells are described in more detail in Chapter 1 of this volume). In accordance with the nomenclature used at the time, these cells have been called the colony-forming unit– agar (CFU-A), colony-forming cell (CFC), or colony-forming unit – culture (CFU-C), and were the first of the in vitro-detected hematopoietic colony-forming cells [5,6]. To produce macrophages from CFU-GM, either macrophage colony-stimulating factor (M-CSF, also called CSF-1) or low concentrations of granulocyte-macrophage colony-stimulating factor (GM-CSF) are required [7]. By mechanisms that are still not understood, the M-CSF receptor, which is identical to the *c-fms* proto-oncogene and is a member of the intrinsic tyrosine-kinase receptor family with autophosphorylating activity, is induced. The M-CSF receptor induces several G-protein intracellular signals, which eventually activate the *fos/jun* proto-oncogenes that encode DNA binding proteins of the macrophage transcription promoter complex. Signaling mechanisms at the nucleus level then induce the monocytic differentiation program. The role of M-CSF, its relationship to other growth factor molecules and the signal transduction process have been reviewed in detail in a recent monograph [8]. As mentioned above, the macrophage can produce a vast array of different growth regulating factors, of which two happen to be M-CSF and GM-CSF. Thus, macrophages can regulate their own production and the expression of the M-CSF gene [8].

As the CFU-GM enters the macrophage lineage, it is progressively transformed into a monoblast, promonocyte and monocyte. Some of the characteristics of these cell populations are given in Table 1. The monoblast and promonocyte are present in the bone marrow (BM). The monocyte is a blood-borne cell. The monocyte enters or is recruited into a specific organ where it undergoes further maturation to a macrophage whose function is determined by the organ or tissue in which it resides (Figure 1). Recruitment of monocytes to become resident macrophages probably occurs under the auspices of chemokines. Monocytes roll along the endothelium, a process controlled by selectins, especially E-selectin. The latter is the ligand to the surface glycoprotein antigen sialyl-Lewis^x (CD15s) present on monocytes [9,10]. This slows down the monocyte so it can interact with adhesion molecules such as VCAM-1, ICAM-1 and ICAM-2 on the pavement endothelium [9]. The corresponding receptors on monocytes are $\alpha_4\beta_1$ integrin, CD11/CD18, and CD14 [11-15]. Once attached, the cells probably follow a chemokine gradient to the tissue or organ [16-18]. Exactly what determines recruitment into a specific organ or tissue is unknown, but it is clear that macrophages are extremely heterogeneous and exist in subpopulations [19-30]. It is this heterogeneity that not only makes the macrophage such an interesting cell to study, but also a very difficult cell to study. As described below, macrophages can exist in different states, and it is important to know what the functional state of the macrophage population is that will be studied, since this will determine the results obtained.

2. TISSUE PROCUREMENT AND PROCESSING

Little is known about the proportion of circulating human monocytes that enter each tissue. In this area, nearly all the studies have been performed in the mouse, where it has been shown that about 56% of all monocytes leaving the circulation become Kupffer cells of the liver, about 8% enter the peritoneal cavity and about 15% become alveolar macrophages. It follows that to study human macrophages in vitro, the source of cells is restricted to certain organs and tissues and will depend on what the questions and goals of the research are going to be. Added to this consideration is the inherent heterogeneity of the macrophage population within an organ or tissue, which also has to be taken into account, since this can affect the type of assay performed.

Table 1. Features of macrophage precursor cells

MONOBLAST	
Size	16μm
Cytoplasmic features	Wright-Giemsa staining: deep blue, clear cytoplasm. Large eccentric, central nucleus. 1-2 large nucleoli. Slight membrane ruffling, but no pseudopods.
Differentiation markers	Usually lacking. Very weak for peroxidase, non-specific esterases, acid phosphatases. No receptors for IgG or complement.
Number of divisions	2
Estimated cell cycle time	19.5h
Estimated time to monocyte	6 days
PROMONOCYTE	
Size	16-18μm
Cytoplasmic features	Wright-Giemsa staining: blue with some neutral red staining. Membrane slightly ruffled, heterogeneous nuclei with indentation. Rough endoplasmic reticulum and active Golgi. Cytoplasm contains filaments. Membrane ruffling with pseudopod production.
Differentiation markers	Peroxidase, fluoride-sensitive non-specific esterases, acid phosphatase.
Number of divisions	1-2
Proportion in bone marrow	Approximately 3%
Estimated concentration	600 x 10^6 cells/kg body weight
MONOCYTE	
Size	15-18μm (largest peripheral blood cells).
Cytoplasmic features	Wright-Giemsa staining: grey-blue cytoplasm with red-purple granules. Indented, central to slightly off-center large nucleus occupying approximately half the cell volume. Ruffled plasma membrane with microvilli.
Differentiation markers	Nonspecific esterases are biochemical markers for monocytes, although monocytes contain an array of different enzymes. Peroxidase production ceases as monocytes complete their maturation process. CD14 positive.
Number of divisions	1
Proportion in blood	3-8%
Concentration	200-800 cells/μl
Estimated monocyte turnover	7 x 10^6 cells/kg/h (cf. Neutrophils: 6 x 10^7 cells/kg/h)

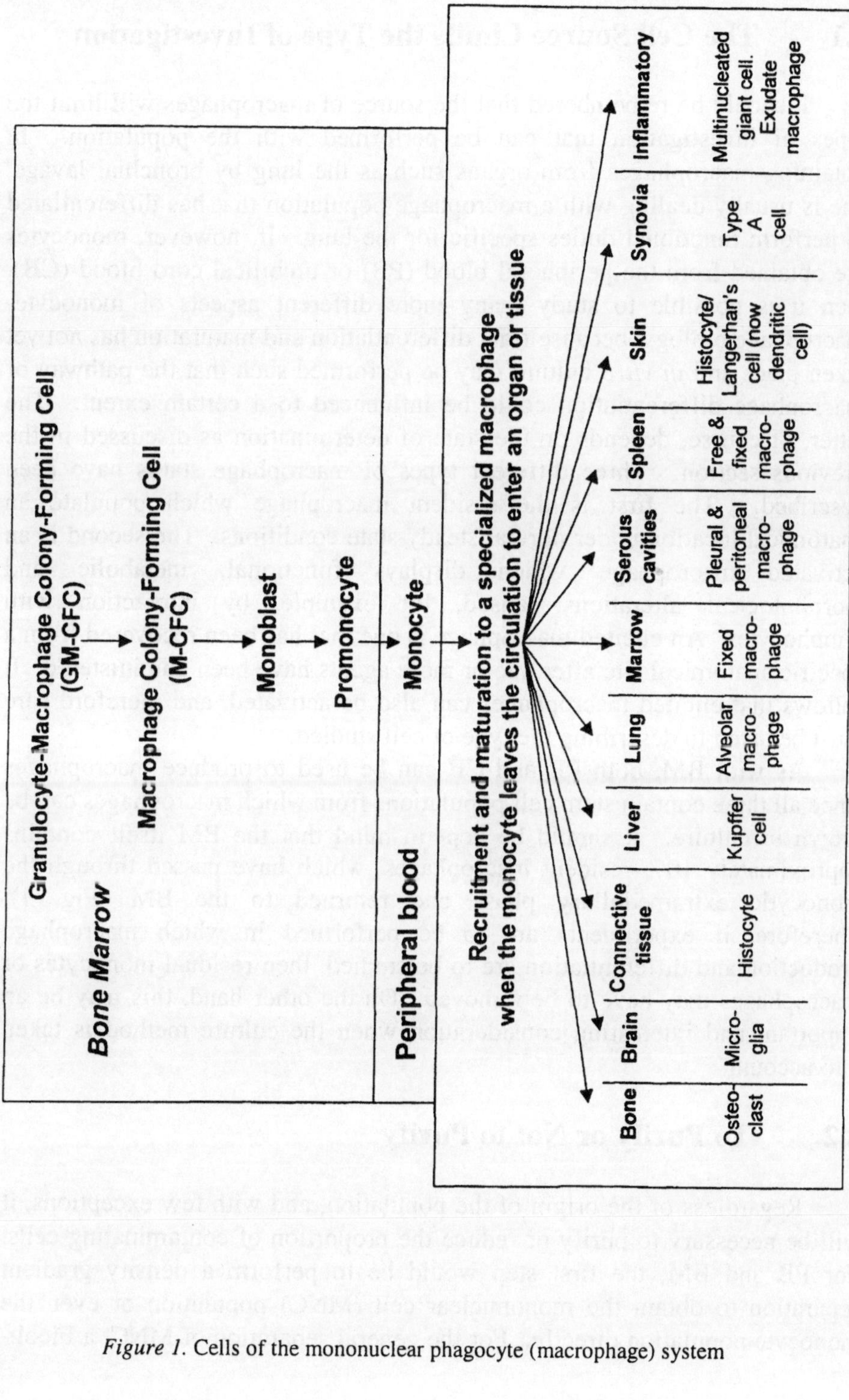

Figure 1. Cells of the mononuclear phagocyte (macrophage) system

2.1 The Cell Source Limits the Type of Investigation

It should be remembered that the source of macrophages will limit the types of investigation that can be performed with the population. In obtaining macrophages from organs such as the lung by bronchial lavage, one is usually dealing with a macrophage population that has differentiated to perform functional duties specific for the lung. If, however, monocytes are obtained from the peripheral blood (PB) or umbilical cord blood (CB), then it is possible to study many more different aspects of monocyte-macrophage biology, because final differentiation and maturation has not yet taken place and *in vitro* culture may be performed such that the pathway of macrophage differentiation could be influenced to a certain extent. The latter, of course, depends on the state of determination as discussed in the previous section. Three different types of macrophage status have been described. The first is the resident macrophage which populate an anatomical location under normal steady-state conditions. The second is an activated macrophage which displays functional, metabolic and morphological alterations caused, for example, by interaction with lymphocytes. An elicited macrophage is one that has been recovered from a specific anatomical site after one or more agents have been administered. It follows that elicited macrophages can also be activated, and therefore care must be taken in describing the type of cell studied.

As with BM, both PB and CB can be used to produce macrophages since all three contain stem cell populations from which macrophages can be grown in culture. It should be kept in mind that the BM itself contains approximately 10% resident macrophages, which have passed through the monocytic extramedullary phase and returned to the BM (Fig. 1). Therefore, if experiments are to be performed in which macrophage production and differentiation are to be studied, then residual monocytes or macrophages may have to be removed. On the other hand, this may be an important and interesting consideration when the culture method is taken into account.

2.2 To Purify or Not to Purify

Regardless of the origin of the population, and with few exceptions, it will be necessary to purify or reduce the proportion of contaminating cells. For PB and BM, the first step would be to perform a density gradient separation to obtain the mononuclear cell (MNC) population or even the monocyte population directly. For the general separation of MNC, a Ficoll-type procedure is used [31-34]. There are several different types of density gradient media on the market. PB monocytes have a lower buoyant density than other cells in the blood and these have overlapping densities when a continuous gradient is used for separation. The result is cross-contamination which is obviously not the object of the exercise. To obviate this problem, a discontinuous gradient can be performed using NycoPrep with a density of 1.068 g/ml. The cell preparation is layered over half the volume of Nycoprep 1.068 and centrifuged at 600 *g* for 15 min at 20 °C. A broad band is obtained below the interface, the top part of which is removed. After careful washing of the cells, a high purity and a reasonable yield of monocytes can be obtained (see also [35-37]). Percoll density gradients have also been used to separate and purify monocytes and macrophages [38,39].

A new method for isolation and purification of monocytes uses a positive-selection magnetic bead process from Miltenyi Biotec (Auburn, CA). The cell suspension is incubated with a panel of non-monocyte antibodies and then magnetically labeled using magnetic beads coupled to an anti-hapten antibody. Non-monocytes are removed by a magnetic field as the cell suspension flows through a column, leaving a highly purified population of monocytes [40-43]. However, macrophages can phagocytose the particles used for separation and this in turn can lead to a change in function of the cells. It should be pointed out that if the cell suspension is contaminated by red blood cells, then a general MNC separation using a density gradient procedure is recommended before a purification process.

Macrophages obtained directly from an organ, such as the lung [44] or liver [45], can be obtained relatively pure either by adhesion to plastic, centrifugal elutriation (providing a sufficient starting volume is available) or by cell sorting. It should be remembered that any procedure involving monocyte or macrophage adhesion invokes a change in functionality of the cell. Therefore, depending on the type of experiment to be performed, adhesion may not be the optimal method of removing other cells from the macrophage population. Centrifugal elutriation is a complicated and initially costly procedure requiring a great deal of time to develop, but can produce a high quality product. Cell sorting is the most expensive procedure and can also be very time consuming, producing a highly purified but low yield product. With the advent of new technology cell sorters (approximately 25,000 events/sec and more), a highly purified, high yield product can be obtained in a shorter period of time.

Production of macrophages from BM can be performed by using one of two basic procedures and these will be described in detail in Section 4. In summary, an initial selection procedure is, in most cases, necessary, even if it only removes erythrocyte contamination.

2.3 To Freeze or Not to Freeze

In general, it is not advisable to freeze selected and/or purified monocyte-macrophage populations. The cells do not withstand the freezing and thawing procedure well. PB, CB and BM can be cryopreserved [46-49]. Cryopreservation is a general procedure in all institutes performing hematopoietic transplantation procedures. After thawing, monocytes-macrophages can be obtained, but the numbers are a limiting factor. In general, approximately half the original cell concentration will be obtained after thawing. However, macrophage (and dendritic cell, see Chapter 7) growth and expansion from the stem cell population can be performed using cryopreserved hematopoietic cells.

If cryopreservation of monocytes-macrophages is to be performed, then certain factors must be taken into account. First, the initial concentration of monocytes-macrophages should be at least 5 x 10^6 cells/ml, since a high proportion of cells are going to die or are lost during the thawing procedure. Second, the serum concentration should be at least 10%, since monocytes and macrophages do not fare well in the absence of serum. The type of material used for cryopreservation should also be considered (e.g. DMSO, glycerol or hetastarch). In addition, the rate of freezing and thawing are exceptionally important parameters, as well as the dilution procedure (if any) of the cryopreservation material used, since this cannot usually remain during subsequent cell culture procedures. Some investigators would dilute the cryopreservative immediately in a large volume (10 ml) of medium supplemented with serum, while others may perform a gradual dilution procedure over a period of 10 min. We have found that for primary cells, the latter procedure provides a better yield than the former.

3. ASSAY TECHNIQUES

3.1 General Characteristics

Macrophages can be detected by several methods. Macrophages are 20-80 μm in size, but rarely, if at all, demonstrate a proliferative response. The nucleus is large and characteristically indented like a kidney and is located to one side of the cell. They adhere avidly to plastic and glass and upon doing so, change their appearance and function. Macrophages contain non-specific esterase, but no myeloperoxidase granules. They contain lysosomes and secrete lysozyme at a steady rate. They actively phagocytose particles (e.g. opsonized latex, iron and ink). They possess Fc and C3b complement receptors as well as lactoferrin receptors on their surface.

Thus, there are numerous assays to detect monocytes and macrophages and the reader is advised to consult the articles listed at the end of this chapter [29,50].

3.2 Flow Cytometry

Monocytes and macrophages can be detected by flow cytometry using monoclonal antibodies against two specific antigens on the cell membrane. One is CD14, which is expressed at high levels on monocytes and macrophages. This antigen is a high affinity receptor for complexes of lipopolysaccharide (LPS) and serum LPS-binding protein. The other is CD64, a high affinity receptor for IgG (FcγRI), a type I transmembrane glycoprotein. CD64 plays a role in phagocytosis and antibody-dependent cell-mediated cytotoxicity (ADCC). In a forward/side scatter dot plot of BM or PB, monocytes are seen as a cluster of events with medium to high forward scatter and low to medium side scatter (Fig. 2). Macrophages usually are large (i.e. high forward scatter) and complex (i.e. high side scatter) cells. The event cluster is usually located above the lymphocyte population, but below the granulocytes. Either $CD14^+$ and/or $CD64^+$ cells can be detected using CD45, a pan-leukocyte antigen, to perform the initial gate procedure. However, as seen in Table 2, there are many different types of CD antigens that can be detected on all cellular components of the mononuclear phagocyte system, and these can be used to enumerate different subpopulations.

3.3 *In Vitro* Production of Macrophages and Detection of Monocyte-Macrophage Progenitor Cells Using the Colony-Forming Technique

This procedure not only detects whether progenitor cells can give rise to macrophages, but can also be used to produce macrophages. Another technique for production and expansion of resident macrophages is given in the next section. This latter technique relies on the growth of progenitor cells to produce macrophages in suspension culture and would therefore not be viewed as a detection technique.

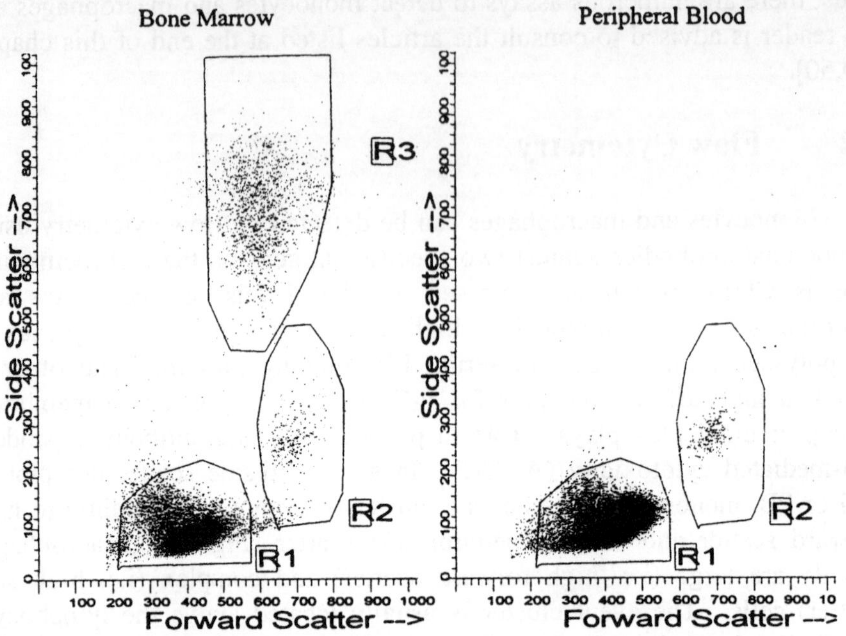

Figure 2. Flow cytometry profile and position of the monocyte population from normal human bone marrow and peripheral blood. R1 = lymphocyte gate. The stem cells are located in this gate. R2 = monocyte gate. As the monocytes mature into macrophages, they will increase both their forward and side scatter. R3 = granulocyte gate. In the peripheral blood sample, there are relatively few granulocytes. The flow cytometry profiles were produced on a Becton Dickenson FACSort and 50,000 events (cells) were acquired.

The production of macrophages is dependent on the presence of either M-CSF or low concentrations of GM-CSF. The presence of such growth factors in conditioned medium and its stimulatory effect on murine BM cells in semi-solid culture was developed independently by Bradley and Metcalf [5] in Australia and Pluznik and Sachs [6] in Israel. Both techniques were reported in 1966 and used agar to immobilize the cells. Although agar is still used, many laboratories have replaced it with water-soluble methylcellulose [51]. In the author's laboratory, the same general method has been used to detect all *in vitro* hematopoietic stem and progenitor cell populations. The major difference occurs in the concentration and type of growth factors used and the length of culture incubation. Otherwise all components remain the same regardless of the cell type being assayed. Furthermore, for the detection assay, a micro-technique is used. If the production of macrophages is required, this same method can easily be scaled-up to accommodate larger cell numbers. The principal components and method is as follows.

Table 2. Membrane antigen (CD) phenotypes of different members of the mononuclear phagocyte system

CFU-GM	M-CFC	Promonocyte	Monocyte	Macrophage	Activated Macrophage
CD13	CD13	CD13	(CD4)	CD11c	CD11c
CD33	CD15	CD14	CD9	CD14	CD14
CD 34	CD33	(CD15)	CD11b,c	CD16	CD16
CD115	CD115	CD33	CDw12	CD26	CD26
CD 116	CD116	CD115	CD13,14,15	CD31	CD31
CDw123	CDw123	CD116	CDw17	CD32	CD32
CD w131	CDw131	CDw123	(CD31)	CD36	CD36
		CDw131	CD32,33,35	CD45RO	CD45RO
			CD36,38	CD45RB	CD45RB
			(CD40)	CD63	CD63
			CD43	CD68	CD68
			(CD45RO)	CD71	CD71
			(CD45RA)	CD74	CD74
			(CD45RB)	CD87	CD87
			CD49b,e,f	CD88	CD88
			CD63,64,65s	CD101	CD101
			CD68	CD119	CD119
			(CD74)	CD121b	CD121b
			CD84,85,86	CD155	CD155
			CD87,89,91	CD156	CD156
			CDw92,93,98		+
			CD101,102		CD23
			CD115,116		CD25
			CD119		CD69
			CDw121B		CD105
			CDW123		
			CD127		
			CDw128,131		
			CD147,155		
			CD156,157		
			CD162,163		
			CD164		

Brackets indicate weak CD antigens

3.3.1 Materials

Petri dishes (35 mm x 10 mm) containing 4 wells (Greiner, Wilmingon, Delaware. No. 627170). These are bacteriological Petri dishes which are NOT sterilized. This appears to be a strange situation, but when plastic Petri dishes are produced, the plastic components are subjected to both heat and pressure which does effectively sterilize them. Plastic is usually sterilized by radiation, which can change the growth surface properties of the plastic and cause cells to preferentially adhere. This results in a change in the make-up of the cell types grown. For this reason, we have found that non-sterilized Petri dishes are far superior to those that have been sterilized by radiation.

1. Plastic tubes, 5 ml, individually wrapped, polystyrene with pushdown cap (Falcon No. 2003)
2. Incubator at 37 °C, gassed with an atmosphere of 5% CO_2, 5% O_2, and 90% nitrogen (Heraeus, South Plainfield, NJ). To achieve low oxygen tension (discussed below), nitrogen is pumped into the incubator chamber to reduce the normal atmospheric oxygen tension from about 21% to 5%.
3. Methylcellulose (MeC) (2% stock solution containing α-thioglycerol (1 x 10^{-2} M stock solution))
4. Human AB serum, heat-inactivated (56 °C for 20 min and sterile filtered)
5. Pure human iron-saturated transferrin (1 x 10^{-10} M stock solution)
6. M-CSF or GM-CSF (R&D Systems, Minneapolis, MN)
7. Iscove's Modified Dulbecco's Medium (IMDM) freshly prepared and sterile filtered

3.3.2 Method

1. Add the components in the Table to sterile plastic tubes. Add cells to fetal bovine serum (FBS) first and MeC should be added last.
2. The MeC is added with a 1 ml syringe without a needle.
3. Once all the components have been added to the tube, make sure all tubes are capped and vortex each tube so that all the components are mixed thoroughly.
4. Label the lid of the Petri dishes with a marker pen.
5. Using an Eppendorf Multipipette or a 1 ml syringe fitted with a 20 gauge needle, withdraw almost all of the mixture in the tube (between 500 and 550 µl) and dispense 100 µl into each of the four wells of the Greiner plate. Using a pipette tip, distribute the mixture around the well so that the surface is completely covered. Thus, each culture is plated in quadruplicate.
6. Put all the plates into a large sterile glass round or oblong dish, stacking the plates on top of each other if necessary, and add a small container filled with distilled water. Cover the dish with aluminum foil.
7. Place the dish containing the plates in a humidified incubator at 37 °C, containing an atmosphere of 5% CO_2 and 5% O_2.
8. Incubate for 14 days to obtain pure macrophage colonies.
9. Count the colonies containing >50 cells/colony under an inverted or preparation microscope.
10. The results are usually given in concentration (i.e. per 10^5 cells plated).

Table 3.

Component	Final concentration	Dilution	Add following in µl
Human AB serum (heat inactivated)	15%		90
Human transferrin (iron-saturated)	1×10^{12}g/ml from a 1×10^{10} g/ml stock	1:100	6
rhGM-CSF or rhM-CSF	2 or 20 ng/ml		6
IMDM			228
Cells	0.5×10^5 cells/ml from a 0.5×10^6 cells/ml stock	1:10	60
MeC in IMDM with α-TG		1:2.857	210
Total			600

3.3.3 Considerations

There are several aspects of this method that are different to many other colony-forming assays reported. The first is that α-thioglycerol is added at the time the MeC is prepared. This addition is not obligatory to the method, but makes it easier. The addition of α-thioglycerol is to maintain many cellular components in the reduced state, thereby reducing free radical formation, which can kill cells and reduce the plating efficiency. This, together with the second aspect, that of culturing cells under reduced oxygen tension, has an added effect in that oxygen toxicity is further reduced. The combined effect is to increase the plating efficiency of the assay, thereby increasing the sensitivity of the system. By increasing the sensitivity, the initial plated cell concentration can also be reduced if necessary. Human AB serum is used for these cultures rather than FBS. Although a final concentration of 15% is given, as with most of the components, this has to be titrated to obtain optimal results. Again, this is not obligatory and the reader may find that a batch of FBS is better than human serum. Although macrophages cannot in general be cultured or maintained under true serum-free conditions, most serum-free media contain FBS substitutes such as bovine serum albumin (BSA). Serum-free media have also been designed especially for the culture of human macrophages (e.g. Macrophage-SFM medium from Life Technologies).

It should be noted that, with the exception of BM, PB and to some extent peritoneal exudate, most organs and tissues have very low numbers of macrophage progenitor cells, so that the above detailed method cannot be used satisfactorily. In this situation, the assay techniques described at the beginning of this section are all applicable.

4. CULTURE TECHNIQUES

In general, macrophages from sources other than the BM, PB and peritoneal cavity cannot be grown, but only maintained in culture. This is because the macrophages from these tissues and organs are resident macrophages, which usually do not have the capacity to proliferate. To maintain macrophages in culture, they can be incubated in plastic or glass dishes or on slides for several days, but will die after a short period of time. Long-term culture of macrophages derived from these sites is usually very difficult. However, using the method described below for production of macrophages from BM, long-term culture over several weeks is possible.

As with all culture techniques, finding the optimal conditions for maintaining monocytes and macrophages is a trial and error process, which takes time and patience. When setting up any culture technique, it goes without saying that the appropriate assay systems have to be on hand to measure the end-points of the method. For macrophages, this may include the assay for non-specific esterase. A histochemical stain can also be combined with another macrophage-specific marker, which is conjugated to a fluorescent label, so that the cells can be viewed and evaluated directly in the culture plate under an inverted fluorescence microscope. Whichever assay system is used, the growth and maintenance of monocytes and macrophages is dependent on several parameters.

4.1 Type of Growth Surface

The growth surface on which cells find themselves is probably one of the most important and underestimated aspects of tissue culture. There are many different types of plastic that can be used for tissue culture, and even if a polystyrene plastic is used, it does not mean to say that it is the same from one manufacturer to another. A culture plate or flask may be denoted tissue culture grade, but there are very large differences in plastic composition, treatment of the plastic after manufacture and the method by which the vessels are sterilized. For example, certain types of radiation can cause a change in the surface properties of the plastic. Ion accumulation can do the same. Of all the different types of plastic tested, we have found that Nuncalon (Nunc) plates have provided the most consistent and reproducible macrophage cultures. Not only is the type of plastic important, but if glass is used, there are also different qualities of glass which may play a role in the maintenance of macrophages *in vitro*. If slide cultures are to be used, a very helpful container in which microscope-format slides can be placed is the Quadriperm plate (Heraeus). This plate has four culture wells in which slides can be placed and cells cultured.

4.2 Serum or Serum-Free

As mentioned above, monocytes survive *in vitro* in the absence of serum only when they adhere to a surface, and rapidly die in suspension cultures in the absence of serum. However, monocytes can be induced to mature in the presence of between 0.5% and 5% serum [52]. As discussed above, for human monocytes and macrophages, human AB serum has been found to be better than batches of FBS, but this has to be determined empirically in each culture system.

4.3 Other Culture Components

Using a single medium for all types of culture results in saving both time and cost. Although many investigators will use different media for different cell types, we have found that for hematopoietic cells, regardless of whether they are stem, progenitor or mature end cells, IMDM is a good general medium to use. We have always purchased the powder form and have prepared the medium fresh or in sufficient quantity to last about one week. In addition, the medium is aliquoted into 20-25 ml bottles and capped tightly with very little air in the bottle. Although this sounds rather extreme, we have found consistent results are always obtained, since not only does the pH remain constant, but also glutamine, vitamins and other nutrients do not have time to deteriorate. We have also found that addition of α-thioglycerol at a final concentration of 1×10^{-4} M, together with low oxygen tension (see below) and a final concentration of 1×10^{-10} M human iron-saturated transferrin are required for optimal growth.

4.4 Growth Factors

As mentioned above, both M-CSF and low concentrations of GM-CSF can be added to cultures to induce macrophage production. As detailed below, this is not an absolute necessity since macrophages produce both M-CSF and GM-CSF and can therefore regulate their own production. It is this property that is taken advantage of in the culture procedure described below, in which BM cells are cultured on nonadherent growth surfaces to produce a relatively pure population of resident macrophages. One should also take into account that addition of either M-CSF or GM-CSF will influence the type of macrophage obtained, since activated macrophages can be produced in the presence of growth factors.

4.5 Low Oxygen Tension

We started to grow hematopoietic cells under low oxygen tension in 1980, soon after it was reported by Bradley [53] that culturing CFU-GM under reduced oxygen tension increased the plating efficiency. We subsequently found that if β-mercaptoethanol, α-thioglycerol, vitamin E or reduced glutathione was added to cultures which were incubated under low oxygen tension, a substantial increase in plating efficiency leading to enhanced assay sensitivity occurred. The reason was simple. Addition of these agents reduced oxidation, and combined with low oxygen tension, oxygen toxicity and free radical formation was decreased [54-57]. Subsequently, α-thioglycerol and low oxygen tension (5%) have been used for all our cultures, including the production of macrophages in culture. The advantageous use of low oxygen tension for growing human hematopoietic cells, including macrophages, has also been well documented [58-61].

4.6 Production of Resident Macrophages on a Nonadherent Growth Surface

This procedure was originally developed for murine macrophages [51,55,62] and has been expanded to include the production of human macrophages. It is used for growing macrophages from BM MNC, and utilizes two distinct properties of these cells. The first is that no exogenous growth factors are added because the small number of resident macrophages originally present produce low concentrations of GM-CSF which causes cells to be induced into the monocyte-macrophage lineage. The second is that the cells are not allowed to adhere, which is in part responsible for the production of resident macrophages rather than activated macrophages. Nonadherence of the cells is achieved by culturing on hydrophobic Teflon foils. For the present method, Petriperm dishes (Heraeus) are used, consisting of 60 mm diameter Petri dishes with a thin, stretched Teflon foil as the growth surface. There are several advantages of using Teflon rather than other plastic types. First, Teflon is non-toxic. Second, one side of the foil has a nonadherent (hydrophobic) surface while the other side has an adherent (hydrophilic) surface. Third, Teflon foil is glass-clear which allows for excellent microscopic examination of the cells. If the cells are to be stained or manipulated, this can also be performed directly in the plate. Fourth, Teflon is gas-permeable, thereby allowing the gas mixture in the incubator to come into direct contact with the cells resting on the growth surface. This is especially important when using low oxygen tension. Gases have a very low solubility, and unless the cells are cultured as a thin film, it will take a long period of time for the low oxygen effect to penetrate to the cells in culture. The disadvantages of this system are that the Petriperm culture plates are relatively expensive and are only manufactured in one size. However, the system can be scaled-up by making Teflon bags using Biofoil or by purchasing Teflon bags from retailers (e.g. Baxter). The basic method is relatively simple.

1. MNC at a concentration of 5-10 x 10^6 cells/3 ml culture are suspended in IMDM supplemented with 10% human AB serum, 1 x 10^{-4} M α-thioglycerol and 1 x 10^{-10} M human iron-saturated transferrin in a Petriperm dish.
2. The culture dishes are incubated between 14 and 21 days at 37 °C in a humidified atmosphere containing 5% CO_2 and 5% O_2. Make sure the incubator trays on which the culture plates are resting have holes so that the gas can enter the dish from below. The plates can be carefully removed during the incubation period and the cells viewed under an inverted microscope.
3. If cells are to be removed, this can usually be performed by swirling the contents of the dish or very gently scraping the cells using a sterile rubber policeman. The cells come away from the surface very easily and can be removed either with a pipette or syringe taking care not to pierce the Teflon foil.

With this relatively easy culture procedure, macrophages are not only produced, but also expand in numbers. Many of the cells contaminating the original preparation will have died out during the culture period. If the supernatant from these cultures is assayed, it will be found that colony-stimulating factors are present. As stated above, the process is dependent on the serum used, which will have to be pre-screened to achieve optimal growth.

5. UTILITY OF THE SYSTEM

This culture system can easily be scaled-up to accommodate the production of large numbers of macrophages. However, the open culture system described here allows assays and tests to be performed *in situ*. Cells can be fixed in the dish if required, and the Teflon foil to which the cells are fixed cut away with a knife or scissors and used for other tests, such as *in situ* hybridization or *in situ* PCR. Using the adherent side of the Teflon foil, other cell types can be cultured (e.g. dendritic cells). Using the stationary culture system described, important aspects of macrophage determination can be studied. In addition, although no growth factors are added to this basic culture procedure, this does not mean that the effect of growth factors on macrophage differentiation, activation and interaction with other cells cannot be investigated using modified culture techniques. This is, in fact, one of the few culture methods in which nearly all aspects of macrophage biology can be investigated. Yet it is not specific for macrophages. Almost any type of cell can be cultured on adherent or nonadherent Teflon foil. Because of the clarity of the image under the inverted microscope, complex image analysis can be performed directly on cells in the plate. If electron microscopy is required, fixation, staining and embedding can also be performed directly in the plate, and the Teflon peeled off very easily from the block after the process has been completed.

6. CONCLUDING REMARKS

Regardless of which culture system is used, the principal aim of all culture methodology is to simulate, as far as possible, the *in vivo* conditions. As with other primary hematopoietic cells, there are many important aspects to culturing macrophages. Some of these may appear trivial in nature, but not taking these aspects into account could mean that valuable time and money is wasted. Macrophages are a heterogeneous and difficult population of cells to study. Due to their versatility in adapting to a multitude of different environmental conditions, small changes in the extracellular milieu can result in changes in macrophage functional capacity. Nevertheless, it is precisely this versatility that makes the macrophage such an interesting cell to study.

ACKNOWLEGMENT

This work was supported by the Center for Cancer Treatment and Research, Palmetto Richland Memorial Hospital and the South Carolina Cancer Center.

REFERENCES CITED

1. Rich IN (1995) Hemopoietic-initiating cells *J Perinat Med* 23:31-38.
2. Rich IN (1995) Primordial germ cells are capable of producing cells of the hematopoietic system in vitro *Blood* 86:463-472.
3. Kelemen E and Jamissa M (1980) Macrophages are the first differentiated blood cells formed in human embryonic liver *Exp Hemat* 8:996-1000.
4. Takahashi K Yamanura F and Naito M (1989) Differentiation, maturation, and proliferation of macrophages in the mouse yolk sac: A light-microscopic, enzyme-cytochemical, immunohistochemical, and ultrastructural study *J Leuko Biol* 45:87-96.

5. Bradley TR and Metcalf D (1966) The growth of mouse BM cells in vitro *Aust J Exp Biol Med* 44:287-287.
6. Pluznik DH and Sachs L (1966) The induction of clones of normal mast cells by a substance from conditioned medium *Exp Cell Res* 43:553-553.
7. Metcalf D (1977) *Hemopoietic Colonies. In Vitro Cloning of Normal and Leukemic Cells* Springer-Verlag Heidelberg.
8. Stanley RE (1994) Colony Stimulating Factor 1 (Macrophage Colony Stimulating Factor), in Thomson AW (ed.) *The Cytokine Handbook.* Academic Press, New York, pp.387-418.
9. Faruqi RM and DiCorleto PE (1993) Mechanisms of monocyte recruitment and accumulation *Br Heart J* 69:S19-29.
10. Walz G Aruffo A Kolanus W Bevilacqua M and Seed B (1990) Recognition by ELAM-1 of the sialyl-Lex determinant on myeloid and tumor cells *Science* 250:1132-1135.
11. Verdegaal EM Beekhuizen H Blokland I and van Furth R (1993) Increased adhesion of human monocytes to IL-4-stimulated human venous endothelial cells via CD11/CD18, and very late antigen-4 (VLA- 4)/vascular cell adhesion molecule-1 (VCAM-1)-dependent mechanisms *Clin Exp Immunol* 93:292-298.
12. Beekhuizen H Blokland I and van Furth R (1993) Cross-linking of CD14 molecules on monocytes results in a CD11/CD18- and ICAM-1-dependent adherence to cytokine-stimulated human endothelial cells *J Immunol* 150:950-959.
13. Beekhuizen H Corsel-Van Tilburg AJ Blokland I and van Furth R (1991) Characterization of the adherence of human monocytes to cytokine- stimulated human macrovascular endothelial cells *Immunology* 74:661-669.
14. Beekhuizen H Blokland I Corsel-Van Tilburg AJ Koning F and van Furth R (1991) CD14 contributes to the adherence of human monocytes to cytokine- stimulated endothelial cells *J Immunol* 147:3761-3767.
15. Springer TA (1990) Adhesion receptors of the immune system *Nature* 346:425-434.
16. Rollins BJ (1996) Monocyte chemoattractant protein 1: a potential regulator of monocyte recruitment in inflammatory disease *Mol Med Today* 2:198-204.
17. Sozzani S Locati M Allavena P Van Damme J and Mantovani A (1996) Chemokines: a superfamily of chemotactic cytokines *Int J Clin Lab Res* 26:69-82.
18. Miller MD and Krangel MS (1992) Biology and biochemistry of the chemokines: a family of chemotactic and inflammatory cytokines *Crit Rev Immunol* 12:17-46.
19. Martinez-Pomares L Platt N McKnight AJ da Silva RP and Gordon S (1996) Macrophage membrane molecules: markers of tissue differentiation and heterogeneity *Immunobiology* 195:407-416.
20. Takahashi K Naito M and Takeya M (1996) Development and heterogeneity of macrophages and their related cells through their differentiation pathways *Pathol Int* 46:473-485.
21. Naito M Umeda S Yamamoto T Moriyama H Umezu H Hasegawa G Usuda H Shultz LD and Takahashi K (1996) Development, differentiation, and phenotypic heterogeneity of murine tissue macrophages *J Leukoc Biol* 59:133-138.
22. Riches DW (1995) Signalling heterogeneity as a contributing factor in macrophage functional diversity *Semin Cell Biol* 6:377-384.
23. Gordon S (1995) The macrophage *Bioessays* 17:977-986.
24. Gordon S Clarke S Greaves D and Doyle A (1995) Molecular immunobiology of macrophages: recent progress *Curr Opin Immunol* 7:24-33.
25. Naito M (1993) Macrophage heterogeneity in development and differentiation *Arch Histol Cytol* 56:331-351.
26. Rutherford MS Witsell A and Schook LB (1993) Mechanisms generating functionally heterogeneous macrophages: chaos revisited *J Leukoc Biol* 53:602-618.

27. Dijkstra CD and Damoiseaux JG (1993) Macrophage heterogeneity established by immunocytochemistry *Prog Histochem Cytochem* 27:1-65.
28. Papadimitriou JM and Ashman RB (1989) Macrophages: current views on their differentiation, structure, and function *Ultrastruct Pathol* 13:343-372.
29. Gordon S Keshav S and Chung LP (1988) Mononuclear phagocytes: tissue distribution and functional heterogeneity *Curr Opin Immunol* 1:26-35.
30. Van Furth R (1992) *Mononuclear Phagocytes: Biology of Monocytes and Macrophages.* Kluwer Academic Publishers
31. Boyum A (1968) Separation of leukocytes from blood and BM. Introduction *Scand J Clin Lab Invest Suppl* 97:7:7-
32. Boyum A (1968) Isolation of leukocytes from human blood. A two-phase system for removal of red cells with methyl cellulose as erythrocyte-aggregating agent *Scand J Clin Lab Invest Suppl* 97:9-29.
33. Boyum A (1968) Isolation of mononuclear cells and granulocytes from human blood. Isolation of mononuclear cells by one centrifugation, and of granulocytes by combining centrifugation and sedimentation at 1 g *Scand J Clin Lab Invest Suppl* 97:77-89.
34. Boyum A (1968) A one-stage procedure for isolation of granulocytes and lymphocytes from human blood. General sedimentation properties of white blood cells in a 1g gravity field *Scand J Clin Lab Invest Suppl* 97:51-76:51-76.
35. Sone S and Key ME (1986) Antitumor and phagocytic activities of rat alveolar macrophage subpopulations separated on a discontinuous gradient of bovine serum albumin *J Biol Response Mod* 5:595-603.
36. Nau P and Peters JH (1986) Human PB accessory cell: isolation by hypotonic density gradient, functional, and phenotypical characterization *Immunobiology* 173:82-97.
37. Sandron D Reynolds HY Laval AM Venet A Israel-Biet D and Chretien J (1986) Human alveolar macrophage subpopulations isolated on discontinuous albumin gradients. Cytological data in normals and sarcoid patients *Eur J Respir Dis* 68:177-185.
38. Vray B and Plasman N (1994) Separation of murine peritoneal macrophages using Percoll density gradients *J Immunol Methods* 174:53-59.
39. Plasman N and Vray B (1993) Mouse peritoneal macrophages: characterization of functional subsets following Percoll density gradients *Res Immunol* 144:151-163.
40. Kossovsky N Hu CK Sponsler EE Chun G and Millett D (1993) High-purity resident tissue macrophage isolates from human synovium and periprosthetic tissues using immunomagnetic techniques *Clin Orthop* 262-268.
41. Lutz DA Chen XM and McLaughlin BJ (1993) Isolation of the phagocytic compartment from macrophages using a paramagnetic, particulate ligand *Anal Biochem* 214:205-211.
42. Ennen J Findeklee H Dittmar MT Norley S Ernst M and Kurth R (1994) CD8+ T lymphocytes of African green monkeys secrete an immunodeficiency virus-suppressing lymphokine *Proc Natl Acad Sci U S A* 91:7207-7211.
43. Bischoff SC Krieger M Brunner T Rot A von T, V Baggiolini M and Dahinden CA (1993) RANTES and related chemokines activate human basophil granulocytes through different G protein-coupled receptors *Eur J Immunol* 23:761-767.
44. Spiteri MA Clarke SW and Poulter LW (1992) Isolation of phenotypically and functionally distinct macrophage subpopulations from human bronchoalveolar lavage *Eur Respir J* 5:717-726.
45. Friedman SL Rockey DC McGuire RF Maher JJ Boyles JK and Yamasaki G (1992) Isolated hepatic lipocytes and Kupffer cells from normal human liver: morphological and functional characteristics in primary culture *Hepatology* 15:234-243.
46. Newton I Charbord P Schaal JP and Herve P (1993) Toward CB banking: density-separation and cryopreservation of CB progenitors *Exp Hematol* 21:671-674.

47. Reiners B Hermann J Fuchs D Prager J Thranhardt H Burger E Kob D and Zintl F (1989) Cryopreservation of marrow, purging and autologous BM transplantation in childhood *Folia Haematol Int Mag Klin Morphol Blutforsch* 116:601-608.

48. Tzeng CH Hsieh RK Chuang MW Liu JH Chen KY Yung CH Wang SY and Chen PM (1988) BM cryopreservation and clinical implications in autologous BM transplantation *Proc Natl Sci Counc Repub China [B]* 12:88-94.

49. Sputtek A Jetter S Hummel K and Kuhnl P (1997) Cryopreservation of PB progenitor cells: characteristics of suitable techniques *Beitr Infusionsther Transfusionsmed* 34:79-83:79-83.

50. Herscowitz HB Holden HT Bellanti JA Ghaffar A (1981) *Manual of Macrophage Methodology. Collection, characterization and function.* Marcel Dekker, Inc. New York.

51. Rich IN Heit W and Kubanek B (1982) Extrarenal erythropoietin production by macrophages *Blood* 60:1007-1018.

52. Andreesen R Brugger W Scheibenbogen C Kreutz M Leser HG Rehm A and Lohr GW (1990) Surface phenotype analysis of human monocyte to macrophage maturation *J Leukoc Biol* 47:490-497.

53. Bradley TR Hodgson GS and Rosendaal M (1968) The effect of oxygen tension on hemopoietic and fibroblast cell proliferation in vitro. *J Cell Physiol* 97:517-522.

54. Rich IN and Kubanek B (1982) The effect of reduced oxygen tension on colony formation of erythropoietic cells in vitro *Brit J Haemat* 52:579-588.

55. Rich IN (1986) A role for the macrophage in normal hemopoiesis II Effect of varying physiological oxygen tensions on the release of hemopoietic growth factors from bone-marrow-derived macrophages in vitro *Exp Hemat* 14:746-751.

56. Rich IN (1987) Erythropoietin production by macrophages: Cellular response to physiological oxygen tensions and detection of erythropoietin gene expression by in situ hybridization, in Rich IN (ed.) *Molecular and Cellular Aspects of Erythropoietin and Erythropoiesis. NATO ASI Series, Vol. H8.* Springer-Verlag, Berlin Heidelberg, pp.291-310.

57. Rich IN (1990) Oxygen sensing mechanisms in the macrophage and erythropoietin production, in Eyzaguirre C Fidone SJ Fitzgerald RS Lahiri S and McDonald D (eds.) *Arterial Chemoreception.* Springer-Verlag, New York,

58. Muench MO Gasparetto C and Moore MA (1992) The in vitro growth of murine high proliferative potential- colony forming cells is not enhanced by growth in a low oxygen atmosphere *Cytokine* 4:488-494.

59. Koller MR Bender JG Papoutsakis ET and Miller WM (1992) Beneficial effects of reduced oxygen tension and perfusion in long-term hematopoietic cultures *Ann N Y Acad Sci* 665:105-16:105-116.

60. Broxmeyer HE Cooper S Lu L Miller ME Langefeld CD and Ralph P (1990) Enhanced stimulation of human BM macrophage colony formation in vitro by recombinant human macrophage colony- stimulating factor in agarose medium and at low oxygen tension *Blood* 76:323-329.

61. Broxmeyer HE Cooper S and Gabig T (1989) The effects of oxidizing species derived from molecular oxygen on the proliferation in vitro of human granulocyte-macrophage progenitor cells *Ann N Y Acad Sci* 554:177-84:177-184.

62. Rich IN (1986) A role for the macrophage in normal hemopoiesis I Functional capacity of BM-derived macrophages to release hemopoietic growth factors *Exp Hemat* 14:738-445.

Chapter 6

Isolation and Cultivation of Osteoclasts and Osteoclast-Like Cells

[1]Philip Osdoby, [1]Fred Anderson, [2]William Maloney and [1]Patricia Collin-Osdoby
[1]Department of Biology and Division of Bone and Mineral Metabolism, Washington University, St Louis, MO 63130 and [2]Department of Orthopedics, Washington University Medical Center, St Louis, MO 63110. Tel: 001-314-935-4044; Fax: 001-314-935-5134; E-mail: osdoby@biodec.wustl.edu

1. INTRODUCTION

Skeletal development, normal bone remodeling, and calcium homeostasis are processes that require the coordinated regulation and integration of elaborate communication networks governing bone formation by osteoblasts and bone resorption by multinucleated osteoclasts. Alterations in osteoclast numbers and/or bone resorptive activity have been implicated as causative factors in pathologies ranging from postmenopausal osteoporosis and hypercalcemia of malignancy to periodontal disease, rheumatoid arthritis, and orthopedic implant loosening [1-6]. Therefore, numerous biomedical investigators interested in basic and clinical aspects of the cellular and molecular biology of bone development and remodeling have sought to develop useful models for such studies [7, 8]. Here, we focus on recent progress in the area of human osteoclast and human osteoclast-like cell culture models.

Osteoclasts are known to originate from hematopoietic precursors related to monocytic cells either found in the circulation or resident in the bone marrow (BM). For this reason, many of the research strategies employed to explore the signals and stages of osteoclastogenesis under normal or pathological conditions have drawn from the field of hematology [for review see 9]. One approach has involved the use of human myelomonocytic cell lines, such as HL-60 [10, 11] and FLG 29.1 [12, 13], which serve as sources of unlimited cell material and can be induced by various developmental promoters to express osteoclast-like characteristics. Unfortunately, regimens to date do not appear to produce significant numbers of cells from these cell lines that express a fully mature osteoclastic phenotype. The key feature of osteoclasts that distinguishes them from all other cell types is their unique ability to excavate resorption pits when cultured on mineralized matrices. Thus, the now classic *in vitro* resorption pit assay [14] is considered the standard by which full osteoclastic development is measured. Other attributes that arise during osteoclast development, and are characteristic of the mature phenotype, are high expression of tartrate-resistant acid phosphatase (TRAP) activity [15], cathepsin K/O [16], $\alpha_v\beta_3$ integrin receptors [17], calcitonin receptors [18], carbonic anhydrase II [19], acid secretion [20], a specific membrane antigen recognized by monoclonal antibody 121F [21], and a set of morphological and ultrastructural changes that include the formation of a polarized ruffled border membrane [22]. The inability to reach high expression levels of these features and a functional bone-resorptive state indicates that studies employing these myelomonocytic cell lines are most informative relative to understanding aspects of early osteoclastogenesis.

Another approach, which generally has been more successful for the *in vitro* formation of osteoclast-like cells, involves the use of primary cultures of BM cells. Originally, feline, rabbit, and avian long-term BM cultures were often used to generate *in vitro* formed multinucleated osteoclast-like cells for study [23, 24]. In recent years, conditions have become well established to also generate osteoclast-like cells from the BM or circulating mononuclear cells (MNC) of murine [25], porcine [26], rat [27], and human species [7, 28-35]. The murine and porcine osteoclastogenic model systems have proven in particular to be the most consistent and reliable sources for the generation of multinucleated osteoclast-like cells, because such cells express most, if not all, of the characteristics of mature osteoclasts, including the key functional attribute of bone pit formation on resorbable substrates. More problematic have been approaches designed to employ human whole BM [7], fractionated human BM [28-32], enriched circulating human MNC [33-35], or MNC obtained from giant cell tumors of bone (or other tissues) [36-39] to consistently generate functional osteoclast-like cells. However, several investigators have recently reported convincing evidence for the formation of functional human osteoclast-like cells from human blood mononuclear or BM cell populations *in vitro* [40-43]. In addition to permitting an investigation of the stages involved in osteoclast development, these human *in vitro* model systems have the added advantage of potentially providing sufficient numbers of such cells for biochemical, molecular, and functional studies. This would not be possible using *in vivo* formed osteoclasts isolated from normal human bone samples, owing to the relatively small number of such cells that are present in normal human bone tissues. Nevertheless, use of these *in vitro* systems may be limited by the lengthy time periods required for generating the osteoclast-like cells (generally 2-4 weeks), the expense of required growth factors, the difficulty in securing sufficient amounts of human BM in a timely and frequent manner, and the uncertainty as to what percentage of the cells in these systems reach the level of a fully mature bone-resorptive functional osteoclast.

Since it has not been easy to either generate large numbers of bone-resorptive osteoclast-like cells *in vitro* or to isolate large numbers of functional *in vivo* formed osteoclasts from normal human bone tissues, some investigators have turned toward using *in vivo* formed human multinucleated cells obtained from human giant cell tumors of bone known as osteoclastomas [44-47]. These cells form *in vivo* in response to excessive osteoclast-inducing developmental signals produced by other cells within the tumor, but are not themselves thought to represent a neoplastic component of the tumor. Besides expressing all of the features described above that are expected for mature functional osteoclasts, these cells readily form resorption pit lacunae on mineralized substrates, and hence are valuable for *in vitro* studies of human osteoclast resorptive activity. However, because the incidence of such tumors is rare and the terminally differentiated osteoclast-like cells do not divide or survive long in culture, it can be difficult to obtain enough material on a regular basis to conduct a series of experiments using these cells. Moreover, due to the unusually high degree of multinucleation in such cells relative to normal human osteoclasts, and the possibility that their hormonal responses may not exactly parallel those of normal human osteoclasts, there may be some limitations associated with the use of these cells.

Despite the scarcity of osteoclasts normally present in human bone, their isolation has been attempted by various investigators over the years owing to the aforementioned lack of a clearly acceptable substitute for such cells. Murrills et al. [48] were one of the first groups to try isolating mature human osteoclasts from fetal tissue for *in vitro* modulator studies of resorptive function. More recently, small numbers of human osteoclasts have been isolated from the craniofacial skeleton by Lambrecht and Marks [49, 50]. As a result of the high sensitivities that it is possible to achieve with many current technologies, and the advent of novel single-cell analyses (including amplified immunodetection, *in situ* hybridization, and confocal microscopy), fewer cells are required these days for many biochemical, molecular, cellular, and functional studies. Therefore, the direct isolation of mature *in vivo* formed human osteoclasts has become an increasingly more attractive and feasible approach for the *in vitro* study of such cells. Our recent work has focused on evaluating potential sources of human bone tissue and suitable methodologies to reproducibly isolate enriched populations of large numbers of human bone-resorptive osteoclasts [51, 52]. Conditions associated with significant localized bone loss were considered good potential candidates for human bone tissue that might yield high numbers of osteoclasts. This included bone obtained as discarded material during total hip replacement (THR) surgery, a procedure currently performed more than 120,000 times annually in the United States and increasing in frequency. Using human femoral heads derived from osteoporotic females or elderly males experiencing a prolonged fracture period prior to THR surgery, or segments of bone removed during surgical resection from sites of implant loosening, we developed procedures to obtain many-fold greater numbers of mature *in vivo* formed bone-resorptive human osteoclasts than had been isolated by investigators previously. Moreover, these isolated human osteoclast populations can be cultured for weeks at a time, express all the characteristics expected of mature osteoclasts, respond to bone-active hormones and other modulators, and provide sufficient material for most biochemical, immunological, physiological, molecular, and functional analyses. Here, we describe the isolation, culture, and characterization of isolated human osteoclasts, as well as human BM-derived osteoclast-like cells, and their uses for various *in vitro* studies.

2. TISSUE PROCUREMENT AND PROCESSING

2.1 Human Osteoclasts (*In Vivo* Formed)

Various sources of human bone material have been screened previously to determine which could serve as suitable sources for the isolation of high numbers of authentic bone-resorptive osteoclasts (hOC) [51, 52]. Generally, femoral heads obtained as discarded surgical material from osteoporotic and/or fracture patients undergoing hip replacement, or segments of bone removed from sites of implant loosening during surgical resection, routinely provided the greatest numbers of hOC. These have been obtained through the cooperation of orthopedic surgeons affiliated with academic institutions and/or teaching/research hospitals. It is usually possible to learn the age, sex, reason for removal of the bone segment, and general condition of the patient for each bone sample. To isolate hOC, human bone is directly placed into (~100 ml or sufficient volume to submerge the bone sample) chilled alpha-minimal essential medium (α-

MEM, GIBCO BRL, Gaithersburg, MD) containing 2% antibiotic/antimycotic (GIBCO BRL) at the time of surgical removal and maintained at 4 °C until it is processed. It is important to process the bone within 30 minutes to several hours of its excision since the viability of isolated hOC is severely compromised after 16-24 hours post-excision. Bone is cut into small blocks (0.5-1 sq in) with a manual bone saw (VWR, Chicago, IL) and then further reduced in size (0.5-1 cm) using a bone cutter (Roboz, Rockville, MD), the pieces are rinsed twice with agitation in sterile Moscona's low bicarbonate buffer, pH 7.4 (MLB, also known as Ca^{2+}, Mg^{2+}-free TBSS and consisting of 137 mM NaCl, 2.7 mM KCl, 0.42 mM $NaHPO_4$, 2.4 mM $NaHCO_3$, and 11.1 mM dextrose) to remove red blood cells, some marrow, and matrix debris, and the bone pieces are then briefly rinsed three more times with MLB before being placed into a T-75 flask (Corning, Fisher Scientific, St. Louis, MO) for enzymatic release of hOC as follows. Two methods that give comparable results relative to hOC yield and viability can be used for this purpose, method 2 having been designed for cases in which bone samples have become available only relatively late in the day.

In method 1, the samples are processed immediately by incubating the bone pieces in a T-75 flask containing 172 ml α-MEM with 2% antibiotic/antimycotic, 2 ml (60 mg) of a 3% stock solution (stored at -20°C) of collagenase (Worthington Biochemical Corp, Lakewood, NJ) in Hanks' balanced salt solution (HBSS, GIBCO BRL) and 8.1 ml of a 1% aqueous stock solution (stored at –20 °C) of trypsin (Worthington Biochemical Corp) in MLB for 90 min at 37 °C in a 95% air, 5% CO_2 moist atmosphere. In method 2, the cut bone pieces are incubated overnight humidified at 37 °C in 95% air, 5% CO_2 in 176 ml of α-MEM containing 2% antibiotic/antimycotic as above, but with less collagenase (1 ml of 3% stock solution in HBSS, 30 mg) and trypsin (4 ml of a 1% stock solution in MLB). At the end of the incubation period for either method, trypsin/collagenase activity is neutralized by the addition of 2 ml of fetal bovine serum (FBS, GIBCO BRL), 0.4 ml heparin is added, and the bones and solution are transferred to a specimen container (Fisher Scientific, St. Louis, MO) for agitation to release hOC from the bones. A custom made rocker arm shaker has been employed (but is not essential) to agitate the bones in chilled MLB once for 0.5 min at a moderate intensity, followed by twice in fresh MLB for 2 min each at a more vigorous level. After each cycle of agitation, the supernatants containing hOC are removed and sequentially filtered to remove small bone particles through 350 μm and 110 μm nylon mesh filters (Nitex, Tetko, Inc., Briarcliff Manor, NY) into containers placed on ice. Thereafter, the filtrates (typically ~200 ml) are combined, the cells collected by centrifugation (210 *g*, 10 min, 4 °C), and the preparations enriched for hOC by Percoll fractionation with or without subsequent immunomagnetic sorting using a specific anti-OC monoclonal antibody (mAb) designated 121F [53] as follows. Other means of enriching for hOC have also been successfully employed (e.g. serum gradient fractionation, rapid attachment to tissue culture dishes or bone substrates) but are not discussed here.

Sterile solutions, supplies, and techniques should be employed from this point onward. For Percoll fractionation, the cells are typically resuspended in 20 ml of a chilled solution of 35% Percoll (Pharmacia, Piscataway, NJ) and 0.2 ml heparin in HBSS, 10 ml is placed into each of two 50 ml tubes held on ice, each tube is gently overlayed with 3 ml chilled HBSS, and the Percoll tubes are centrifuged in a swinging bucket rotor at 400 *g* for 20 min at 4 °C. However, if the initial human bone sample is relatively large, it is better to resuspend the cells in 40 ml Percoll and set up four tubes to centrifuge. Following centrifugation, the top layer and interface are aspirated together from each tube individually, transferred to new tubes, diluted with HBSS to 50 ml, inverted to wash, and then pelleted at 300 *g* for 10 min at 4 °C. Further density gradient fractionation on 6% or 8% Percoll gradients, while valuable for the enrichment of chicken osteoclasts, has not worked as well with hOC which are frequently of smaller size. Yields and cell viability are determined by mixing 0.1 ml of the hOC (typically resuspended in 3 ml HBSS) with 0.1 ml of a sterile solution of trypan blue (0.4% in 0.9% NaCl) in a microfuge tube, immediately placing a drop on a hemocytometer slide, gently lowering the coverslip on top so as not to dislodge large hOC, and counting the total number of hOC and viable (non-blue) hOC in a microscope according to standard procedures. Cells can be centrifuged and resuspended for culture (see below) at this point, or subjected to an additional round of enrichment for hOC via immunomagnetic capture as follows. Magnetic polystyrene beads (4.5 µm in diameter), supplied covalently conjugated with affinity-purified sheep anti-mouse IgG (Dynal Inc, New Hyde Park, NY), are placed (100 µl of a 50% slurry as supplied) into a microcentrifuge tube and washed three times. This is accomplished by the repeated addition of 1 ml phosphate-buffered saline (PBS), inversion or gentle pipetting to mix, and the application of a magnet held to the side of the tube (~1 min) for the removal of washes. mAb 121F (30 µg, available upon request from this laboratory) in a volume of 250 µl is added to the washed beads, the beads are gently resuspended, and the reaction mix is incubated for 1-4 hr with gentle end-over-end mixing or periodic finger tapping at room temperature until needed for hOC enrichment. At that point, the reacted beads are magnetically sorted, washed three times with PBS, and resuspended in a small volume (~200 µl) for addition to hOC. mAb-coupled beads should be prepared while the isolation of hOC is underway so that they will be ready to use once the hOC have been 35% Percoll enriched. Following 35% Percoll fractionation, the hOC are immediately resuspended in a 50 ml tube in 6 ml of phenol red-free medium 199 Earle's salts supplemented with 8.3 mM $NaHCO_3$, 100 mM HEPES (pH 6.8), 5% FBS (GIBCO BRL), and 2.5% antibiotic/antimycotic (OC medium). After addition of the 121F mAb-coupled magnetic beads, the 50 ml tube is swirled several times to rapidly mix the cells and beads, the tube is placed into a container of ice so that it is set slanted at a 45^0 angle and the bead/cell mixture is visible, and the container is placed on a rotary shaker adjusted to slowly mix the beads and cells for 30 minutes. In a sterile hood, the bead-bound hOC are magnetically sorted (5 min per cycle due to the larger volumes) while held on ice, washed gently by inversion and magnetic sorting three times with 40 ml chilled HBSS, and either resuspended for culture (Section 3) or extracted for RNA, etc. (Section 4). The supernatants from the magnetic sortings are again magnetically sorted to recapture any lost bead-bound hOC, and these are added back to the main sample. In addition, since these final supernatants often still contain numerous hOC, they can be centrifuged (300 *g*, 10 min) to capture such cells, and the pellet resuspended for culture. Although the yield of isolated hOC varies between individual human bone samples, good preparations will typically yield 10^5-10^7 hOC, 70-90% of which should be viable hOC. Enrichments of at least 40% on a per cell basis (>80% on a per nuclear basis) are routinely achieved for hOC using 35% Percoll fractionation, and are frequently as high as 90% on a per cell basis (>98% on a per nuclear basis) following 121F immunomagnetic affinity capture of hOC. Their identification as bone-resorbing hOC is verifed as described in Section 4.

2.2 Human Osteoclast-Like Cells (*In Vitro* Formed)

In general, there is insufficient BM in bone samples obtained for hOC isolation to culture for the formation of human BM-derived osteoclast-like cells (hOCL), although this source has occasionally proven suitable for this procedure. Most often, bones for this purpose have been obtained as either discarded surgical specimens of human rib bone from thoracic surgery or as long bones from accident victims. BM should be removed as soon as possible after obtaining the bone sample, and the preparation immediately processed using sterile techniques to ensure good results. BM is flushed from bones using a sterile solution (~50-100 ml) of 2.8% heparin sodium solution (from beef lung, 1000 U/ml, Pharmacia and Upjohn Co., Kalamazoo, MI) in HBSS and a sterile 30 ml hypodermic syringe fitted with an 18-gauge needle. The flushed BM is then passed sequentially through 350 μm and 110 μm nylon mesh filters (Nitex, Tetko, Inc.) into 50 ml tubes set in ice, followed by centrifugation of the filtrates at 175 *g* for 5 min to collect the cells. Meanwhile, two 50 ml tubes are prepared and set aside that each contain 0.5 ml heparin sodium solution and 15 ml Ficoll-Hypaque (Pharmacia Biotech), and a check is performed on the pH of a Moscona's high bicarbonate buffer (MHB, prepared as for MLB except that 12 mM instead of 2.4 mM $NaHCO_3$ is used) to ensure that it is within the range of pH 7.2 to 7.4 (adjust with sterile 1N HCl, if necessary). Following centrifugation, the lipid pads and supernatants overlying the loose cell pellets are slowly withdrawn and discarded, the cells in each tube are gently resuspended in 6 ml MHB and combined into a new 50 ml tube containing 0.5 ml heparin sodium solution, and the volume is brought to 15 ml with additional MHB as needed. An equal volume (15 ml) of FBS is then added, the cells are mixed by gentle pipetting, and half of this cell suspension (15 ml) is very gently layered by slow release against the side of the tube onto each of the two prepared tubes containing Ficoll and heparin. The gradients are centrifuged at 400 *g* for 30 min at room temperature, the supernatants overlying the fluffy MNC layers are removed and discarded, and the fluffy layer from each gradient is withdrawn into a new 50 ml tube containing ~25 ml MHB. The volume of each tube is subsequently increased to 45-50 ml with MHB, the cells are shaken vigorously, the cell suspension is centrifuged at 175 *g* for 5 min at 25 °C, and the pellets are each resuspended in ~1.5 ml MHB. These are combined into one tube, the volume increased to 20-30 ml with MHB, and sufficient FBS is added for a final concentration of 10%. After mixing, an aliquot is taken (0.1 ml), mixed with an equivalent volume of trypan blue (0.4% in 0.9% NaCl), and both the total number and viable (non-blue) number of MNC (excluding erythrocytes) is immediately determined using a hemocytometer according to standard procedures. The human BM MNC are cultured to form hOCL multinucleated cells and their properties evaluated as described below.

3. CULTURE TECHNIQUES

3.1 Preparation of Sterile Devitalized Bone or Ivory Discs for Resorption Studies

Ivory obtained either through donation from a local zoo or the Federal Department of Fish and Wildlife Services, or a segment of bovine cortical bone obtained from a local slaughterhouse and thoroughly washed, can serve as suitable substrates for resorption pit analyses. The bone or ivory is sliced by hand with a jigsaw into small chunks, then reduced to pieces of 0.4 mm thickness and further cut into smaller 5 mm^2 circular or rectangular slices using a low speed Isomet saw (Buehler, Lake Bluff, IL). After washing in deionized water at 4 °C overnight, the discs are soaked repeatedly in 70% ethanol in sterile tubes, and are then routinely stored in 70% ethanol at –20 °C. For experimental use, the required number of discs are aseptically removed from the tube (using alcohol soaked tweezers), transferred to a new sterile 50 ml Falcon tube, rinsed extensively by inversion at least three times with ~40 ml sterile HBSS, and then transferred into culture wells or dishes containing HBSS prior to the plating of cells. Since ethanol inhibits osteoclast bone resorption, it is important that the bone or ivory slices be well rinsed before cells are plated onto them.

3.2 Culture of *In Vivo* Formed hOC

Percoll-fractionated hOC can be cultured on tissue culture plastic, glass coverslips, and/or calcified resorptive substrates (typically bone or ivory) depending upon what they are to be used for. For many purposes, it is convenient to set up a 24-well plate with a sterile circular glass coverslip in each well (placed into the well via alcohol sterilized tweezers), and 2-6 small devitalized bovine cortical bone or ivory discs directly placed on top of each coverslip (similarly transferred with sterilized tweezers). These are kept moist in sterile HBSS or medium until the hOC are ready to plate into the wells. At that time, the hOC are resuspended in OC medium, and 250,000-400,000 hOC in 0.5 ml are plated per well of the 24-well plate. Within several hours, hOC attachment and spreading on tissue culture wells or glass coverslips should become evident and many large, irregularly shaped, multinucleated cells can be readily observed by light microscopy. After overnight incubation (for sufficient numbers of hOC to have attached) at 37 °C in a moist atmosphere of 95% air, 5% CO_2, the bone or ivory slices can be individually moved out of these wells into new single wells of a 48-well plate (one disc per well) containing fresh OC medium, and their incubation continued with feeding every four days (0.3-0.5 ml OC medium), plus or minus various modulators (e.g. growth factors, hormones, other test agents), for several days or up to 2-3 weeks. The samples can then be harvested for resorption pit analysis, cytochemical/immunocytochemical staining, or electron microscopic evaluation as described below. Resorption pit analyses are best conducted after at least 7-10 days of incubation to allow for the slow resorption kinetics of hOC. No pits are evident within the first 3-4 days of culture, but single, non-lobulated excavations are typically observed by ten days of hOC culture on bone or ivory. hOC cultured on glass coverslips are suitable for enzymatic, cytochemical, immunocytochemical, other single-cell analyses, or *in situ* molecular hybridization analyses (some of which are described below), and may be harvested for these procedures within 1-5 days of their initial plating. Alternatively, hOC cultures can be harvested for protein and subjected to various biochemical or electrophoretic/Western blot analyses. Culture medium (note exact volumes and times of incubation) can be saved at the time of feeding, briefly centrifuged to remove cell debris, and the supernatants frozen at –80 °C until needed for analysis of released products (e.g. growth factors, lipid metabolites, nitric oxide radicals). It is important to have wells without cells, but containing medium, to collect as controls for such measurements. If desired, the release of substances can be normalized for the amount of protein or RNA extracted from the same wells (and used for other purposes). For analyzing whether hOC express mRNA for various known or unknown markers, it is preferable to use the mAb 121F immunomagnetically purified cell preparations in order to optimize the purity of the sample population. Sufficient RNA for Northern blot analysis is usually difficult to come by, and so, more sensitive molecular techniques such as reverse transcriptase polymerase chain reaction (RT-PCR) or ribonuclease protection assay (RPA) should be employed. Despite their purity, magnetically isolated hOC preparations are usually not cultured for experimental study since hOC readily phagocytose the magnetic beads and mAb 121F inhibits osteoclast bone resorption (and therefore expression of TRAP, etc.) [53, 54]. Therefore, Percoll-fractionated isolated hOC can be cultured with or without modulators in order to assess regulatory effects on mRNA expression levels for various known or unknown markers. This is best achieved by culturing a larger number of cells together in 35 mm or 60 mm tissue culture dishes, or pooling 2-4 similarly treated wells from a 24-well plate to obtain sufficient RNA for analysis by RT-PCR or RPA as described below. Single-cell analytical methods (e.g. immunostaining, *in situ* hybridization) can be used in parallel cultures if it is important to document whether the hOC present in these populations undergo the regulatory changes detected by molecular analysis for specific markers.

3.3 Culture of *In Vitro* Formed hOCL

Following their inductive formation in culture (on glass coverslips, if desired), hOCL can be further cultured with or without modulators and the cells, protein, RNA, or released substances subjected to the same sorts of analyses as employed for hOC. The one notable exception to this is bone pit resorption studies since hOCL make few, if any, pit excavations on mineralized matrices. The hOCL formed in culture can be enriched by the selective removal of stromal cells, as well as themselves be released from the tissue culture dishes in order to plate the hOCL in other wells for study, according to the following procedure. The tissue culture medium is withdrawn from the culture dish, the cells are rinsed once briefly and once again for 15 min at 37 °C with MHB (5 ml and 8 ml per 100 mm dish, respectively), and the stromal cells are then subjected to mild trypsinization by incubation of the dish for several minutes in a digestion solution (7 ml per 100 mm dish) comprised of 30 ml MHB plus 15 ml 1% EDTA in MLB and 0.6 ml of a 1% stock solution of trypsin (in MLB, stored at –20 °C, Worthington Biochemical Corp.). After 3-4 minutes of incubation at 37 °C, the dish is tapped sharply once or twice, the detached cells (primarily stromal cells) removed, the adherent cells (primarily hOCL) rinsed once with MHB, and the remaining cells fed with fresh medium for continued culture. If hOCL are also to be removed from the dish, then a fresh aliquot of digestion solution is added to the dish instead of culture medium, the cells are incubated for a further 25-30 minutes at 37 °C, FBS is added to a final concentration of 10% in order to inhibit trypsin activity, and the cells are gently dislodged by pipetting up and down a few times, followed by transfer to a 50 ml tube. The dish is rinsed with ~5 ml MHB, any remaining cells are scraped off into this rinse using a rubber policeman, and the rinse is combined with cells in the 50 ml tube. The cells are then collected by centrifugation at 210 *g* for 5 min, and resuspended in medium for culture at 0.6 x 10^6 cells per well of a 6-well plate.

4. ASSAY TECHNIQUES

4.1 Morphology and Ultrastructure

Standard protocols can be used to evaluate the morphological and ultrastructural characteristics of hOC and hOCL for determination of an OC phenotype. By light microscopy, hOC typically appear as large multinucleated cells of varying sizes and shapes having a grainy cast, and often one or more pseudopodial extensions per cell (Fig. 1, A and C). By comparison, hOCL are generally larger and contain more nuclei than hOC, exhibit less of a grainy texture, and frequently resemble flattened pancakes (Fig. 1E). Immunomagnetically-isolated hOC are typically decorated with multiple beads per cell and can be so thoroughly coated with antibody-conjugated beads that they appear as a ball of beads (Fig. 1B). When cultured on plastic, glass, or bone/ivory, immunomagnetically-isolated hOC spread out and internalize the beads, rather than shed them as do non-phagocytic cells. For transmission electron microscopic (TEM) visualization, pelleted cells or those on bone/ivory slices are rinsed briefly with HBSS, fixed in 2.5% glutaraldehyde in HBSS (pH 7.4), rinsed, postfixed on ice in 1% osmium tetroxide, washed, dehydrated in a graded series of alcohols (70-100%), embedded in Epon, and thin sections cut with a diamond knife. The sections are mounted on colloidin coated-carbon stabilized copper grids, stained with uranyl acetate and lead citrate, and examined and photographed using a transmission electron microscope. Multiple nuclei (often clustered within the cell and varying in number between cells), abundant mitochondria, numerous vesicles, extensive vacuolation, well-developed perinuclear Golgi complexes, prominent rough endoplasmic reticulum, free polysomes, and ruffled border membrane and clear zone domains characteristic of OC should be evident. For scanning electron microscopy (SEM), the cells on glass coverslips or bone/ivory are rinsed, fixed for at least 24 hours in 2.5% glutaraldeyde in HBSS (pH 7.4), rinsed, exposed to osmium vapors, alcohol dehydrated, mounted on grids, critical point dried, gold sputter coated, and viewed and photographed using a scanning electron microscope. By SEM, hOC cultured on plastic appear as large cells having a complex morphology, many fine filopodial projections, microvilli and membrane blebs over the cell surface, and a peripheral cytoplasmic skirt (Fig. 2A). If cultured on bone/ivory for less than seven days, well-defined resorption lacunae are generally not yet evident in association with hOC (Fig. 2B). However, hOC cultured for more extended periods of ten days or longer exhibit resorption pits having characteristically exposed collagen fibrils, occasionally manifested as a resorption track (Fig. 2C), but more often appearing as a unilobular cavity adjacent to and partially underlying the hOC actively engaged in resorption (Fig. 2D). Typically, ~10% of the hOC population have formed resorption pits within ten days. These pits can be viewed more clearly if desired by removal of the hOC from the bone/ivory, either initially before the gold coating step or after viewing the sample and then recoating with gold to visualize the pits alone.

4.2 Cytochemical Staining

As a general differential stain, Difquik (Criterion Sciences, Riverdale, NJ) is a quick and easy method (eosin Y, azure A, and methylene blue) to discriminate nuclear and cytoplasmic detail in cultured and fixed hOC (Fig. 3A) and hOCL. In addition, although not fully specific for OC, high TRAP activity is a characteristic property of OC that is upregulated during OC development and is important for their resorption of bone. Consequently, TRAP activity is monitored in most OC developmental model systems (Fig. 1, D and F; Fig. 3, B and D). TRAP levels should significantly increase in hOCL cultures induced by developmental promoters in comparison with either uninduced multinucleated cells or the BM precursor cell population from which they were derived (Fig. 1F). TRAP levels should be even higher for *in vivo* formed isolated hOC in comparison with hOCL (Fig. 1D and Fig.

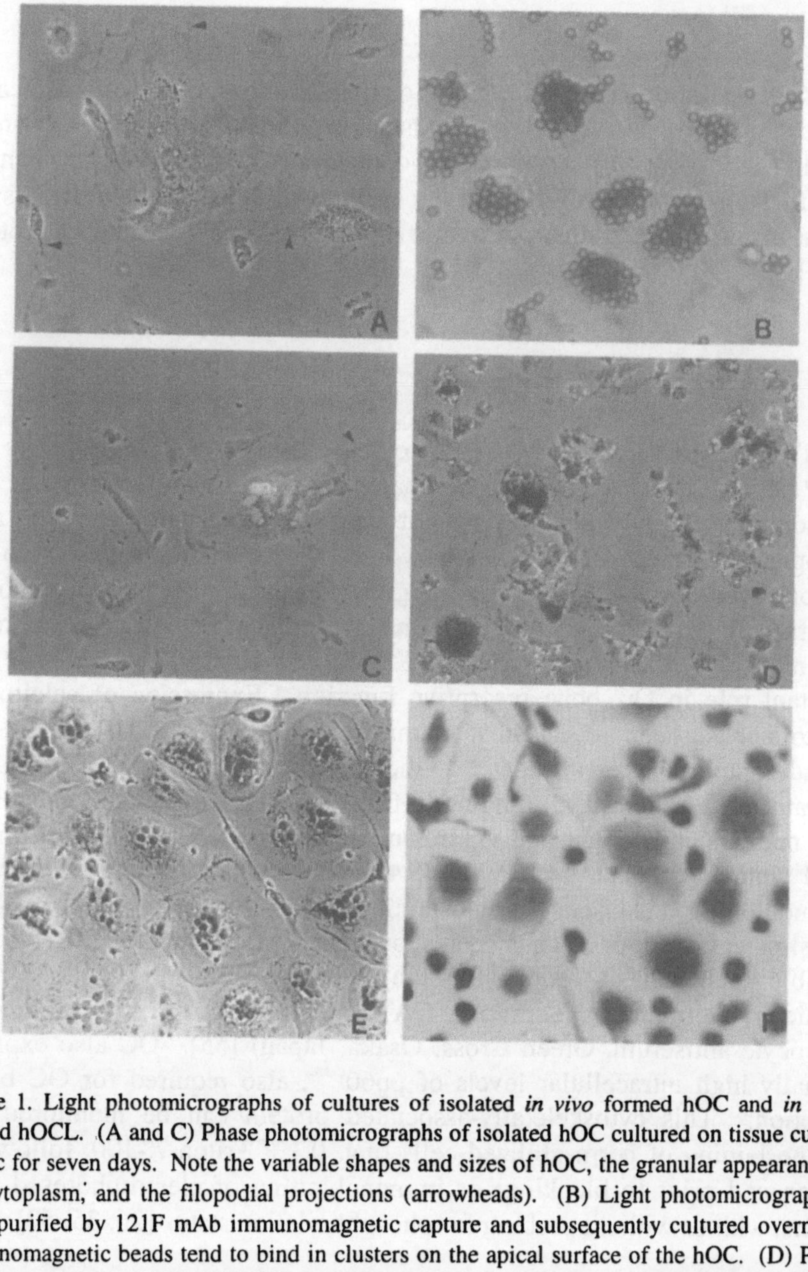

Figure 1. Light photomicrographs of cultures of isolated *in vivo* formed hOC and *in vitro* formed hOCL. (A and C) Phase photomicrographs of isolated hOC cultured on tissue culture plastic for seven days. Note the variable shapes and sizes of hOC, the granular appearance of the cytoplasm, and the filopodial projections (arrowheads). (B) Light photomicrograph of hOC purified by 121F mAb immunomagnetic capture and subsequently cultured overnight. Immunomagnetic beads tend to bind in clusters on the apical surface of the hOC. (D) Phase photomicrograph of isolated hOC partially purified by 35% Percoll enrichment, cultured overnight, and stained (dark red) for TRAP activity. Both large and small TRAP positive hOC are evident. (E) Phase photomicrograph of BM-derived *in vitro* formed hOCL cultured for 16 days. Note that relative to hOC, these cells appear as larger, more uniformly sized, rounded, flattened cells having multiple nuclei located toward the central region of the cytoplasm. (F) Bright field photomicrograph of hOCL cultured for 16 days and stained for TRAP activity. Original magnification in A-F was 280X.

3B). Cells cultured on plastic, glass, or bone/ivory can be cytochemically stained for relative TRAP activity levels using published procedures in order to evaluate individual cells within the population [15], or cell extracts can be quantified for levels of TRAP activity using a microplate enzymatic assay and the measured enzymatic activities normalized for cell extract protein levels [24].

4.3 Antigenic Profile

Together with specific morphological features and high levels of TRAP activity, hOC exhibit certain characteristic surface markers that are commonly monitored since they are not expressed in related monocytes and macrophages or macrophage polykaryons. These include expression of calcitonin receptors (CTR), $\alpha_v\beta_3$ integrin (vitronectin receptor), carbonic anhydrase II (CA II), and the 121F mAb-reactive OC-specific membrane antigen (Fig. 3, C and D). Each of these has been shown to have an important role in OC bone resorptive function. Expression of calcitonin receptors has been classically monitored at the cellular level via autoradiographic detection of ^{125}I-labeled salmon calcitonin (Peninsula Laboratories, Belmont, CA) binding to fixed (and TRAP stained) cells [35]. Other markers can be detected via immunological methods such as enzyme-linked immunosorbent assay (ELISA) [24], gel electrophoresis and immunoblotting [21], or immunostaining [24] of cells either on glass coverslips or bone/ivory as has been shown for the 121F antigen (121F mAb available upon request through this lab), $\alpha_v\beta_3$ integrin (e.g. LM609 mAb, Chemicon International, Temecula CA) [52], or CA II (goat anti-human erythrocyte antiserum, Green Cross, Osaka, Japan) [55]. OC also express unusually high intracellular levels of pp60$^{c\text{-}src}$, also required for OC bone resorption. This cytoskeletally-associated protein can be monitored by immunostaining of permeabilized cells (e.g. 0.1% Triton X-100, following fixation and prior to blocking) or immunoblotting of electrophoresed cell extracts, with or without probing for phosphorylation status [26, 35, 52].

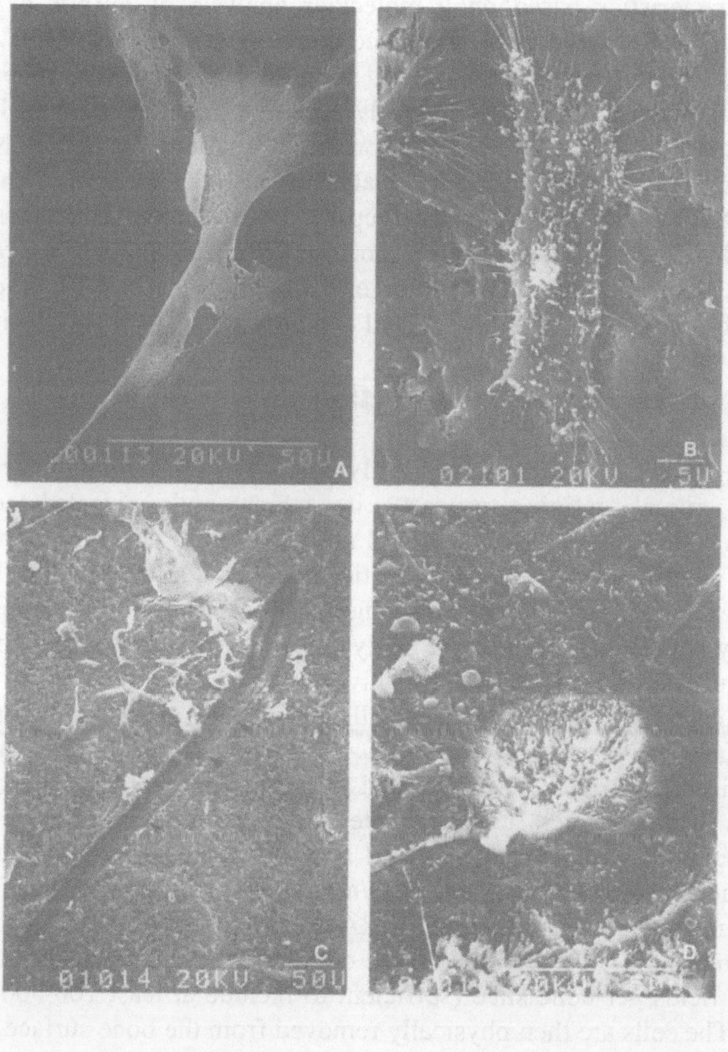

Figure 2. Scanning electron micrographs of isolated and cultured hOC. (A) hOC cultured on tissue culture plastic for three days. Note the long filopodial extensions emanating from the large cell body (bar = 50 μm). (B) hOC cultured on bone for three days. Many cytoplasmic blebs and microfilopodia can be observed stretching from the plasma membrane to contact the bone matrix at multiple sites. No resorption activity is apparent (bar = 5 μm). (C) hOC cultured on ivory for 14 days. Several small and one large hOC (arrows) are observed in association with a long resorption tract (arrowheads) (bar = 50 μm). (D) A single oval resorption pit displaying a network of exposed collagen fibers is evident in hOC cultured on ivory for 14 days. This resorption pit has been visualized after removal of hOC from the bone (bar = 50 μm).

4.4 Molecular Profile

With the advances in technology that have occurred in recent years, it is becoming more common these days to monitor the expression of various OC characteristics based on a molecular analysis of mRNA expression levels for markers like those described above (e.g. TRAP, CTR, CAII, and $\alpha_v\beta_3$ integrin). Because only small amounts of RNA can generally be obtained, especially from hOC, highly sensitive methods, such as RT-PCR or RPA, are usually employed for routine evaluations instead of Northern blot analysis [51, 52]. Molecular analysis is also the method of choice for demonstrating the expression of cathepsin O (also known as cathepsin K or O2) in hOC or hOCL, a cysteine protease which is uniquely expressed in OC and represents the major and essential degradative enzyme responsible for OC breakdown of non-mineralized components of the bone matrix [56].

4.5 Bone Resorptive Function

This is the key attribute of a fully-developed OC, and therefore serves as the critical defining property by which multinucleated cells are considered to be OC. Although OC are capable of resorbing particles of bone in culture, so are other phagocytic cells, despite the fact that the latter cannot excavate resorption pits on bone/ivory. Consequently, the resorption pit assay is valuable not only for analyzing the effects of various treatments or modulators on OC bone-resorptive function, but also as a way of confirming that the multinucleated cells under study are OC in nature. This is exemplified by noting that hOC will produce resorption lacunae on mineralized substrates, whereas hOCL generally do not, and, at most, only etch the surface of such substrates. Resorption pit analysis of hOC cultured on bone/ivory is routinely performed after 10-14 days of culture (to allow for sufficient pit formation since few/no pits form within the first few days of culture) by rinsing, fixing, and TRAP staining hOC as described above. The number of TRAP stained OC is determined for a constant number of random fields per bone slice (sufficient to include at least 100-300 OC per slice). The cells are then physically removed from the bone surface, and the numbers and areas of pits formed are quantified within these same exact fields using an ocular reticle on the microscope and a computer-linked darkfield reflective light image analysis system [54, 57]. Resorption measures are subsequently normalized to express the results as mean area of bone resorbed per OC, mean number of pits per OC, and mean area of individual pits. Several trials, each having 3-5 replicates for control and treated conditions, are generally performed to achieve statistically significant results.

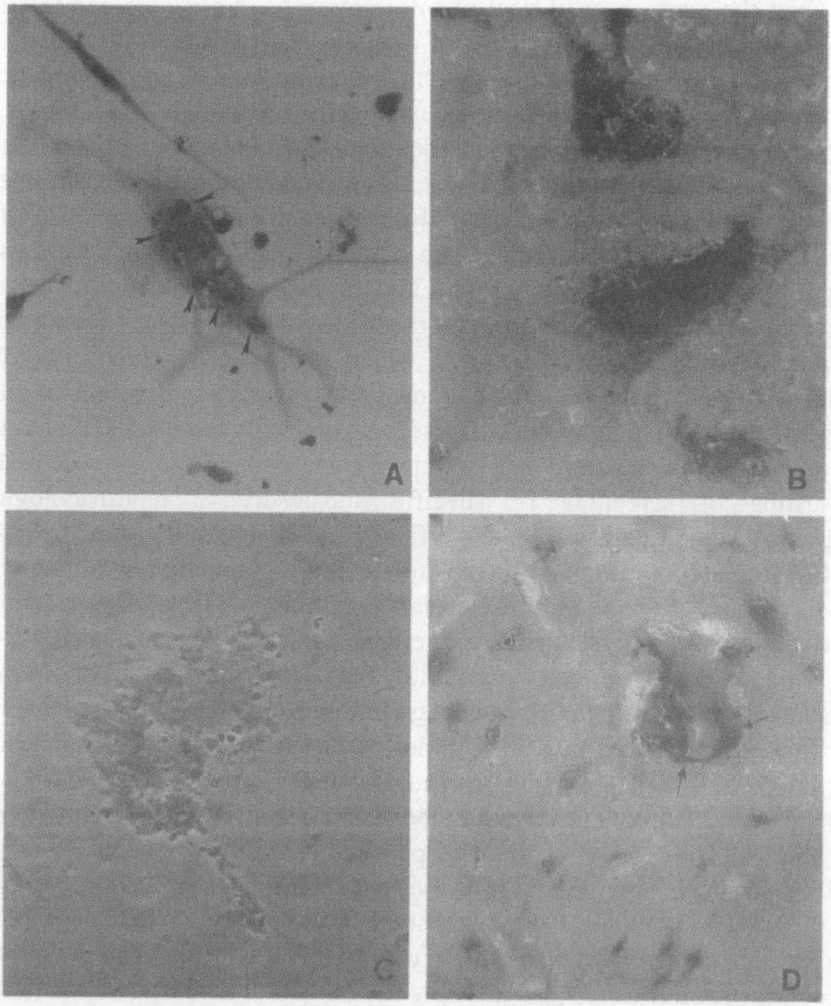

Figure 3. Light photomicrographs of hOC following cytochemical or immunocytochemical staining. (A) hOC cultured for seven days on plastic and stained with Difqik reveals at least five nuclei (arrowheads) and many pseudopodial processes (original magnification was 280X). (B) Bright field reflective light photomicrograph of hOC cultured on bone for ten days and stained for TRAP (original magnification was 224X). (C) Phase contrast photomicrograph of hOC cultured on plastic for seven days and immunoreacted with mAb 121F, followed by a biotinylated secondary antibody, streptavidin β-galactosidase conjugate, and X-gal detection (blue-green reaction product). A single large multinucleated hOC is shown that is well spread out and exhibits positive reactivity (original magnification was 280X). (D) Bright field reflective light photomicrograph of hOC cultured on bone for 14 days and doubly stained by reaction first with mAb 121F (as in panel C), followed by TRAP staining (as in panel B). Many hOC appear reactive both for binding of 121F (dark bluish stain) and for TRAP activity (red staining). Also note the large resorption pit affiliated with the double stained hOC (arrows, original magnification was 224X).

5. UTILITY OF SYSTEM

Over the years, a great deal of information has been obtained relative to our understanding of the process of osteoclastogenesis and the regulation of osteoclast-mediated bone resorption, based on studies employing non-human osteoclast and osteoclast precursor cells. Although there appear to be many similarities in osteoclast development, behavior, and function, between human and non-human species, and valuable therapeutic regimens currently in use have already derived from research initially focused on the latter, there is also evidence that bone modeling and remodeling processes are not fully equivalent across species [58-60]. Consequently, certain aspects of human osteoclast formation, differentiation, resorptive activity, and regulation may differ from that which has been observed for other species. Access to these human bone cell systems is therefore a great advantage both for delineating precise, and possibly unique, cellular and molecular mechanisms involved in human bone physiology, as well as for the testing or screening of specific anti-resorptive pharmacological compounds and strategies for their efficacy in human cells. An important benefit associated with the ability to isolate and manipulate any human cell also relates to the inherent limitations that exist relative to performing experimental studies on humans. Therefore, these two human cell types, hOC and hOCL, represent valuable tools for the study of human osteoclastogenesis and mature osteoclast function. Although BM-derived *in vitro* formed hOCL and giant cell tumor of bone-associated multinucleated cells have been in use for experimental studies for more than a decade, similar studies employing large numbers of normal *in vivo* formed bone-resorbing hOC have only just become possible using the protocols presented herein. Thus, it is now possible for the first time for researchers to perform studies with sufficient numbers of authentic bone-resorptive hOC to explore diverse questions relating to hOC from a biochemical, immunological, molecular, physiological, pharmacological, and functional standpoint, that were previously difficult or not feasible to investigate *in vitro*. In addition to the multiple experimental approaches described in Sections 3 and 4 above, most of the standard techniques and assays used for investigating other isolated and cultured cells can also be applied to hOC and hOCL. In particular, this includes single cell analyses (e.g. motility, ion fluxes, signal transduction, cytoskeletal and cell shape modulation), microscopy (e.g. confocal, optical sectioning), biochemical analyses (e.g. enzymatic activities, Western blot studies), and functional assays (e.g. resorption modulation by hormones, factors, or other agents and production of degradative ions or enzymes).

The ability to isolate numerous hOC also enhances the inherent value of the hOCL developmental model system for osteoclastogenesis since similarities and differences between the *in vitro* partially differentiated hOCL and the mature *in vivo* formed hOC can now be directly assessed, as well as conditions explored for optimizing the promotion of a fully mature OC phenotype *in vitro*. Moreover, using mature hOC, it may be possible to identify previously unknown novel genes that are expressed only in the mature functional OC or under particular physiological conditions, and to identify factors important for the final stages of osteoclastogenesis in the hOCL system. Ultimately, the availability of such tools may facilitate the development of new diagnostic or therapeutic strategies for alleviating the osteopenia associated with various bone disorders such as osteoporosis, rheumatoid arthritis, tumor-associated osteolysis, and periodontal disease.

6. CONCLUDING REMARKS

In this chapter, we have tried to convey the significance of two valuable human cell culture model systems, the human BM-derived *in vitro* hOCL osteoclastogenesis model system, and the culture of isolated *in vivo* formed normal bone-resorptive hOC, for use in a wide range of studies encompassing developmental, functional, and regulatory issues pertaining to OC-mediated bone resorption. In particular, our new found ability to isolate large numbers of normal bone-resorptive hOC makes it possible, for the first time, for researchers to conduct individual cell and population studies with sufficient numbers of authentic hOC to answer important questions that have been unresolved, and perhaps difficult to address to date. It is likely that future research will increasingly employ these isolated bone-resorptive hOC as a substitute for, or possibly a complement to, work utilizing *in vitro* formed partially-differentiated hOCL, preosteoclastic leukemic or transformed cell lines, and multinucleated giant cells obtained from osteoclastoma bone tumors. Therefore, the hOCL and hOC culture systems together should provide researchers an opportunity to more completely probe fundamental processes underlying OC development and function in normal human bone modeling and remodeling, as well as to gain valuable new insights into the mechanisms responsible for bone loss associated with various pathological conditions, an understanding that could lead to the development of improved diagnostic or therapeutic approaches to combat such osteopenia.

ACKNOWLEDGEMENTS

The authors wish to thank Dr. Gary Miller for providing additional human bone, and Dr. Teresa Sunyer, Ms. Linda Rothe, Ms. Janice Weihe, and Ms. Aurora Fausto for their valuable participation in studies related to the hOC and hOCL cells described here. This work was supported in part by NIH grants AG15435, AR32087, and AR42715 to P.O., and DK46547 to. P.C.O.

LITERATURE CITED

1. Kanis JA (1995) Bone and cancer: pathophysiology and treatment of metastases. *Bone* 17: 101S-105S.
2. Taube T, Elomaa I, Blomqvist C, Beneton MN, and Kanis JA (1994) Histomorphometric evidence for osteoclast-mediated bone resorption in metastatic breast cancer. *Bone* 15: 161-6.
3. Mundy GR (1993) Cytokines and growth factors in the regulation of bone remodeling. *J Bone Miner Res* 8: S505-10.
4. Suda T, Udagawa N, Nakamura I, Miyaura C, and Takahashi N (1995) Modulation of osteoclast differentiation by local factors. *Bone* 17: S 87-S 91.
5. Wiebe SH, Hafezi M, Sandhu HS, Sims SM, and Dixon SJ (1996) Osteoclast activation in inflammatory periodontal diseases. *Oral Diseases* 2: 167-80.
6. Manolagas SC, Bellido T, and Jilka RL (1995) New insights into the cellular, biochemical, and molecular basis of postmenopausal and senile osteoporosis: Roles of IL-6 and gp130. *Intl J Immunopharmacol* 17: 109-16.
7. MacDonald BR, Takahashi N, McManus LM, Holahan J, Mundy GR, and Roodman GD (1987) Formation of multinucleated cells that respond to osteotropic hormones in long term human BM cultures. *Endocrinol* 120: 2326-33.
8. Helfrich MH, Thesingh CW, Mieremet RH, and van Iperen-van Gent AS (1987) Osteoclast generation from human fetal BM in cocultures with murine fetal long bones. A model for *in vitro* study of human osteoclast formation and function. *Cell Tiss Res* 249: 125-36.
9. Roodman GD (1996) Advances in bone biology - The osteoclast. *Endocrine Revs* 17: 308-332.
10. Yoneda T, Alsina MM, Garcia JL, and Mundy GR (1991) Differentiation of HL-60 cells into cells with the osteoclast phenotype. *Endocrinol* 129: 683-9.
11. Somerman MJ, Berry JE, Khalkhali-Ellis Z, Osdoby P, and Simpson RU (1995) Enhanced expression of alpha v integrin subunit and osteopontin during differentiation of HL-60 cells along the monocytic pathway. *Exp Cell Res* 216: 335-41.
12. Gattei V, Bernabei PA, Pinto A, Bezzini R, Ringressi A, Formigli L, Tanini A, Attadia V, and Brandi ML (1992) Phorbol ester induced osteoclast-like differentiation of a novel human leukemic cell line (FLG 29.1). *J Cell Biol* 116: 437-47.
13. Khalkhali-Ellis Z, Collin-Osdoby P, Li L, Brandi ML, and Osdoby P (1997) A human homolog of the 150 kD avian osteoclast membrane antigen related to superoxide dismutase and essential for bone resorption is induced by developmental agents and opposed by estrogen in FLG 29.1 cells. *Calcif Tiss Intl* 60: 187-93.
14. Boyde A and Jones SJ (1991) Pitfalls in pit measurement. *Calcif Tiss Intl* 49: 65-70.

15. Minkin C (1982) Bone acid phosphatase: tartrate-resistant acid phosphatase as a marker of osteoclast function. *Calcif Tiss Intl* 34: 285-90.
16. Littlewoodevans A, Kokubo T, Ishibashi O, Inaoka T, Wlodarski B, Gallagher JA, and Bilbe G (1997) Localization of cathepsin K in human osteoclasts by in situ hybridization and immunohistochemistry. *Bone* 20: 81-86.
17. Horton MA, Dorey EL, Nesbitt SA, Samanen J, Ali FE, Stadel JM, Nichols A, Greig R, and Helfrich MH (1993) Modulation of vitronectin receptor-mediated osteoclast adhesion by Arg-Gly-Asp peptide analogs: a structure-function analysis. *J Bone Miner Res* 8: 239-47.
18. Rathod H, Malcolm AJ, Gillespie JI, Berry V, Pooley J, Piggott NH, and Datta HK (1994) Characterization of a subtype of primary osteoclastoma: extracellular calcium but not calcitonin inhibits aggressive HLA-DR-positive osteoclastoma possessing 'functional' calcitonin receptors. *J Pathol* 174: 293-9.
19. Zheng MH, Fan Y, Wysocki S, Wood DJ, and Papadimitriou JM (1993) Detection of mRNA for carbonic anhydrase II in human osteoclast-like cells by in situ hybridization. *J Bone Miner Res* 8: 113-8.
20. Chatterjee D, Chakraborty M, Leit M, Neff L, Jamsa-Kellokumpu S, Fuchs R, Bartkiewicz M, Hernando N, and Baron R (1992) The osteoclast proton pump differs in its pharmacology and catalytic subunits from other vacuolar H(+)-ATPases. *J Exp Biol* 172: 193-204.
21. Oursler MJ, Li L, and Osdoby P (1991) Purification and characterization of an osteoclast membrane glycoprotein with homology to manganese superoxide dismutase. *J Cell Biochem* 46: 219-33.
22. Gay CV (1996) Role of microscopy in elucidating the mechanism and regulation of the osteoclast resorptive apparatus. *Microsc Res Tech* 33: 165-70.
23. Fuller K and Chambers T (1989) Bone matrix stimulates osteoclastic differentiation in cultures of rabbit BM cells. *J Bone Miner Res* 4: 179-183.
24. Collin-Osdoby P, Oursler M, Rothe L, Webber D, Anderson F, and Osdoby P (1995) Osteoclast 121F antigen expression during osteoblast conditioned medium induction of osteoclast like cells *in vitro*: Relationship to calcitonin responsiveness, tartrate resistant acid phosphatase levels, and bone resorptive activity. *J Bone Miner Res* 10: 45-58.
25. Akatsu T, Tamura T, Takahashi N, Udagawa N, Tanaka S, Sasaki T, Yamaguchi A, Nagata N, and Suda T (1992) Preparation and characterization of a mouse osteoclast-like multinucleated cell population. *J Bone Miner Res* 7: 1297-306.
26. Galvin RJ, Bryan P, Horn JW, Rippy MK, and Thomas JE (1996) Development and characterization of a porcine model to study osteoclast differentiation and activity. *Bone* 19: 271-9.
27. Kukita A, Kukita T, Shin J, and Kohashi O (1993) Induction of mononuclear precursor cells with osteoclastic phenotypes in a rat BM culture system depleted of stromal cells. *Biochem Biophys Res Commun* 196: 1383-1389.
28. Flanagan AM, Horton MA, Dorey EL, Collins DA, Evely RS, Moseley JM, Firkin FC, Chambers TJ, Helfrich MH, and Martin TJ (1992) An assessment of the ability of human BM cultures to generate osteoclasts. *Intl J Exptl Pathol* 73: 387-401.
29. Demulder A, Suggs SV, Zsebo KM, Scarcez T, and Roodman GD (1992) Effects of stem cell factor on osteoclast-like cell formation in long-term human marrow cultures. *J Bone Miner Res* 7: 1337-44.
30. Roux S, Quinn J, Pichaud F, Orcel P, Chastre E, Jullienne A, and De Vernejoul MC (1996) Human cord blood monocytes undergo terminal osteoclast differentiation *in vitro* in the presence of culture medium conditioned by giant cell tumor of bone. *J Cell Physiol* 168: 489-98.

31. Connor JR, Dodds RA, James IE, and Gowen M (1995) Human osteoclast and giant cell differentiation: the apparent switch from nonspecific esterase to tartrate resistant acid phosphatase activity coincides with the in situ expression of osteopontin mRNA. *J Histochem Cytochem* 43: 1193-201.
32. Kassem M, Mosekilde L, Rungby J, Mosekilde L, Melsen F, and Eriksen EF (1991) Formation of osteoclasts and osteoblast-like cells in long-term human BM cultures. *Apmis* 99: 262-8.
33. Fujikawa Y, Sabokbar A, Neale S, and Athanasou NA (1996) Human osteoclast formation and bone resorption by monocytes and synovial macrophages in rheumatoid arthritis. *Annals Rheum Diseases* 55: 816-822.
34. Pierelli L, Scambia G, Donofrio G, Ciarli M, Fattorossi A, Bonanno G, Menichella G, Battaglia A, Panici PB, Tommasi M, Mancuso S, and Leone G (1997) Generation of multinuclear tartrate-resistant acid phosphatase positive osteoclasts in liquid culture of purified human peripheral blood CD34(+) progenitors. *Brit J Haematol* 96: 64-69.
35. Matayoshi A, Brown C, DiPersio JF, Haug J, Abu-Amer Y, Liapis H, Kuestner R, and Pacifici R (1996) Human blood-mobilized hematopoietic precursors differentiate into osteoclasts in the absence of stromal cells. *Proc Natl Acad Sci USA* 93: 10785-90.
36. Zambonin Zallone A, Grano M, Colucci S, Zigrino P, De Bellis M, Zambonin G, and Serra M (1995) Human osteoclast-like cells from giant cell tumors of bone: a new tool for investigating bone resorption and osteoclast biology. *Calcif Tiss Intl* 56: S24.
37. Grano M, Colucci S, De Bellis M, Zigrino P, Argentino L, Zambonin G, Serra M, Scotlandi K, Teti A, and Zambonin Zallone A (1994) New model for bone resorption study *in vitro*: human osteoclast-like cells from giant cell tumors of bone. *J Bone Miner Res* 9: 1013-20.
38. James IE, Dodds RA, Lee-Rykaczewski E, Eichman CF, Connor JR, Hart TK, Maleeff BE, Lackman RD, and Gowen M (1996) Purification and characterization of fully functional human osteoclast precursors. *J Bone Miner Res* 11: 1608-18.
39. Ito M, Hsu CT, Naito S, Matsuo T, Onizuka S, Sekine I, Fujii H, and Matsuoka Y (1992) Osteoclast-like giant cell tumour of the gallbladder. *Virchows Archiv - A, Pathol Anat Histopathol* 420: 359-66.
40. Mbalaviele G, Orcel P, Morieux C, Nijweide PJ, and de Vernejoul MC (1995) Osteoclast formation from human cord blood MNCs co-cultured with mice embryonic metatarsals in the presence of M-CSF. *Bone* 16: 171-7.
41. Purton LE, Lee MY, and Torok-Storb B (1996) Normal human peripheral blood MNCs mobilized with granulocyte colony-stimulating factor have increased osteoclastogenic potential compared to nonmobilized blood. *Blood* 87: 1802-8.
42. Higuchi S, Tabata N, Tajima M, Ito M, Tsurudome M, Sudo A, Uchida A, and Ito Y (1998) Induction of human osteoclast-like cells by treatment of blood monocytes with anti-fusion regulatory protein-1/CD98 monoclonal antibodies. *J Bone Miner Res* 13: 44-49.
43. Sarma U and Flanagan AM (1996) Macrophage colony-stimulating factor induces substantial osteoclast generation and bone resorption in human BM cultures. *Blood* 88: 2531-40.
44. Oursler MJ, Pederson L, Fitzpatrick L, Riggs BL, and Spelsberg T (1994) Human giant cell tumors of the bone (osteoclastomas) are estrogen target cells. *Proc Natl Acad Sci USA* 91: 5227-31.
45. Li YP, Chen W, and Stashenko P (1996) Molecular cloning and characterization of a putative novel human osteoclast-specific 116-kDa vacuolar proton pump subunit. *Biochem Biophys Res Commun* 218: 813-21.

46. Goldring S, Roelke M, Petrison K, Bhan A (1987) Human giant cell tumors of bone. Identification and characterization of cell types. *J Clin Investig* 79: 483-491.
47. Galvin RJ, Cullison JW, Avioli LV, and Osdoby PA (1994) Influence of osteoclasts and osteoclast-like cells on osteoblast alkaline phosphatase activity and collagen synthesis. *J Bone Miner Res* 9: 1167-78.
48. Murrills RJ, Shane E, Lindsay R, and Dempster DW (1989) Bone resorption by isolated human osteoclasts *in vitro*: effects of calcitonin. *J Bone Miner Res* 4: 259-68.
49. Lambrecht JT (1990) Cell cultures of human osteoclasts for testing biomaterials. *Deutsche Zahnarztliche Zeitschrift* 45: 82-6.
50. Lambrecht JT and Marks SC, Jr. (1996) Human osteoclast-like cells in primary culture. *Clin Anat* 9: 41-5.
51. Osdoby P, Tsay A, Rothe L, Collin-Osdoby P, and Goldring S (1997) IL-8 expression is elevated in human osteoclasts associated with arthritic and Paget's disease as well as in human osteoclast-like cells exposed to inflammatory cytokines. *J Bone Miner Res* 12: S 199.
52. Collin-Osdoby P, Anderson F, Sunyer T, Rothe L, Weihe J, Maloney W, Miller G, and Osdoby P (1998) Isolation of bone-resorptive human osteoclasts: Culture, characterization, and hormonal regulation *in vitro*. (Submitted).
53. Collin-Osdoby P, Oursler MJ, Webber D, and Osdoby P (1991) Osteoclast-specific monoclonal antibodies coupled to magnetic beads provide a rapid and efficient method of purifying avian osteoclasts. *J Bone Miner Res* 6: 1353-65.
54. Collin-Osdoby P, Li L, Rothe L, Anderson F, Kirsch D, Oursler MJ, and Osdoby P (1998) Inhibition of avian osteoclast bone resorption by monoclonal antibody 121F: A mechanism involving the osteoclast free radical system. *J Bone Miner Res* 13: 67-78.
55. Toyasawa S, Ogawa Y, Chang C, Hong S, Yagi T, Kuwahara H, Wakasa K, and Sakurai M (1991) Histochemistry of tartrate-resistant acid phosphatase and carbonic anhydrase isoenzyme II in osteoclast-like giant cells in bone tumors. *Virchows Archiv A Pathol Anat* 418: 255-261.
56. Drake F, Dodds R, James I, Connor J, Debouck C, Richardson S, Lee-Rykaczewski E, Coleman L, Rieman D, Barthlow R, Hastings G, and Gowen M (1996) Cathepsin K, but not cathepsins B, L, or S, is abundantly expressed in human osteoclasts. *J Biol Chem* 271: 12511-12516.
57. Kasten T, Collin-Osdoby P, Patel N, Osdoby P, Krukowski M, Misko T, Settle S, Currie M, and Nickols A (1994) Potentiation of osteoclast bone resorptive activity by inhibition of nitric oxide synthase. *Proc Natl Acad Sci USA* 91: 3569-3573.
58. Murrills RJ, Stein LS, Horbert WR, and Dempster DW (1992) Effects of phorbol myristate acetate on rat and chick osteoclasts. *J Bone Miner Res* 7: 415-23.
59. Pacifici R (1996) Estrogen, cytokines, and pathogenesis of postmenopausal osteoporosis. *J Bone Miner Res* 11: 1043-1051.
60. Gay C (1991) Avian osteoclasts. *Calcif Tiss Intl* 49: 153-154.

Chapter 7

Isolation and Culture of Dendritic Cells

[1]Michael A Morse and [2]H Kim Lyerly
Departments of [1]Medicine and [2]Surgery, Duke University Medical Center, Box 2606, Durham NC 27710. Tel: 001-919-681-8350; Fax: 001-919-681-7970; E-mail: morse004@mc.duke.edu

1. INTRODUCTION

Dendritic cells (DC) appear to be the most potent antigen presenting cells. Because of their ability to initiate $CD4^+$ T-helper (T_h), $CD8^+$ cytotoxic T-lymphocyte (CTL) [1-3], and T-cell-dependent antibody responses, there is considerable interest in obtaining adequate numbers of DC for use in immunotherapy strategies and for studying auto-immunity, HIV infection, transplantation, graft-versus-host disease, and induction of tolerance. The intended application often dictates the type of DC required or the conditions under which the DC are obtained. This chapter will discuss the various types of DC and their proposed developmental pathways as a prelude to describing methods for their isolation, culture, and manipulation.

In order to appreciate the diverse methodologies for obtaining DC, it is necessary to understand their heterogeneity. The name "dendritic cell" has been given to several developmentally related and unrelated populations of cells possessing morphologic, phenotypic, and functional similarities. The term dendritic cell has been applied to Langerhans cells (LC) of the epidermis, veiled cells in the afferent lymphatics and thoracic duct, interstitial DC in the connective tissue of solid organs, blood DC, interdigitating DC of lymphoid organs (paracortical region of lymph nodes, spleen, Peyer's patches), germinal center DC, follicular DC, and thymic DC. DC may be found in virtually every tissue of the body, and while some of these cells are of the same lineage, differing only in location or phase of development (immature or mature), others appear to be of different lineages.

At least three pathways of DC development have been suggested by *in vitro* culture experiments, although the *in vivo* correlates of the intermediates observed *in vitro* are still a matter of speculation. Steinman [4] has called the three pathways the "sentinel/non-lymphoid," "migratory/myeloid," and "toleragenic/lymphoid." These pathways have different progenitors and intermediates and result in DC with different phenotype and function. They also can be recreated *in vitro* with specific cytokines and culture conditions.

The sentinel/non-lymphoid DC which express CD1a, the Langerhans cell marker, and reside in the epidermis, mucosa, and possibly the interstitium of organs [4], appear to be able to arise directly from $CD34^+$ progenitors [5,6]. The migratory/myeloid DC arise from bone marrow (BM) or cord blood (CB) $CD34^+$ hematopoietic progenitors with the potential to differentiate towards granulocytes, monocytes, or DC [7-10]. One intermediate in this pathway is a $CD14^+$ myelomonocytic cell with the potential to differentiate into either macrophages or DC, depending on the cytokine milieu [5]. It has also been observed that peripheral blood (PB) $CD14^+$ monocytes may be driven to DC by the addition of exogenous granulocyte-macrophage colony-stimulating factor (GM-CSF) and interleukin-4 (IL-4), with or without tumor necrosis factor- α (TNF-α) [11, 12], but it is not known whether these $CD14^+$ monocytes are related to the bipotential $CD14^+$ myelomonocytic intermediates described for the CB and BM progenitor-derived DC.

In contrast to myeloid-derived DC, lymphoid-derived DC develop from very immature thymic T-cell precursors and $CD34^+$ BM precursors with the potential to give rise to T-, B-, and NK cells, but not myeloid cells [13-15]. A subset of these thymic DC (defined by CD8 α expression) has high Fas ligand expression and the capacity to kill Fas^+ $CD4^+$ T-cells, suggesting an immunomodulatory role [16]. Another lymphoid-derived population is the plasmacytoid T cell with a $CD4^+CD11c^-$ phenotype found in the T-cell areas of lymphoid organs [17] which on exposure to IL-3 and CD40-ligand differentiates into an interdigitating DC.

It is also necessary to comment on a phenotypically and functionally distinct population of cells bearing the name "dendritic cell," but entirely unrelated to other pathways, the follicular DC (FDC) found in germinal centers. FDC, which trap and present antigen in the form of immune complexes to B cells [18], are likely derived from mesenchymal fibroblasts and not hematopoietic cells, and possess a phenotype ($CD4^-CD11c^-$ $CD14^+CD21^+$) not characteristic of other DC [18]. In contrast, another population of DC called germinal center DC [19] are phenotypically similar to, and may derive from, myeloid lineage $CD4^+CD11c^+$ blood DC. They present antigen to T-cells, but, unlike FDC, cannot stimulate B cells.

In addition to different developmental pathways, DC may be classified by their level of maturity. Although the precursors for DC do not have distinguishing features, mature DC have many similarities, appearing as large cells with elongated, stellate processes or veils and are defined by lack of lineage-specific markers such as CD3 (T-cell), CD14 (monocytes), CD19 (B-cell), and CD56 (NK cell), and by high expression of HLA-DR, HLA class I, CD11c, CD18, CD40, CD54 (ICAM-1), and CD58 [1]. The expression of CD83 [20], p55 (fascin) [21], the co-stimulatory molecules CD80 (B7.1) and CD86 (B7.2) [1], and CMRF-44 [22] are upregulated in mature DC. Maturation is also marked by a decreasing ability to acquire and process antigen, but an increasing potency of T-cell stimulation [23-25].

For the successful utilization of DC, it is important to match the DC type, maturational phase, and functional attributes with the desired application and to develop the most effective strategy for obtaining those particular DC. The two major methods for obtaining DC are direct isolation from the organ or tissue which is most accessible or *in vitro* culture from precursors.

2. ISOLATION OF DENDRITIC CELLS

Historically, DC were obtained by isolation procedures beginning with the work of van Voorhis [26]. Currently, isolation is mainly performed to study DC from a specific location (e.g. thymic DC or germinal center DC). Procedures (reviewed in [27] and [28]) have been described for isolating DC from virtually every organ or tissue including skin [29-35], synovial tissue [36], interstitial tissues [37], liver [38], thymus [39], germinal centers of lymph nodes (tonsils) [19, 40-42], BM, and most frequently, PB [43-50]. PB and BM are often obtained from volunteers or healthy donors for allogeneic transplants, while germinal center material is generally obtained from tonsillar tissue removed because of recurrent infections, and thymus is obtained from children undergoing cardiovascular surgery [39]. It is important to distinguish between those procedures that isolate truly fresh DC and those that utilize a short (18-36 hr) incubation period and that would be better referred to as short-term cultured DC.

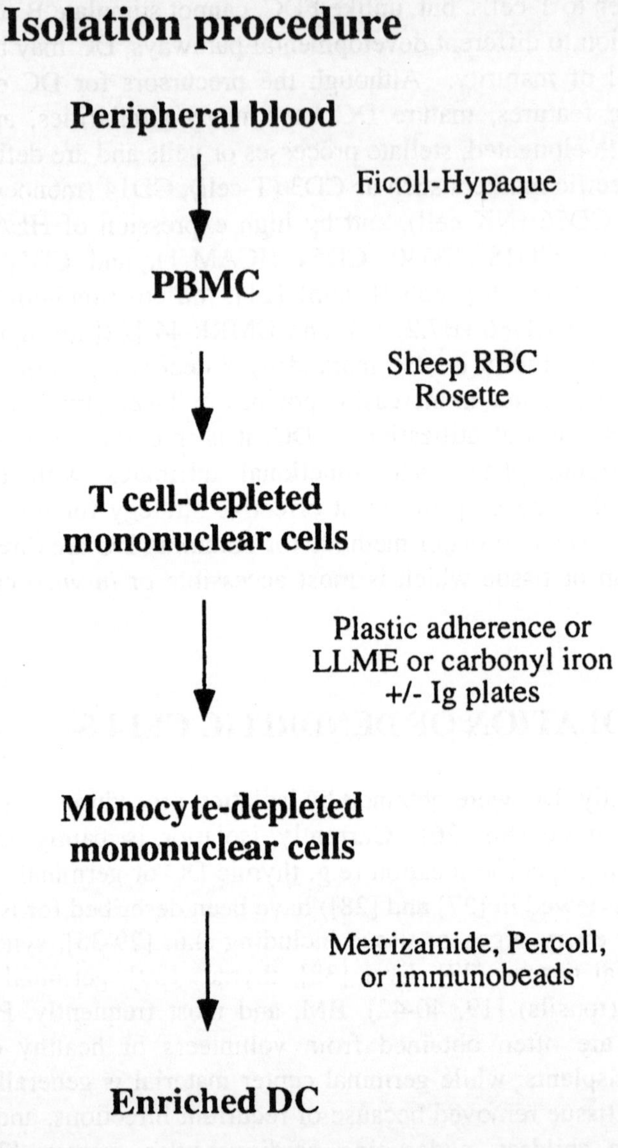

Figure 1. Schema for the isolation of DC from PB. Depletion of contaminating cells is performed by a combination of density gradient centrifugation, plastic adherence, panning, and immunobeads.

As the most readily available source, PB is the most common starting point for isolation procedures (see Fig. 1). Because the frequency of PB DC is low (0.15% of circulating mononuclear cells (MNC)) [20] and because there is no single marker which accurately identifies DC, multiple steps are required to deplete contaminating cells and select for DC. Many protocols are a variation of the short-term culture technique described by Freudenthal and Steinman [43] in which MNC are separated from PB or buffy coats by density gradient centrifugation over Ficoll-Hypaque. The T-cells are removed by sheep red blood cell rosetting, and the depleted cell population is allowed to adhere for 36 hours on plastic (to remove some monocytes and allow the temporarily adherent DC to detach from the plates). The nonadherent cells are layered over a human IgG coated plate to remove more monocytes (by plastic adherence and Fc receptor binding). Alternatively, contaminating phagocytic cells have been removed using carbonyl iron [47, 48] or L-leucyl L -leucine O-methyl ester (LLME) [49] which is metabolized to a product toxic to monocytes. The remaining cells are passed over a density gradient column and the buoyant (low density) cells are collected, being now enriched for DC. The choice of material for the density gradient column has included bovine serum albumin (BSA) [26], metrizamide [43, 44, 50], Percoll [45-47, 51], and Nycodenz [52]. Flow cytometry [49, 53] or immunomagnetic beads [39, 54] can be used to deplete DC of other lineages, yielding greater purity, but lower cell numbers. With these methods, 0.1-1.0% of the starting number of PB MNC are remaining and are 90-98% pure DC.

Subsequent study has raised questions about whether some of the isolation steps may alter the phenotype or function of DC or of other cells present in the heterogeneous populations such that they resemble DC. For example, Thomas [49] has observed that some subpopulations of DC are firmly adherent and would be lost during procedures which use plastic adherence as a depletion step. Furthermore, metrizamide gradients may alter monocytoid cells so that they resemble DC [55]. LLME is toxic to LC [56] and thus could not be used to purify them. Finally, in those procedures where a 24-36 hour incubation is performed [45, 46, 51], maturation of the DC occurs, and therefore, it is not possible to study DC precursors or unactivated DC. Others have designed protocols for isolating fresh DC using sheep red blood cell rosetting and flow cytometry to deplete cells containing markers not expressed by DC but found on other lineages [53]. These cells appear to differ in morphology and phenotype from DC isolated using extended incubation [49, 57], but acquire an activated phenotype during the culture period. Because of concern regarding phenotypic changes, some have used non-adhesive or Teflon-coated materials for handling DC during isolation [46, 51].

The tissue source chosen for the isolation of DC may also affect the phenotype observed. For example, tonsillar tissue is generally only available after tonsillectomy performed for recurrent inflammation, which may theoretically lead to variability in the phenotypic and functional qualities of the isolated cells. Some tissues such as the spleen have regional differences in the phenotype and function of DC. The periarteriolar DC share similarities with interdigitating DC in T-cell areas of lymph nodes (lacking lysozyme, non-specific esterase, Fc receptors and CD11b and weakly expressing CD4 and CD14), whereas peripheral zone DC have a macrophage-like phenotype [58].

2.1 Isolation of Specific Types of Dendritic Cells: Langerhans Cells

The isolation of epidermal LC [29-35] has served as a prototype for isolating DC from non-hematopoietic tissues. After the epidermis is separated from the dermis by incubation in trypsin, mechanical disruption is used to create epidermal suspensions. While most protocols have used an 18-24 hour incubation at 37 ^{o}C in a serum-containing medium to permit adherence of keratinocytes and enrichment of LC in the nonadherent fraction [29, 30, 33], fresh LC have also been isolated using density gradient centrifugation over BSA columns [33] or Ficoll-metrizoate [29, 31]. Because LC express CD1a, they may also be selected by monoclonal antibody and immunomagnetic bead selection [30, 34], or flow cytometric sorting [35]. As is the case with hematopoietic tissue-derived DC, there are morphologic, phenotypic, and functional differences between fresh and short-term incubated LC consistent with maturation. The freshly isolated human LC possess short dendrites and Birbeck granules, express MHC class II and CDla, and have the capability of acquiring and processing antigen. But after 48-72 hours of incubation, they develop numerous veils, more prominent dendrites, and greater levels of MHC class II, ICAM-1, and BB1/B7. Although freshly isolated LC are initially capable of acquiring endocytic antigens, incubated LC lose this property [33] but have greater T-cell stimulatory activity. In contrast, there are decreases in Birbeck granules and Fc and complement receptors, and loss of CD1a [33].

2.2 Other Methods to Increase the Number of Isolated Dendritic Cells

Because the major drawback to isolating DC remains their low frequency in most tissues, a method for increasing the number of DC present in that tissue would be attractive. Recently, flt-3 ligand (FL), a cytokine with stimulatory effects on BM progenitors, has been observed to dramatically increase the number of DC in tissues such as the murine liver, spleen, lymph nodes, and PB [59]. The DC from the spleen could be separated into at least five phenotypically distinct populations based on expression of CD11c, CD11b, and CD8α. Because the mature splenic monocytes were absent, it was suggested that FL may cause terminal differentiation of a precursor(s) towards DC, rather than mobilization of mature DC. FL has now been administered to normal human volunteers at doses of 10-100 μg/kg/d for 14 consecutive days and was well tolerated [60]. The white blood cell count, and in particular the monocyte component, increased after FL administration. Most importantly, the number of DC in the PB increased 30-fold [61]. It may therefore be possible to administer FL to expand PB DC, thereby increasing the yield from an isolation procedure. The functional nature of these DC remains to be determined.

3. GENERATION OF DENDRITIC CELLS *IN VITRO*

In vitro generation of DC from precursors has a number of advantages over isolation, the most important being the less cumbersome methodologies and higher yields. *In vitro* generation also permits more precise manipulations to achieve the desired phenotypic and functional properties, and limits negative influences which may be present in the donor. For example, mature DC isolated from tumor-bearing hosts have a reduced ability to stimulate antigen-specific CTL [62], limiting their application to tumor immunotherapy. DC generated *in vitro* from BM precursors of tumor-bearing hosts, though, retain their ability to stimulate antigen-specific CTL [63], suggesting that a factor such as vascular endothelial growth factor (VEGF) elaborated in the tumor-bearing host can suppress DC function [64]. Generation of DC from precursors *in vitro* would therefore serve to overcome the suppression of DC maturation and function observed in the tumor-bearing host. Nonetheless, there have been few direct comparisons of the two methodologies for obtaining human DC. In mice, DC cultured from BM with GM-CSF and IL-4 possess similar allostimulatory and secondary T-cell responses as DC isolated from the spleen [65].

In this section, the term "generation" will be used in preference to "grow" to indicate the fact that most procedures do not involve proliferation of DC but rather differentiation of precursors into DC. While some procedures result in proliferation of DC precursors, the cells actually identifiable as DC do not appear to proliferate.

Precursor cells for the *in vitro* generation of DC are available in PB (obtained by phlebotomy or leukapheresis, with or without cytokine mobilization) [66-69], BM (obtained from donors undergoing BM harvests for allogeneic BM transplants) [6, 7, 8, 66, 70-72] and neonatal CB (from discarded placentas) [5, 9, 73]. The precursors derived from these sources include CD34⁺ hematopoietic progenitors, the cells remaining after CD34 selection, CD14⁺ monocytes, lymphocyte-depleted PB MNC [74], and plastic adherent cells. The typical procedure for obtaining CD34⁺ hematopoietic precursors requires separation of MNC by density gradient centrifugation over Ficoll-Hypaque, depletion of monocytes by plastic adherence, and selection for CD34⁺ cells by immunomagnetic beads (e.g. Dynabeads™ (Dynal, Oslo, Norway) or MACS™ (Miltenyi Biotech, Bergish Gladbach, Germany) [5, 8, 67, 70], immunoaffinity columns (e.g. Ceprate™ (CellPro, Inc., Bothell, WA) [66, 69], or flow cytometric sorting [70]. CD14⁺ monocytes have been obtained by plastic adherence of PB MNC, followed by recovery of the adherent cells with cold saline and depletion of contaminating lymphocytes with metrizamide [11]. Plastic-adherent cells are obtained by allowing PB MNC to adhere to plastic for two hours [75, 76] and gently removing the nonadherent cells.

3.1 Generation of DC from $CD34^+$ Cells

$CD34^+$ cells have been chosen as the starting point for DC generation by some groups in order to study DC developmental pathways or to attempt to increase the yield of DC by promoting proliferation of precursors. DC generation from $CD34^+$ precursors has been successfully accomplished with a number of different cytokine combinations, but there is no consensus as to the optimal combination. Ye [72] cultured $CD34^+$ cells in medium containing GM-CSF and c-kit ligand (KL), and found that the addition of TNF-α during the first five days yielded a higher number and proportion of DC, but after five days there was no effect on the number, proportion, or expression of costimulatory molecules on DC. When IL-4 or CD40 ligand (CD40L) replaced TNF-α from day 5 to 10 of culture, the absolute number, proportion, and expression of CD1a and CD80 increased compared with maintaining TNF-α. This result is expected, as CD40L cross-links CD40 on DC, increasing DC survival, maturational level, and expression of accessory molecules [77]. The total number of cells in culture was reduced, though, possibly reflecting an inhibitory effect on other cell types. By selecting for $CD1a^+$ cells with immunomagnetic beads at day five, and subculturing the enriched cells with GM-CSF, KL, IL-4, and CD40L, most of the cells at day ten were HLA-DR^+, $CD80^+$, $CD86^+$, $CD54^+$, $CD4^+$, though half were CD1a negative. Thus, Ye concluded that the optimal conditions for growth and differentiation of DC from CD34$^+$ precursors was GM-CSF, KL, and TNF-α for the first five days followed by selection for CD1a$^+$ cells and then subculture with GM-CSF, KL, IL-4, and CD40L in the last five days. Interestingly, in a different study, CD34$^+$ CB progenitor cells incubated with CD40L were found to proliferate and differentiate, independent of GM-CSF, into DC that expressed high HLA-DR, but no CD1a or CD40 [78].

Other cytokines have been found to increase the yield of DC in the CD34$^+$ system. Young [6] and Szabolcs [8] demonstrated that the addition of KL (to GM-CSF and TNF-α) effected a 50- to 100-fold increase in the DC precursor they referred to as a colony-forming unit (CFU)-DC. Sienna [67] reported that the addition of KL, FL, or both to the standard condition of GM-CSF and TNF-α increased the yield of DC by 2.5-fold for each cytokine alone and 5-fold for the combination. Similarly, Maraskovsky [79] showed that the addition of KL (to GM-CSF, TNF-α and IL-4) increased the number of DC generated from human CD34$^+$ BM cells 5-fold, and FL increased the number 6-fold, but the combination effected an 11-fold increment in CD1a$^+$ DC. Although it has not been reported, calculations based on these results suggest that massive numbers of DC could be generated *in vitro*. Sienna [67] calculated that if CD34$^+$ cells derived from an entire mobilized PB (mPB) leukapheresis product were used, 0.6 x 10^9 DC/kg (or 40 x 10^9 DC for an adult patient) would be obtained. One must be cautious of these assumptions, because the conditions present when small-scale studies are performed are not always reproducible when processing is done on a clinical-scale. Furthermore, in regard to clinical studies, many of the cytokines described above are not available in clinical grade, are expensive, and add to the complexity of the procedures. Therefore, a new technique for generating DC from CD34$^+$ precursors on thymic epithelium [80] is particularly interesting. Using thymic epithelium, it is possible to generate mature DC without adding exogenous cytokines. These DC are phenotypically and functionally equivalent to DC generated *in vitro* with the use of cytokines.

3.2 Generation of DC from CD14$^+$ and Plastic-Adherent Cells

The other commonly employed technique for culturing DC from PB is by culturing either CD14$^+$ cells [11, 12], T- and B-cell depleted PB MNC [74, 81, 82], or the fraction of PB MNC adherent to plastic after a two hour incubation [74, 76, 82, 83] (see Fig. 2). The starting population of cells is cultured in medium containing GM-CSF (800 U/ml) and IL-4 (500-1000 U/ml), with or without TNF-α for one week, and the DC-containing population is then harvested by washing the cells from the flask or plate.

The purity of the resulting product depends on the purity of the starting population. Romani has commented on the difficulty in standardizing the method of obtaining two hour plastic-adherent precursors [74], because the purity depends on the vigorousness of the attempts to wash away the nonadherent cells. Overall, DC yield from these methods is approximately 6-8% [12, 76] of the starting MNC population. This technique has now been applied in clinical-scale DC generation. In our current phase I studies of infusions of DC pulsed with either the CEA peptide CAP-1 or RNA encoding CEA in patients with metastatic CEA-expressing adenocarcinomas, patients undergo a 3-4 hour leukapheresis (to collect approximately 1.5-2.0 x 10^8 cells/kg) and the PB MNC are separated from the product on an automated cell separator. The PB MNC are allowed to adhere to plastic flasks for two hours and the nonadherent fraction is removed. The adherent cells are cultured in AIM-V™ serum-free medium containing GM-CSF and IL-4 for seven days. At the completion of the culture period, the cells are harvested and depleted of T- and B-cells using a magnetic column, pulsed with the peptide or RNA, and tested for efficacy *in vitro* using the functional assays to be discussed below. The typical yield of DC, equal to 5-7 % of the original starting number of PB MNC (Morse, unpublished observations), is adequate for clinical studies.

Whether $CD34^+$ progenitors or plastic adherent and $CD14^+$ selected PB MNC represent the optimal starting point is debatable. Certainly, the $CD34^+$ progenitor-derived cells require two weeks of culture for the generation of DC [67, 68], whereas only one week is required for the other precursors. Also, it appears that unless selected for CD1a during the culture, $CD34^+$ progenitor-derived DC have a lower percentage of CD1a and appear to have slightly lower allostimulatory activity in the mixed lymphocyte reaction (MLR) assay [84]. In contast, the yield of cells from $CD34^+$ precursors is generally 4- to 8-fold higher because of their ability to proliferate [83]. Nonetheless, with additional manipulations, augmentation of the yields from $CD34^-$ cells is possible. An approximately 3-fold increase in the number of DC derived from $CD34^-$ PB precursors was achieved by adding IL-3 and TNF-α to the standard IL-4 and GM-CSF-containing medium [84]. Finally, at least one study [85] has found that DC generated from $CD34^+$ cells are more potent than those generated from monocytes at inducing tumor antigen-specific CTL *in vitro* from the PB of patients with low tumor antigen-specific CTL precursor frequencies. Therefore, the ideal method and source of precursors for generating DC still requires considerable research.

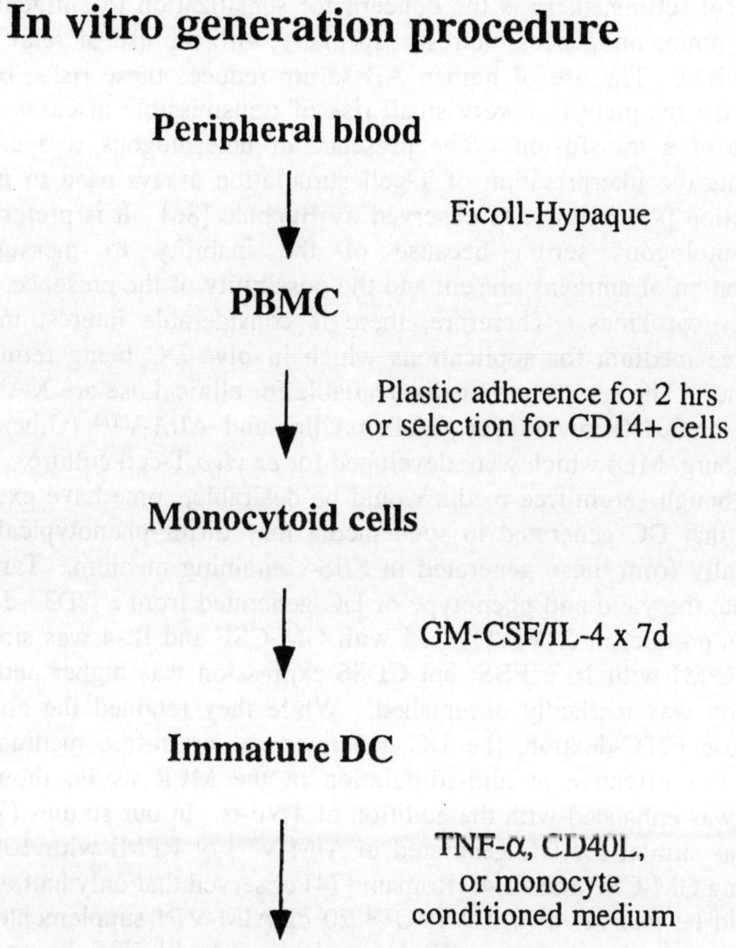

Figure 2. Schema for the generation of DC from PB. Plastic adherent cells, $CD14^+$ monocytes, or lymphocyte-depleted MNC are incubated in medium containing GM-CSF and IL-4. Maturation is effected by the addition of CD40L, TNF-α, or monocyte-conditioned medium.

3.3 Choice of Medium

Many of the procedures for generating DC *in vitro* use heterologous serum-containing medium, but this is undesirable, particularly for clinical applications of DC, but also for basic research studies of DC function. In the clinical setting, there is the concern for sensitization to antigens other than the tumor antigen of interest, especially with the use of fetal bovine serum (FBS). The use of human AB serum reduces these risks, but still exposes the recipient to a very small risk of transmissible diseases akin to the risks of a transfusion. The presence of heterologous sera can also complicate the interpretation of T-cell stimulation assays used to monitor DC function (see below), as observed by Buchele [86]. It is preferable to avoid autologous serum because of the inability to measure the concentration of antigens present and the possibility of the presence of DC-inhibitory cytokines. Therefore, there is considerable interest in using serum-free medium for applications which involve DC being returned to their donor. The serum-free media available for clinical use are X-VIVO™ (10, 15, or 20) (Biowhittaker, Walkersville) and AIM-V™ (Gibco BRL, Gaithersburg, MD) which were developed for *ex vivo* T-cell culture.

Although serum-free media would be desirable, some have expressed concern that DC generated in such media may differ phenotypically and functionally from those generated in FBS-containing medium. Tarte [84] found that the yield and phenotype of DC generated from a CD34-depleted apheresis product in X-VIVO™ 15 with GM-CSF and IL-4 was similar to that in RPMI with 10% FBS, but CD86 expression was higher and CD1a expression was markedly diminished. While they retained the ability to endocytose FITC-dextran, the DC generated in serum-free medium were slightly less effective at allo-stimulation in the MLR assay, though the activity was enhanced with the addition of TNF-α. In our studies [76], the MLR was similar for DC generated in AIM-V™ or RPMI with 10% FBS containing GM-CSF and IL-4. Romani [74] observed that only half as many DC could be generated in X-VIVO™ 20 or AIM-V™ supplemented with 1% autologous plasma than in RPMI supplemented with FBS due to greater adherence of DC generated in X-VIVO™ 20. X-VIVO™ 10 and 15 produced "unsatisfactory" yields. What constituent of serum is responsible for the slight differences is not clear. One study suggested that TGF-β could replace serum [87] when generating DC from $CD34^+$ cells in GM-CSF, TNF-α, and KL in serum-free medium.

Another important group of questions regarding the use of DC generated in serum-free medium containing GM-CSF and IL-4 centers on their level of maturity and the stability of their phenotype and function. Romani [74] demonstrated that DC generated under these conditions are immature, based on their lack of expression of CD83, but have the capacity to stimulate resting T-cells and process and present tetanus toxoid to antigen-specific T-cells. We have observed that these immature DC are capable of stimulating a tumor antigen-specific CTL response *in vitro* from naive T-cells [88]. Nonetheless, Romani [74] and Bender [82] point out that these DC are not stable. When cultured in the absence of cytokines, they revert to the phenotype and functional activity of monocytes that have much less potent T-cell stimulatory capacity. Exposure for three additional days to a monocyte-conditioned medium (produced by culturing autologous monocytes on Ig-coated plates) resulted in mature DC that exhibited phenotypic stability even after removal of cytokines [74, 82]. It remains to be determined whether immature DC will mature *in vivo* when exposed to the cytokine milieu of the patient, or whether maturational agents will be required during the *in vitro* generation of the DC vaccine. From a regulatory standpoint, it is preferable if defined, standardized materials are used in DC processing, and thus, further work should be directed at defining the necessary cytokines found in monocyte-conditioned medium.

Other culture conditions that remain to be more completely evaluated are the role of cell density and the presence of contaminating cells. Removal of contaminating red blood cells and platelets has been suggested [83] because they contain immunosuppressive factors such as gangliosides and TGF-β that might inhibit DC development, though this has not been experimentally proven. DC generation and maturation may be dependent on cell density [84] as well. High cell density (1 x 10^6 cells/cm^2 and 5 x 10^6 cells/ml) during the first overnight incubation promoted faster and more reliable generation of DC than lower densities (0.3 x 10^6 cells/ml) at the start of culture. In contrast, the expression of CD83 and upregulation of CD86 after the addition of TNF-α required low cell densities (0.5 x 10^6 cells/ml) and was abolished at high densities (2 x 10^6 cells/ml). Although the culture surfaces have received little attention, Mahnke [89] has demonstrated that murine DC cultured on collagen coated plates (but not exposed to soluble collagen or fibronectin) developed greater T-cell stimulatory activity.

3.4 Methods for Loading DC with Antigen *In Vitro*

For applications of DC involving immunotherapy, it has become important to establish conditions for delivering antigens to DC. Animal models have suggested several efficacious approaches for delivering antigen to DC including peptide pulsing, co-incubation with DNA [90] or RNA encoding the antigen of interest [91], pulsing with tumor extracts [92], fusion with tumors [93], and infection with viral vectors [94]. While it is beyond the scope of this chapter to discuss the advantages and drawbacks of these different methods, the method chosen may affect the culture condition utilized for DC development. For example, different methods of delivering antigen may require different maturational stages of the DC.

4. ASSAYS FOR ASSESSING DENDRITIC CELLS

Because of the lack of a single defining characteristic for DC and the wide range of phenotype and function observed even for DC of the same lineage, a combination of different assays is typically required to characterize a group of cells believed to be DC. The appearance of large cells with dendrites or veils is the typically described DC morphology, but depending on the culture conditions, monocytes may demonstrate a similar morphology [81]. Phenotypic staining pattern has been the most extensively studied. As described above, different types of DC have somewhat different phenotypes. For myeloid-derived DC which are the most frequently generated, high expression of HLA-DR and CD40, low expression of CD14, and the presence of variable amounts of CD1a, CD80, CD86, and CD54 are typical. CD14 expression interestingly has been found to vary with the antibody used. With Leu-M3, expression is frequently undetectable on cultured DC but with My4 it is reduced but often still present [81]. The state of maturity of the DC is reflected by the expression of CD80, CD83, and CMRF-44.

The standard functional assay for DC is the MLR [95] in which graded numbers of irradiated stimulator cells are co-cultured in triplicate with allogeneic responder cells for 96 hours, after which ^{3}H-thymidine is added for 18 hours before harvesting the responders to determine their proliferation as detected by thymidine incorporation into their DNA. DC are the most potent stimulators of responder cells in the MLR, exceeding monocytes and PB MNC. The activity in the MLR is a reflection of the high expression of HLA class II molecules, though others have found MLR activity of DC also correlates with CD1a expression [81]. Although it is the most frequently performed functional assay, the MLR does not detect an activity of DC normally exhibited *in vivo*; nonetheless, it is relatively easy to perform.

Proliferation of autologous T-cells in response to DC loaded with recall antigens such as tetanus toxoid or influenza proteins or peptides [95] can also be measured by thymidine uptake, and reflects an activity actually exhibited by DC *in vivo*. Because the interest in generating DC has more recently been focused on their use in immunotherapy protocols, assays which measure the ability of the generated DC to stimulate naive T-cell responses against tumor or virus-specific epitopes are particularly important. The standard assay for measuring the induction of CTL responses is the microcytotoxicity assay [95], in which HLA-matched target cells labeled with a radionuclide (e.g. Chromium-51) or other measurable agent (e.g. Europium) are co-incubated with graded numbers of T-cells for four hours and the amount of released label is used to calculate the percentage of target cells lysed. DC loaded with tumor or viral antigens are generally able to stimulate a potent CTL response as measured by the microcytotoxicity assay. Another assay relying on microcytotoxicity is the limiting dilution assay which quantifies the antigen-specific CTL precursor frequency in a population of responder cells. DC loaded with antigen are able to stimulate a detectable number of antigen-specific CTL precursors after only two cycles of stimulation. Unfortunately, the microcytotoxicity assay is cumbersome and time consuming (21 days from the start of the first T-cell stimulation with antigen-loaded DC), and therefore newer assays are now being evaluated.

The ELISPOT [96] is an ELISA-based assay for measuring the number of T-cells which respond to the presentation of antigen by releasing cytokines such as IL-2, interferon-γ, TNF-α, and IL-4. The ELISPOT is performed in five basic steps: (1) coating a nitrocellulose-backed microtiter plate with purified cytokine-specific antibody; (2) blocking the plate to prevent nonspecific absorption of random proteins; (3) incubating the cytokine secreting cells (e.g. T-cells stimulated by antigen-loaded DC) at several different dilutions; (4) adding a labeled second antibody; and (5) detecting the antibody-cytokine complex. The final step is usually an enzyme/substrate reaction that generates a purple colored product which represents a permanent "footprint" that can be quantitated microscopically, visually, or electronically. Each purple spot represents one single cell secreting the cytokine of interest. The CTL precursor frequency provided by ELISPOT assay is comparable to that obtained by limiting dilution analysis. The FastImmune™ (Becton-Dickinson, San Jose, CA) test [97] is a flow cytometry-based assay for measuring the number of T-cells which make cytokine in response to antigen stimulation. As with the ELISPOT, antigen-loaded DC are used to stimulate T-cells, after which Brefeldin A is added to prevent cytokine release from the cells. After permeabilization, the cells are stained with anti-CD4, anti-CD8, anti-CD69 (activation marker), and monoclonal antibody specific for the cytokine of interest and examined by flow cytometry analysis. Because the ELISPOT and FastImmune™ are rapid assays with quantitative results, they are likely to become the basis for monitoring DC-based studies.

5. FUTURE DIRECTIONS

The considerable interest in using DC for immunotherapy strategies [98-101] has provided the impetus to continue development of techniques for obtaining them in large quantities. The availability of FL to increase the number of DC in PB makes isolation procedures a consideration again. As the number of markers and described subtypes of DC increases [102] and the cytokine requirements for them are elucidated, it will be possible to tailor strategies to obtain the precise DC type desired for a particular purpose. Interest in streamlining and automating the procedures for generating DC to make them less labor intensive and more reliable may make them readily available for widespread clinical use.

LITERATURE CITED

1. Steinman RM (1991) The dendritic cell system and its role in immunogenicity. *Ann Rev Immunol* 9:271-296.
2. Inaba K, Young JW, Steinman RM (1987) Direct activation of $CD8^+$ cytotoxic T lymphocytes by dendritic cells. *J Exp Med* 166: 182-194.
3. Young JW and Steinman RM (1990) Dendritic cells stimulate primary human cytolytic lymphocyte responses in the absence of $CD4^+$ helper T cells. *J Exp Med* 171:1315-1332.
4. Steinman RM, Pack M, Inaba K (1997) Dendritic cell development and maturation. In Ricciardi-Castagnoli P, ed, Dendritic Cells in Fundamental and Clinical Imunology, New York, Plenum Press, pp 1-6.
5. Caux C, Vandervliet B, Massacrier C, et al (1996) $CD34^+$ hematopoietic progenitors from human cord blood differentiate along two independent dendritic cell pathways in response to GM-$CSF^+TNF\alpha$. *J Exp Med* 184:695-706.
6. Young JW, Szabolcs P, Moore MA (1995) Identification of dendritic cell colony-forming units among normal human $CD34^+$ BM progenitors that are expanded by c-kit-ligand and yield pure dendritic cell colonies in the presence of granulocyte/macrophage colony-stimulating factor and tumor necrosis factor alpha. *J Exp Med* 182:1111-1120, 1995.
7. Szabolcs P, Avigan D, Gezelter S, et al (1996) Dendritic cells and macrophages can mature independently from a human BM-derived, post-colony forming unit intermediate. *Blood* 87:4520-30.
8. Szabolcs P, Moore MAS, Young JW (1995) Expansion of immunostimulatory dendritic cells among the myeloid progeny of human $CD34^+$ BM precursors cultured with KL, granulocyte-macrophage colony-stimulating factor, and TNF-α. *J Immunol* 154:5851-5861.
9. Santiago-Schwarz F, Belilos E, Diamond B, et al (1992) TNF in combination with GM-CSF enhances the differentiation of neonatal cord blood stem cells into dendritic cells and macrophages. *J Leuk Biol* 52:274-281.
10. Inaba K, Inaba M, Deguchi K, et al (1993) Granulocytes, macrophages, and dendritic cells arise from a common major histocompatibility complex class II-negative progenitor in mouse BM. *Proc Natl Acad Sci USA* 90:3038-3042.
11. Zhou L-J, Tedder TF (1996) $CD14^+$ Blood monocytes can differentiate into functionally mature $CD83^+$ dendritic cells. *Proc Natl Acad Sci USA* 93:2588-2592.
12. Kiertscher SM and Roth MD (1996) Human $CD14^+$ leukocytes acquire the phenotype and function of antigen-presenting dendritic cells when cultured in GM-CSF and IL-4. *J Leukoc Biol* 59:208-218.
13. Saunders D, Lucas K, Ismaili J, et al (1996) Dendritic cell development in culture from thymic precursor cells in the absence of granulocyte-macrophage colony stimulating factor.*J Exp Med* 184:2185-2196.
14. Galy A, Travis M, Cen D, Chen B (1993) Human T, B, natural killer, and dendritic cells arise from a common BM progenitor cell subset. *Immunity* 3:459-473.

15. Wu L, Antica M, Johnson GR, et al (1991) Developmental potential of earliest precursor cells from adult thymus. *J Exp Med* 174:1617-1627.
16. Suss G, Shortman K (1996) A subclass of dendritic cells kills CD4 T cells via Fas/Fas-ligand-induced apoptosis. *J Exp Med* 183:1789-1796.
17. Grouard G, Rissoan MC, Filguiera L, et al (1997) The enigmatic plasmacytoid T cells develop into dendritic cells with interleukin (IL)-3 and CD40-ligand. *J Exp Med* 185:1101-1111.
18. Liu YJ, Grouard G, de Boutteillier O, Banchereau J (1996) Follicular dendritic cells and germinal centers. *Int Rev Cytol* 166:139-179.
19. Grouard G, Durand I, Filgueira L, et al (1996) Dendritic cells capable of stimulating T cells in germinal centers. *Nature* 384:364-367.
20. Zhou LJ and Tedder TF (1995) Human blood dendritic cells selectively express CD83, a member of the immunoglobulin superfamily. *J Immunol* 154:3821-3835.
21. Mosialos G, Birkenbach M, Ayehunie S, et al (1996) Circulating human dendritic cells differentially express high levels of a 55-kd actin-bundling protein. *Am J Path* 148:593-600.
22. Hock BD, Starling GC, Daniel PB, Hart DNJ (1994) CMRF-44, a novel monoclonal antibody to an activation antigen expressed by the allostimulatory cells within PB, including dendritic cells. *Immunol* 83:573-581.
23. Austyn JM (1992) Antigen uptake and presentation by dendritic leukocytes. *Semin Immunol* 4: 227-36.
24. Sallusto F and Lanzavecchia A (1994) Efficient presentation of soluble antigen by cultured human dendritic cells is maintained by granulocyte/macrophage colony-stimulating factor plus interleukin 4 and downregulated by tumor necrosis factor alpha. *J Exp Med* 179:1109-1118.
25. Caux C, Massacrier C, Vanbervliet B, et al (1994) Activation of human dendritic cells through CD40 cross-linking. *J Exp Med* 180:1263-1272.
26. Van Voorhis WC, Hair LS, Steinman RM, Kaplan G (1982) Human dendritic cells. Enrichment and characterization from PB. *J Exp Med* 155:1172-87.
27. Williams LA, Egner W, Hart DNJ (1994) Isolation and function of human dendritic cells. *Int Rev Cytol* 153:41-103.
28. Jaffe R (1993) Review of human dendritic cells: isolation and culture from precursors. *Pediatric Path* 13:821-837.
29. Cohen PJ and Katz SI (1992) Cultered human Langerhans cells process and present intact protein antigens. *J Invest Dermatol* 99:331-336.
30. Morris J, Alaibac M, Jia MH, Chu T (1992) Purification of functional active epidermal Langerhans cells: A simple and efficient new technique. *J Invest Dermatol* 99:237-240.
31. Teunisen MBM, Wormmeester DJ, Krieg SR, et al (1990) Human epidermal Langerhans cells undergo profound morphologic ancl phenotypical changes during in-vitro culture. *J Invest Dermatol* 94:166-173.
32. Vedel J, Vincendeau P, Bezian JH, Taieb A (1992) Flow cytometry analysis of adhesion molecules on human Langerhans cells. *Clin Exp Dermatol* 17:240-245.
33. Romani N, Lenz A, Glassel H, et al (1989) Cultured human Langerhans cells resemble lymphoid dendritic cells in phenotype and function. *J Invest Dermatol* 93:600-609.
34. Hanau D, Schmitt DA, Fabre MF, Cazenave JP (1988) A method for the rapid isolation of human epidermal Langerhans cells using immunomagnetic microspheres. *J Invest Dermatol* 91:274-279.
35. Inaba K, Schuler G, Witmer MD, et al (1986) Immunologic properties of purified epidermal Langerhans cells. *J Exp Med* 164:605-613.

36. Knight SC, Fryer P, Griffiths S, et al (1989) Class II antigens on dendritic cells from synovial fluids of patients with inflammatory arthritis. *Clin Exp Immunol* 78:19-25.
37. Hart DNJ, Prickett TCR, McKenzie JL, et al (1989) Characterization of interstitial dendritic cells in human tissues. *Transplant Proc* 21:401-403.
38. Prickett TCR, McKenzie JL, Hart DNJ (1988) Characterization of interstitial dendritic cells in human liver. *Transplantation* 46:754-761.
39. LaFontaine M, Landry D, Montplaisir S (1997) Human thymic dendritic cells. *Microsc Res Tech* 38:267-275.
40. Hart DNJ, McKenzie JL (1988) Isolation and characterization of human tonsil dendritic cells. *J Exp Med* 168:157-170.
41. King PD and Katz DR (1989) Human tonsillar dendritic cell-induced T cell responses: analysis of molecular mechanisms using monoclonal antibodies. *Eur J Immunol* 19:581-587.
42. Prickett TCR, McKenzie JL, Hart DNJ (1992) Adhesion molecules on human tonsil dendritic cells. *Transplantation* 53: 483-490.
43. Freudenthal PS and Steinman RM (1990) The distinct surface of human blood dendritic cells as observed after an improved isolation method. *Proc Natl Acad Sci USA* 87:7698-702.
44. Brooks CF and Moore M (1988) Differential MHC class II expression on human PB monocytes and dendritic cells. *Immunology* 63:303-311.
45. Young JW and Steinman RM (1988) Accessory cell requirements for the MLR and polyclonal mitogens, as studied with a new technique for enriching blood dendritic cells. *Cell Immunol* 111:167-182.
46. Xu H, Friedrichs U, Gieseler RK, et al (1992) Human blood dendritic cells exhibit a distinct T-cell stimulating mechanism and differentiating pattern. *Scand J Immunol* 36:689-696.
47. Vakkila J, Lehtonen E, Koskimies S, Hurme M (1987) Dendritic cells in human PB: effective enrichment from the nonadherent cells. *Immunol Lett* 15:229-236.
48. Kuntz-Crow M and Kunkel HG (1982) Human dendritic cells: Major stimulators of autologous and allogenic MLR. *Clin Exp Immunol* 49:, 338-346.
49. Thomas S, Davis LS, Lipsky PE (1993) Isolation and characterization of human PB dendritic cells. *J Immunol* 150:821-34.
50. Knight SC, Farrant J, Bryant A, et al (1986) Nonadherent, low density cells from human PB contain dendritic cells and monocytes both with veiled morphology. *Immunology* 57:595-603.
51. Markowicz S and Engleman EC (1990) Granulocyte-macrophage colony-stimulating factor promotes differentiation and survival of human PB dendritic cells in vitro. *J Clin Invest* 85:955-61.
52. Hart DNJ, Starling G C, Calder VC, and Fernando NJ (1993) B7/BB-1 is a leucocyte differentiation antigen on human dendritic cells induced by activation. *Immunology* 79:616-620.
53. Egner W, Andreesen R, Hart DNJ (1993) Allostimulatory cells in fresh human blood: heterogeneity in antigen presenting cell populations. *Transplantation* 56:945-50.
54. Karhumaki E, Viljanen ME, Cottler-Fox M, et al (1993) An improved enrichment method for functionally competent, highly purified dendritic cells and its application to HIV-infected blood samples. *Clin Exp Immunol* 91: 482-488.
55. Kabel PJ, de Haan-Meulman M, Voorbij AM, et al (1989) Accessory cells with a morphology and marker pattern of dendritic cells can be obtained by elutriator-purified blood monocyte fractions. An enhancing effect of metrizamide in this differentiation. *Immunobiology* 179:395-411.

56. Simon JC, Thiele DL, Schopf E, Sontheimer RD (1992) Effects of the immunosuppressive dipeptide L-leucyl L-leucine O-methyl ester on epidermal Langerhans cells. *J Invest Dermatol* 99:80s-82s.
57. Hart DNJ, Starling G C, Calder VC, and Fernando NJ (1993) B7/BB-1 is a leucocyte differentiation antigen on human dendritic cells induced by activation *Immunology* 79:616-620.
58. Buckley PJ, Smith MR, Braverman MF, Dickson SA (1987) Human spleen contains phenotypic subsets of macrophages and dendritic cells that occupy discrete microanatomic locations. *Am J Pathol* 128:505-519.
59. Marakovsky E, Brasel K, Teepe M, et al (1996) Dramatic increase in the numbers of functionally mature dendritic cells in flt3 ligand-treated mice: multiple dendritic cell subpopulations identified. *J Exp Med* 184:1953-1962.
60. Lebsack ME, McKenna JA, Hoek R, et al (1997) Safety of Flt3 ligand in healthy volunteers. *Blood* 90:abstract #751.
61. Marakovsky E, Roux E, Teepe M, et al (1997) Flt3 ligand increases PB dendritic cells in healthy volunteers. *Blood* 90:abstract #2585.
62. Gabrilovich DI, Ciernik IF, Carbone DP (1996) Dendritic cells in antitumor immune responses. I. Defective antigen presentation in tumor-bearing hosts. *Cell Immunol* 170:101-110.
63. Gabrilovich DI, Nadaf S, Corak J, et al (1996) Dendritic cells in antitumor immune responses. II Dendritic cells grown from BM precursors, but not mature DC from tumor bearing mice, are effective antigen carriers in the therapy of established tumors. *Cell Immunol* 170:111-119.
64. Gabrilovich DI, Chen HL, Girgis K, et al (1996) Production of vascular endothelial growth factor by human tumors inhibits the functional maturation of dendritic cells. *Nature Med* 10:1096-1103.
65. Garrigan K, Moroni-Rawson P, McMurray C, et al (1996) Functional comparison of spleen dendritic cells and dendritic cells cultured *in vitro* from BM precursors. *Blood* 88:3508-3512.
66. Bernhard H, Disis ML, Helmfeld S et al (1995) Generation of immunostimulatory dendritic cells from human $CD34^+$ hematopoietic progenitor cells of the BM and PB. *Cancer Res* 55:1099-1104.
67. Siena S, Di Nicola M, Bregni M, et al (1995) Massive ex vivo generation of functional dendritic cells from mobilized $CD34^+$ blood progenitors for anti-cancer therapy. *Exp Hematol* 23:1463-1471.
68. Strunk D, Rappersberger K, Egger C, et al (1996) Generation of human dendritic cells/Langerhans cells from circulating $CD34^+$ hematopoietic progenitor cells. *Blood* 87:1292.
69. Mackensen A, Herbst B, Kohler G, et al (1995) Delineation of the dendritic cell lineage by generating large numbers of Birbeck granule-positive Langerhans cells from human PB progenitor cells in vitro. *Blood* 86:2699.
70. Rosenzwajg M, Canque B, Gluckman JC (1996) Human dendritic cell differentiation pathway from $CD34^+$ hematopoietic precursor cells. *Blood* 87:535-544.
71. Reid CDL, Stackpoole A, Meager A, et al (1992) Interactions of tumor necrosis factor with granulocyte-macrophage colony-stimulating factor and other cytokines in the regulation of dendritic cell growth in vitro from early bipotent $CD34^+$ progenitors in human BM. *J Immunol* 149:2681-2688.
72. Ye Z, Gee AP, Bowers WE, et al (1996) *In vitro* expansion and characterization of dendritic cells derived from human BM $CD34^+$ cells. *Bone Marrow Transplant* 18:997-1008.

73. Caux C, Dezutter-Dambuyant C, Schmitt D, et al (1992) GM-CSF and TNF-alpha cooperate in the generation of dendritic Langerhans cells. *Nature* 360:258-261.
74. Romani N, Reider D, Heuer M, et al, Generation of mature dendritic cells from human blood. An improved method with special regard to clinical applicability. *J Immunol Methods* (1996) 196:137-151.
75. Romani N, Gruner S, Brang D, et al (1994) Proliferating dendritic cell progenitors in human blood. *J Exp Med* 180:83-93.
76. Morse MA, Zhou L-J, Tedder TF, et al (1997) Generation of dendritic cells *in vitro* from peripheral blood mononuclear cells with granulocyte-macrophage-colony-stimulating factor, interleukin-4, and tumor necrosis factor-alpha for use in cancer immunotherapy. *Ann Surg* 226:6-16.
77. Caux C, Massacrier C, Vanbervliet B, et al (1994) Activation of human dendritic cells through CD40 cross-linking. *J Exp Med* 180:1263-1272.
78. Flores-Romo L, Bjork P, Duvert V, et al (1997) CD40 ligation on human cord blood $CD34^+$ hematopoietic progenitors induces their proliferation and differentiation into functional dendritic cells. *J Exp Med* 185:341-349.
79. Maraskovsky E, Roux M, Tepee HJ, et al (1995) The effect of flt3 ligand and/or KL on the generation of dendritic cells from human $CD34^+$ BM. *Blood* 86:420a.
80. Miralles GD, Smith CA, Whichard LP, Morse MA, Haynes BF, Patel DD (1998) $CD34^+$CD38-lin- cord blood cells develop into dendritic cells in human thymic stromal monolayers and thymic nodules. *J Immunol* (In press).
81. Chapuis F, Rosenzwajg M, Yagello M, et al (1997) Differentiation of human dendritic cells from monocytes *in vitro*. *Eur J Immunol* 27:431-441.
82. Bender A, Sapp M, Schuler G, et al (1996) Improved methods for the generation of dendritic cells from nonproliferating progenitors in human blood. *J Immunol Methods* 196:121-35.
83. Thurnher M, Papesh C, Ramoner R, et al (1997) In vitro generation of $CD83^+$ human blood dendritic cells for active tumor immunotherapy *Exp Hematol* 25:232-237, 1997.
84. Tarte K, Lu ZY, Fiol G, et al (1997) Generation of virtually pure and potentially proliferating dendritic cells from non-CD34 apheresis cells from patients with multiple myeloma. *Blood* 90:3482-3495.
85. Mortarini R, Anichini A, Di Nicola M, et al (1997) Autologous dendritic cells derived from $CD34^+$ progenitors and from monocytes are not functionally equivalent antigen-presenting cells in the induction of Melan-A/Mart-1_{27-35}-specific CTLs from the PB lymphocytes of melanoma patients with low frequency of CTL precursors. *Cancer Res* 57:5534-5541.
86. Buchele S and Reinhard H (1997) Presentation of tetanus toxoid to autologous T cells by dendritic cells generated from human blood. In Ricciardi-Castagnoli P, ed, Dendritic Cells in Fundamental and Clinical Imunology, New York, Plenum Press, pp 1-6.
87. Strobl H, Riedl E, Scheinecker C, et al (1996) TGF-beta1 promotes in vitro development of dendritic cells from $CD34^+$ hemopoietic progenitors. *J Immunol* 157:1499-1507.
88. Wong C, Morse M, Nair SK (1997) Induction of primary, human antigen-specific cytotoxic T lymphocytes in vitro using dendritic cells pulsed with peptides. *J Immunother* (in press).
89. Mahnke K, Bhardwaj RS, Luger TA, et al (1996) Interaction of murine dendritic cells with collagen up-regulates allostimulatory capacity, surface expression of heat stable antigen, and release of cytokines *J Leukoc Biol* 60:463-472.
90. Alijagic S, Moller P, Artuc M, et al (1995) Dendritic cells generated from PB transfected with human tyrosinase induce specific T cell activation. *Eur J Immunol* (25:3100-3107.

91. Boczkowski D, Nair S, Snyder D, Gilboa E (1996) Dendritic cells pulsed with RNA are potent antigen presenting cells in vitro and in vivo. *J Exp Med* 184:465-472.
92. Nair SK, Snyder D, Rouse BT, Gilboa E (1997) Regression of tumors in mice vaccinated with professional antigen-presenting cells pulsed with tumor extracts. *Int J Cancer* 70:706-715.
93. Gong J, Chen D, Kashiwaba M, Kufe D (1997) Induction of antitumor activity by immunization with fusions of dendritic and carcinoma cells. *Nature Med* 3:558-61.
94. Bhardwaj N, Bender A, Gonzalez N, et al (1994) Influenza virus-infected dendritic cells stimulate strong proliferative and cytolytic responses from human $CD8^+$ T cells. *J Clin Invest* 94:797-807.
95. James ST (1991) Measurement of basic immunologic characteristics of human MNCs. In Coligan JE, Kruisbeek AM, Margulies DH, et al, eds, Current Protocols in Imunology, New York, John Wiley & Sons, pp 7.10-1-7.10.10.
96. Miyahira Y, Murata K, Rodriguez D, et al (1995) Quantification of antigen specific $CD8^+$ T cells using an ELISPOT assay. *J Immunol Methods* 181:45-54.
97. Maino VC, Suni MA, Ruitenberg J (1995) Rapid flow cytometric method for measuring lymphocyte subset activation. *Cytometry* 20:127-33.
98. Mukherji B, Chakraborty NG, Yamasaki S, et al (1995) Induction of antigen-specific cytolytic T cells *in situ* in human melanoma by immunization with synthetic peptide-pulsed autologous antigen presenting cells. *Proc Natl Acad Sci USA* 92:8078-8082.
99. Hsu FJ, Benike C, Fagnoni F, et al (1996) Vaccination of patients with B-cell lymphoma using autologous antigen-pulsed dendritic cells. *Nature Med* 2:52-58.
100. Reichardt V, et al (1996) Idiotypic vaccination using dendritic cells for multiple myeloma patients after autologous PB stem cell transplantation. *Blood* 88: 481 (Abstract).
101. Murphy G, Tjoa B, Ragde H, et al (1996) Phase I clinical trial: T-cell therapy for prostate cancer using autologous dendritic cells pulsed with HLA-A0201-specific peptides from prostate-specific membrane antigen. *The Prostate* 29:371-380.
102. Hart DNJ, Clark GJ, Dekker JW, et al (1997) Dendritic cell surface molecules: A proliferating field. In Ricciardi-Castagnoli P, ed, Dendritic Cells in Fundamental and Clinical Imunology, New York, Plenum Press, pp 439-442.

Chapter 8

In Vitro Proliferation and Differentiation of $CD34^+$ Cells to Neutrophils

James G Bender
Nexell Therapeutics Inc, 9 Parker Street, Irvine, CA 92618. Tel: 001-949-470-6642; Fax: 001-949-470-6644; E-mail: Jbender@nexellinc.com

1. INTRODUCTION

Neutrophils represent the predominant leukocyte present in the blood of humans, and function in host defense by moving to inflammatory sites where they engulf and kill microorganisms. They are produced in bone marrow (BM) from the proliferation and differentiation of a stem cell compartment identified by expression of the CD34 antigen. The BM produces between 60-400 x 10^7 neutrophils per day, and greater than 90% (about 10^{11} neutrophils) are held in reserve in the BM [1]. From this compartment, they are released into the blood and circulate for 6-9 hours, after which they marginate into the capillaries and migrate into tissues where they survive for another 12-48 hours, and then die by apoptosis and are subsequently removed by the reticuloendothelial system [2].

The importance of this lineage has been established through the study of conditions in which neutrophils are missing or dysfunctional. Mature neutrophils have a complex function requiring locomotion, the ability to sense chemotactic gradients, adherence, phagocytosis and a variety of killing mechanisms. Hereditary or acquired defects in many of these functions can become life-threatening. One example of a hereditary condition is chronic granulomatous disease in which a defect in the oxidase enzyme that produces oxygen radicals to kill bacteria results in impaired neutrophil function [3]. Patients with this disease suffer from recurrent infections and, without treatment, die from multiple infections within 5-7 years [3]. Another situation, which emphasizes the importance of neutrophils in host defense, is the life-threatening neutropenia associated with high dose chemotherapy for the treatment of cancer. Risk analysis studies show that the longer these patients remain neutropenic, the higher their risk of infection and mortality, in spite of isolation measures and antibiotic support [4].

The opportunity to grow neutrophils *ex vivo* began with the discovery of molecules that regulate the production of neutrophils. These discoveries began in the 1960s and were prompted by development of a method to grow colonies of neutrophils from BM progenitors in a semisolid matrix in contact with a feeder layer of cells producing diffusible factors [5]. This colony assay technique ultimately led to identification of the neutrophil regulatory cytokines granulocyte colony-stimulating factor (G-CSF) and granulocyte-macrophage colony-stimulating factor (GM-CSF), as well as others. In these early studies, however, minimal characterization was performed on the granulocytes produced in colony assays.

As more cytokines were discovered and the progenitor population that gives rise to neutrophils was better characterized, it became possible to prepare *in vitro* derived neutrophils from liquid cultures of progenitors stimulated with recombinant cytokines. This chapter focuses on the methods and culture systems for producing neutrophils *in vitro*, as well as the characterization of these cell products and their potential for therapeutic use.

2. TISSUE PROCUREMENT AND PROCESSING

2.1 Obtaining $CD34^+$ Cells

$CD34^+$ cells are the recommended starting cell population for preparation of *in vitro* derived neutrophils. This population is present in BM at about 1% and in umbilical cord blood (CB) at <1%. Only rare $CD34^+$ cells are present in normal peripheral blood (PB) [6], although their frequency can be increased significantly by mobilization using a variety of techniques (reviewed in [7]). A commonly used method of mobilization is to administer daily doses of G-CSF in a steady-state situation, and then harvest blood mononuclear cells (MNC) by apheresis on days 4-6. This technique is suitable for allogeneic stem cell donors and cancer patients in remission. Alternatively, patients can be mobilized by giving G-CSF or GM-CSF after myelosuppression, usually with cyclophosphamide, and MNC are harvested when the leukocyte counts begin to recover. These mobilization regimens can provide $CD34^+$ cells in the 1-4% range, and have been increasingly used clinically because of the more rapid neutrophil and platelet engraftment observed with mobilized PB (mPB) stem cells [7]. Large numbers of mobilized $CD34^+$ cells (from 1-8 x 10^8) can be harvested from a single leukapheresis, providing adequate cells for laboratory or clinical studies. It is, however, important to note that the quality of $CD34^+$ cells from different sources may vary by phenotype and function [8].

Further purification may be required depending on the type and size of the culture (discussed in section 4). Enrichment of the $CD34^+$ population can be done with immunoselection methods using monoclonal antibodies (MAbs) attached to magnetic particles (Baxter Isolex® 50 or Miltenyi Biotec MACS) or acrylamide beads (CellPro Ceprate®). These systems are commercially available for laboratory use and enrich $CD34^+$ cells to 70-99% purity.

2.2 Preparation of Mature Neutrophils

Mature neutrophils from PB can be isolated by a combination of density gradient separation and sedimentation. MNC can be removed from blood by Ficoll-Hypaque separation [9]. The cell pellet containing neutrophils and red cells is then resuspended in a buffered gelatin solution such as Plasmagel® [10] or a dextran solution [11]. The red cells sediment after 30-60 minutes, and the leukocyte-rich supernatant is then aspirated and the cells are washed by centrifugation. Residual red cells can be removed by hypotonic lysis in distilled water, resulting in cell preparations that are >90% neutrophils.

2.3 Maintenance of Mature Cells *In Vitro*

Neutrophils are end stage cells that become non-proliferative as they mature, and then die within hours or days of entering tissues. Because of this, timing and culture conditions become important in maintaining these cells *in vitro*. Freshly isolated circulating neutrophils have a short half-life *in vitro* of about 24 hours. This *in vitro* survival can be enhanced by the addition of G-CSF [12], GM-CSF or interferon-γ (IFN- γ) [13], as well as by culturing neutrophils on a fibroblast monolayer [14]. In contrast, *in vitro* derived neutrophils can be maintained for longer periods (up to 35 days) [15], provided that the medium contains serum and the appropriate growth factors [16,17]. This suggests that the removal of neutrophils *in vivo* is an active process, and *in vitro* systems to differentiate neutrophils provide a model for the acquisition of function and the mechanism of neutrophil clearance.

3. ASSAY TECHNIQUES

3.1 CD34$^+$ Cells

Flow cytometry techniques have been developed to quantitate CD34$^+$ cells in transplant products. A simple technique was originally described by Siena et al. [18] which displayed CD34 positive cells versus side-scatter, and a cluster of cells in the low side scatter region was quantitated. This method is widely used and has been standardized in a Nordic workshop [19]. A more complex protocol was developed by Sutherland et al. [20] using two-color CD45 and CD34 staining and a multidimensional gating strategy to identify a cluster of CD34$^+$ cells. More recently, automated methods to enumerate CD34$^+$ cells have been developed using flow cytometry, as well as microvolume fluorimetry [21]. In general, trained technologists starting with good quality cell samples will get equivalent results using any of these methods. The technique for performing CD34 analysis will therefore depend upon training and experience of the technologists and the instrumentation that is available.

3.2 Measurement of Neutrophil Differentiation and Function

3.2.1 Defining differentiation stages by morphology

Morphological studies using Wright–Giemsa staining were initially used to define the intermediate stages as neutrophils proliferate and differentiate into mature cells. These morphological stages were further grouped into mitotic (proliferative) and maturation (non-proliferative) compartments [22].

The mitotic compartment includes the myeloblast, promyelocyte and myelocyte stages. The myeloblast stage with an undifferentiated blast morphology represents >3% of BM [23] and has a transit time of about 18 hours. The promyelocyte stage is the first morphological stage of neutrophils, and is defined by the appearance of primary granules that contain myeloperoxidase. It represents 2-4% of BM [23] with a transit time of 24 hours. The myelocyte stage denotes onset of the formation of secondary (or azurophilic) granules, has a transit time of 104 hours, representing 3-4 cell divisions, and comprises approximately 13% of BM nucleated cells [23].

The maturation compartment includes the metamyelocyte, band stage and mature segmented neutrophil, and accounts for approximately 40% of nucleated cells in BM [23]. These non-dividing cells show progressive condensation of nuclear chromatin resulting in a lobular nuclear morphology, typically with three lobes. Morphologically, these stages define the appearance of tertiary granules containing acid phosphatase and sulfated glycosaminoglycans. The transit time through this compartment is about 4-6 days, giving a total transit time from myeloblast to mature neutrophil of 10-12 days. Interestingly, this represents the time frame of neutropenia following high dose chemotherapy, in which the proliferating cells are eliminated from BM by anti-proliferative chemotherapy drugs.

This same morphological approach to classification can be applied to *in vitro* differentiated neutrophil populations, with a few caveats. Culture-derived neutrophil precursors generally have different nuclear morphology and nucleus to cytoplasm ratio than their textbook counterparts [15]. These differences can make it difficult to do a "textbook" differential with cultured cells, and some ingenuity is required in defining the morphological categories observed in culture.

3.2.2 Defining differentiation stages by antigen expression

More recently, surface antigen expression in combination with flow cytometry has been used to characterize the differentiation stages of neutrophils. This approach was developed with the discovery and classification of the leukocyte surface antigens and has become a powerful way to provide rapid information on cell composition in culture.

Studies of neutrophil differentiation were initially done on normal BM, indicating that CD15 was one of the first antigens expressed as cells differentiate from $CD34^+$ myeloblasts to promyelocytes [24]. Other antigens such as CD64 and CD67 are also expressed at this time, with CD64 being present on an earlier $CD34^+$ CFU-GM phenotype [25]. Two-color staining with CD15 and CD11b was then used to define the differentiation of neutrophils and monocytes [24]. $CD34^+$ cells are largely CD15 and CD11b negative. As they differentiate into the neutrophil lineage, they express CD15 at the promyelocytic and myelocytic stage, and later acquire CD11b as they continue differentiating into metamyelocytes, bands and segmented neutrophils. In contrast, $CD34^+$ cells differentiating into the monocytic lineage express CD11b first, and later CD15 at a lower level than that observed on neutrophils. The final maturation into fully mature bands and segmented neutrophils occurs with the expression of CD16 [24].

We and others have used antigen expression to recapitulate this maturation sequence and monitor neutrophil differentiation in cultures of MNC or selected $CD34^+$ cells stimulated with various combinations of factors [15,26,27]. Isolated $CD34^+$ cells express very little CD15, do not express CD11b, and are not highly proliferative in short term cultures of 5-7 days [15]. $CD34^+$ cells can be expanded in the granulocytic lineage when cultured with interleukin-3 (IL-3), c-kit ligand (KL), G-CSF, and GM-CSF. In this pathway, $CD34^+$ cells first acquire the CD15 antigen and later CD11b [15,24,27] (Fig. 1). After ten days in culture, a $CD15^+CD11b^-$ phenotype is present which is highly proliferative, and when sorted this population contains primarily promyelocytes and myelocytes (Fig. 2). By 21 days in culture, the cells become doubly positive, expressing both CD11b and CD15, and represent the mature stages of metamyelocytes, bands and segmented neutrophils (Fig. 1).

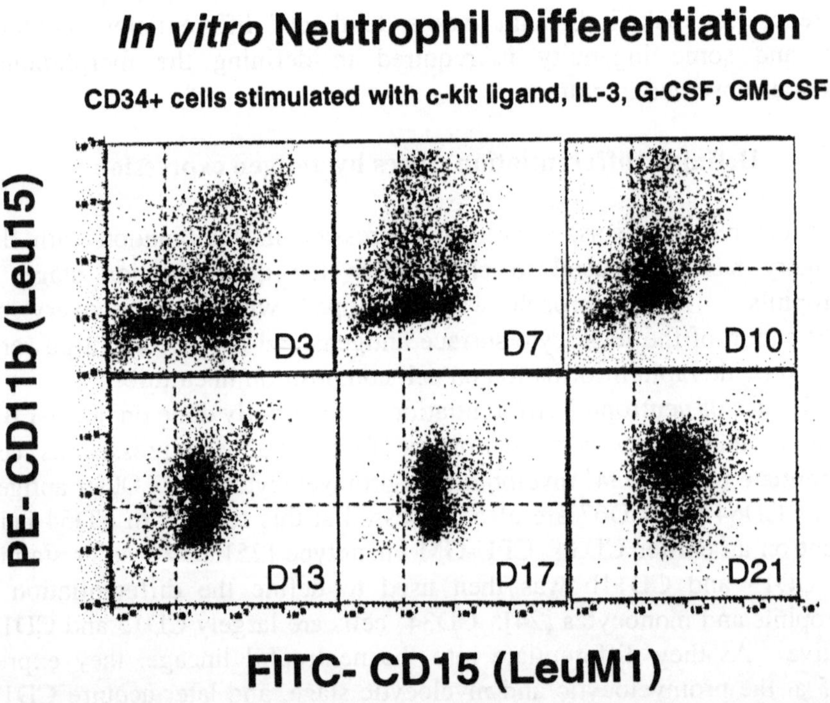

Figure 1. Characterization of CD11b/CD15 expression on $CD34^+$ cells differentiated *in vitro* to neutrophils. Shown are dot plots at different time points from a culture of $CD34^+$ cells stimulated with KL, IL-3, GM-CSF, and G-CSF in human long-term culture medium containing 25% animal serum.

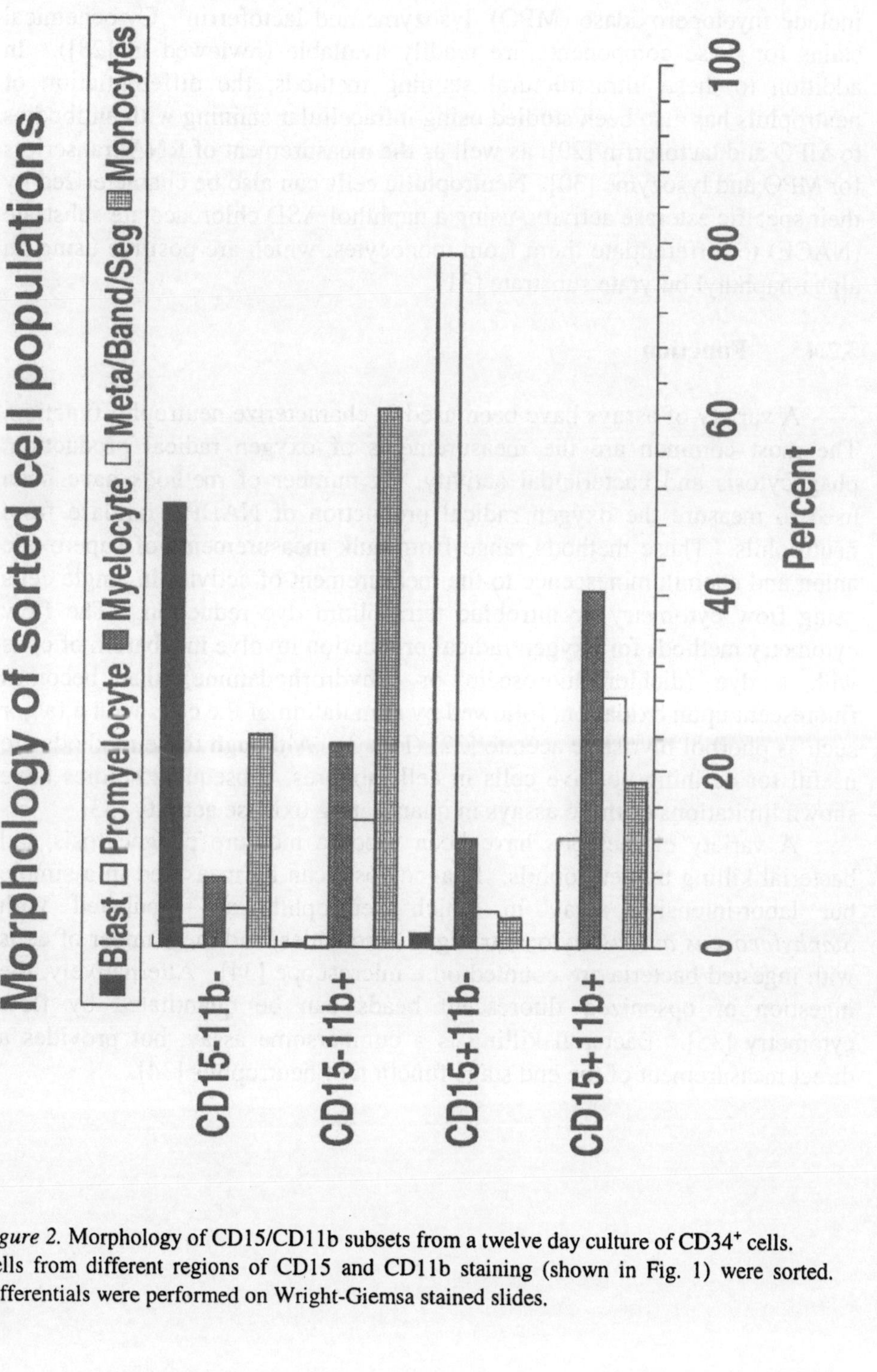

Figure 2. Morphology of CD15/CD11b subsets from a twelve day culture of CD34$^+$ cells. Cells from different regions of CD15 and CD11b staining (shown in Fig. 1) were sorted. Differentials were performed on Wright-Giemsa stained slides.

3.2.3 Granule constituents

The granule markers relevant to the study of neutrophil differentiation include myeloperoxidase (MPO), lysozyme and lactoferrin. Cytochemical stains for these components are readily available (reviewed in [28]). In addition to these ultrastructural staining methods, the differentiation of neutrophils has also been studied using intracellular staining with antibodies to MPO and lactoferrin [29], as well as the measurement of RNA transcripts for MPO and lysozyme [30]. Neutrophilic cells can also be characterized by their specific esterase activity, using a naphthol ASD chloroacetate substrate (NACE) to differentiate them from monocytes, which are positive using an alpha-naphthyl butyrate substrate [31].

3.2.4 Function

A variety of assays have been used to characterize neutrophil function. The most common are the measurements of oxygen radical production, phagocytosis and bactericidal activity. A number of methods have been used to measure the oxygen radical production of NADPH oxidase from neutrophils. These methods range from bulk measurements of superoxide anion and chemiluminescence to the measurement of activity in single cells using flow cytometry or nitroblue tetrazolium dye reduction. The flow cytometry methods for oxygen radical production involve incubation of cells with a dye (dichlorofluoroscein or dihydrorhodamine) that becomes fluorescent upon oxidation, followed by stimulation of the cells with a factor such as phorbol myristate acetate [32] (Fig. 3). Although these methods are useful for defining positive cells in cell mixtures, subsequent studies have shown limitations of these assays in quantitating oxidase activity [33].

A variety of methods have been used to measure phagocytosis and bacterial killing by neutrophils. Phagocytosis can be measured in a simple but labor-intensive assay in which neutrophils are incubated with *Staphylococcus aureus*, cytocentrifuged onto slides, and the number of cells with ingested bacteria are counted on a microscope [34]. Alternatively, the ingestion of opsonized fluorescent beads can be quantitated by flow cytometry [35]. Bacterial killing is a cumbersome assay, but provides a direct measurement of the end stage function of neutrophils [34].

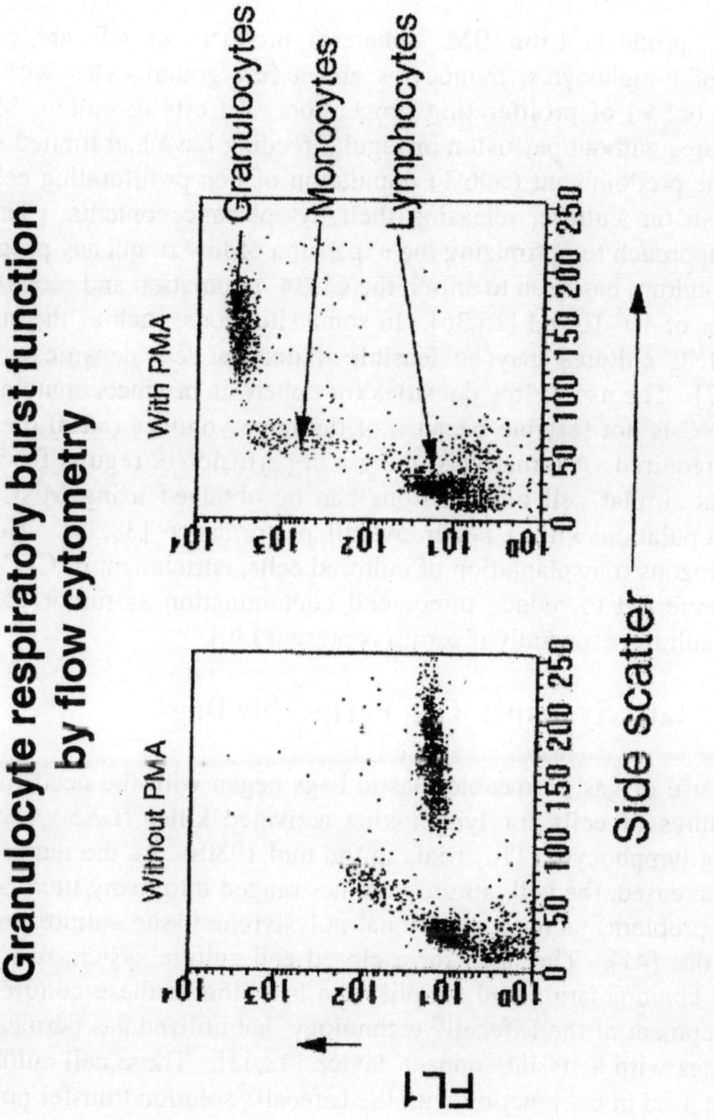

Figure 3. Measurement of granulocyte respiratory burst function by flow cytometry. Leukocytes from normal PB were isolated by lysing the red blood cells, and were then incubated in 100 μM dihydrorhodamine (DHR) for 30 minutes with 5 U of catalase. They were stimulated with 100 ng/ml of phorbol myristate acetate (PMA) for 30 minutes and analyzed by flow cytometry. Shown are the leukocyte populations defined by side scatter and the fluorescence of oxidized DHR in FL1.

4. CULTURE TECHNIQUES

4.1 System Concepts

MNC products from BM, apheresis products or CB are complex mixtures of lymphocytes, monocytes and a few granulocytes with a low frequency (<1%) of proliferating progenitors. Efforts to culture MNC in static cultures without perfusion or regular feeding have had limited success because the predominant (>96%) population of non-proliferating cells dies out early in the culture, releasing their cytoplasmic contents. The most effective approach to optimizing the expansion of low frequency progenitors in a static culture has been to enrich the $CD34^+$ population and seed the cells at densities of 10^4-10^5/ml [15,36]. In some situations, such as the culturing of CB, MNC cultures may be feasible if done at low densities of 1-2 x 10^5/ml [37]. The use of low densities for apheresis products containing 1-6 x 10^{10} MNC is not feasible because of the huge volumes (>100 liters) that would be required. Alternatively, the use of perfusion or regular feeding has shown that similar cell compositions can be obtained using MNC as the starting population with a better overall performance [38,39]. However, with autologous transplantation of cultured cells, enrichment of $CD34^+$ cells may be preferred to reduce tumor cell contamination as tumor cells may persist in culture, especially if serum is present [40].

4.2 Static Systems: Gas Permeable Bags

The use of gas permeable plastic bags began with the need for large-scale cultures of cells for lymphokine activated killer (LAK) and tumor infiltrating lymphocyte (TIL) trials in the mid 1980s. As the required dose of cells increased, the cell culture volumes ranged into many liters, creating logistical problems using conventional polystyrene tissue culture flasks and roller bottles [41]. The need for a closed cell culture system to minimize microbial contamination and simplify the handling of these cultures led to the development of the Lifecell® technology that utilized gas permeable cell culture bags with a sterile connect device [42,43]. These cell culture bags have been used in conjunction with the Lifecell® solution transfer pump, and provide an easy way to handle large cultures with a reduced risk of contamination to the cells, as well as reduced exposure of biohazards to the medical staff.

4.3 Cell Culture Bag Materials

Two companies that provide commercial bags for cell culture are Nexell Therapeutics, Inc. (Irvine, CA, distributed through Baxter Healthcare Corp.) and American Fluoroseal Corp. (Silverspring, MD). Nexell's culture bag is sold under the trade name of Lifecell®(PL732) while American Fluoroseal markets Teflon® cell culture bags. In addition, bags marketed through Nexell as Cryocyte bags (PL269) have also been used for cell culture [44].

An important consideration is the gas permeability of the material. A limited number of bag materials are available that will provide the appropriate gas exchange necessary for cell cultures. Culture performance depends on the degree to which oxygen (and carbon dioxide) can diffuse through the material. Cellular metabolism can vary greatly between cells in a culture as well as at different points during the culture period, resulting in variations in nutrient and oxygen requirements. Progenitor cell cultures have demonstrated better cell proliferation in static cultures under low oxygen conditions (5%), with 5% carbon dioxide [39].

Other factors need to be considered when using plastic bags for cell culture. Though gas permeability is an important consideration, surface characteristics and other properties are equally important, with MNC cultures being less sensitive to material effects than enriched $CD34^+$ cells [45]. Recently, the surface properties found in polystyrene flasks have been incorporated into a gas permeable material using a co-extrusion process that provides a thin layer of polystyrene on the interior surface of the PL732 material used in Lifecell® bags. This material (known as PL2417) has better gas permeability properties than PL732, performance equivalent to Teflon, and has been scaled up for clinical culture of $CD34^+$ cells [46,47].

4.4 Optimal Volumes and Cell to Surface Area Ratio

Utilizing the optimal culture medium depth is important for proper gas diffusion and delivery of oxygen to cells lying on the surface. Maximum bag capacities are much greater than the optimal culture volume. In general, optimal culture volumes are approximately 0.25-1.0 ml/cm^2 of bag surface area.

$CD34^+$ cells proliferate better when initiated at 1-10 x 10^4 cells/ml and the volume to surface area ratio is less than 0.5 ml/cm^2. In general, static bag cultures require feeding after 5-7 days due to the depletion of nutrients such as glucose, pyruvate, amino acids and lipids, and the build up of waste products such as lactate and ammonium ion. Bag cultures can be initiated at low volumes and then fed during the culture period by simply adding more medium and recombinant growth factors. CD34$^+$ cells from apheresis products were cultured in PL2417 prototype bags with a surface area of 200 cm^2 at an initial cell concentration of 1 x 10^4 cells/ml, and at volumes of either 50 ml (0.25 ml/cm^2) or 100 ml (0.5 ml/cm^2). In three experiments, after twelve days in culture without feeding, cells started at 50 ml proliferated better than those started at 100 ml (54x vs. 33x, respectively). Other studies have demonstrated that better cell proliferation occurs when CD34$^+$ cells receive one or more feedings of fresh culture medium and recombinant growth factors during a twelve day culture period [17]. The value of dilutional feeding with CD34 cultures was shown using a bag with a surface area of 750 cm^2, started at a low volume of 200 ml (0.3 ml/cm^2), and then expanded to 800 ml (1.1-1.3 ml/cm^2) by adding 200 ml every three days during a twelve day culture period. These cultures generated twice the number of cells when compared to a culture where the cells were diluted in 800 ml at the beginning of the culture and not fed [48].

4.5 Perfusion Systems

Perfusion systems for cell culture are widely used to produce soluble proteins and other biologics. Only recently have there been efforts to develop perfusion systems to produce therapeutic cell products. There are a variety of design issues when the cells in a culture become the desired product. These include providing a homogeneous environment and providing for easy inoculation and retrieval of the cells. Approaches that address these issues include flat plate chambers with radial or rectangular geometries [38,49] and stirred suspension chambers [50,51]. Flat plate chambers have used membranes [49] or smooth surfaces [52] to provide for stromal development and thereby retain the cells. Alternatively, parallel grooves that are perpendicular to the fluid flow have been successfully used to retain and perfuse nonadherent cells [38]. More traditional designs such as hollow fiber cartridges have been successfully used to culture lymphocytes [53], but have not been widely used for CD34 cultures because of biocompatibility issues associated with the materials used, as well as issues of maintaining a homogeneous environment.

Beyond the fundamental issues of environment and cell retrieval is the question of whether perfusion derives the same cell population obtained with a static method, or with a method using a fractionated population such as the CD34 selected preparation. Studies comparing MNC cultured with perfusion have been shown to result in neutrophil precursor preparations similar to those obtained using CD34$^+$ cells in static culture [38,54].

4.6 Growth Factors and Medium

Many studies have demonstrated the *in vitro* proliferation of $CD34^+$ cells stimulated with recombinant cytokines, and there are a number of excellent reviews of this area [55-57]. A primary observation is that combinations of cytokines are integral for optimally differentiating $CD34^+$ cells to neutrophils *in vitro* [26,27,34,58]. The best combinations include early-acting factors such as KL and flt-3 ligand (FL) in combination with multilineage factors such as IL-3 [34,36,59], and finally the more neutrophil-specific G-CSF and GM-CSF [26]. Studies of factor combinations to expand neutrophilic precursors indicate that a limited combination can give adequate results [60].

Medium is the other important component of the culture system. With growing regulatory concern for disease transmission and immunization to foreign proteins, the use of animal serum in cell culture medium for clinical use is coming under greater scrutiny. These concerns have led to the development of serum-free and defined media for hematopoietic cell culturing. A variety of formulations have been described [61], and several commercial media are available from BioWhittaker (Walkersville, MD), Life Technologies (Gaithersburg, MD), and Quality Biologicals (Gaithersburg, MD). These media are proprietary and usually have complex base formulations such as McCoy's or Iscove's Modified Dulbecco's Medium (IMDM), and add human serum albumin as a protein source. Although these media support the expansion of progenitors [61], terminal maturation of neutrophils does not occur in the absence of serum. An alternative to the use of serum-free medium or the addition of animal serum has been to add autologous plasma or pooled AB serum [34,62,63]. The logistics of preparing and utilizing pooled serum components as culture supplements makes these reagents a difficult choice for wide-scale implementation of a therapy. Autologous plasma, however, is readily available from the apheresis process, and therefore is a potential medium supplement. However, significant variation has been observed in the effects of adding autologous plasma to cultures of $CD34^+$ cells, possibly because of the variation in mobilization regimens and patient variability [64].

We have used combinations of growth factors, with or without the presence of serum, to examine the distribution of neutrophil precursor phenotypes based upon the expression of CD11b and CD15 [15]. In these studies, the distribution of cells expressing CD15 and CD11b was affected by the growth factor combination, as well as the presence of serum. Cultures containing G-CSF resulted in populations with a higher proportion of double positive ($CD15^+CD11b^+$) phenotypes, indicating that G-CSF drives terminal differentiation (Fig. 4). Likewise, serum-free cultures only partially support the late phenotypes of bands and segmented neutrophils, suggesting that additional factors are required for terminal differentiation. The addition of KL and either IL-6 or G-CSF significantly enhanced the total cell numbers that can be obtained compared to using PIXY321 (IL-3/GM-CSF fusion protein, Immunex, Seattle, WA).

4.7 Characterization of *In Vitro* Derived Neutrophils

A major difference observed with *in vitro* derived neutrophils was the low expression of CD16 (less than 20%) normally expressed on mature granulocytes [15,27]. Likewise, differences have been observed in phenotypes when different growth factor combinations are used [65], as well as when different media formulations are used. Significant differences are seen when G-CSF is compared with GM-CSF, with G-CSF stimulating more terminal differentiation of function as well as antigen expression [26,66]. Serum-free media typically do not sustain neutrophil differentiation to mature granulocytes [34], and the density of some antigens may be reduced [47]. Together, these data suggest that additional cytokines or other nutrients in serum are required to drive terminal differentiation of granulocytes *in vitro*. One example of a cytokine that may be involved is IFN-γ. It has been reported to act on the progenitor stage to affect the differentiation and function of mature cells [67].

We have characterized the growth factor requirements of a $CD15^+CD11b^-$ population cultured from $CD34^+$ cells selected from an apheresis product. When single clone-sorted $CD15^+CD11b^-$ cells are cultured in a four growth factor combination: IL-3, KL, G-CSF, GM-CSF, about 15% are capable of high proliferation, greater than eight cells/well. When grown in two growth factors, the G-CSF/GM-CSF combination outperforms the IL-3/KL combination. Further, when grown in a single growth factor, G-CSF supports the most proliferation.

We have also studied the proliferative capacity of CD15/CD11b phenotypes from serum-free cultures [16]. To estimate the ability of these phenotypes to proliferate upon re-infusion, sorted populations of the CD15/CD11b phenotypes from serum-free cultures were sorted into a secondary culture containing 25% animal serum and IL-3, GM-CSF, G-CSF and KL. Although a ten-fold increase in cell numbers was observed from the double negative phenotype ($CD15^-CD11b^-$), the single positive population ($CD15^+CD11b^-$) accounted for the majority of the proliferation for this cell product, proliferating an additional thirty-five fold in the secondary culture containing maturing neutrophil phenotypes. These findings were confirmed by clone sorting studies in which individual cells from these phenotypes were deposited into single wells and the number of progeny counted after seven days.

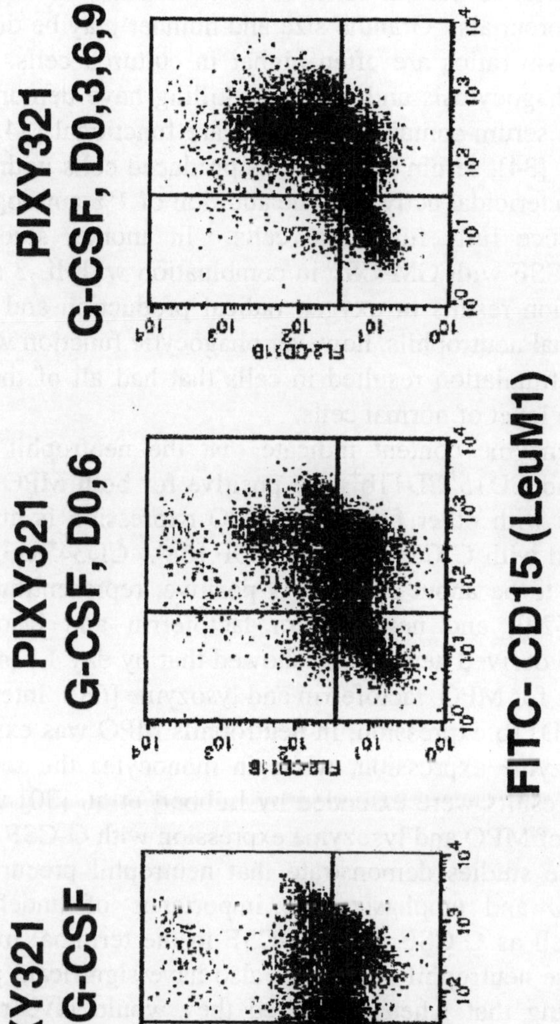

Figure 4. Effect of G-CSF and feeding schedules on the phenotype of cells produced from cultured CD34⁺ cells. CD34⁺ cells from G-CSF mobilized normal volunteers were cultured in X-VIVO™ 10 serum-free medium with 1% human serum albumin, 100 ng/ml PIXY321, with or without feeding with 6 ng/ml final concentration of G-CSF as indicated for twelve days, and the CD15/11b phenotype pattern was determined.

Besides surface antigen expression, morphology and function have been used to characterize cultured granulocytic cells. In general, the morphology is similar to normal cells, but not identical, causing confusion in performing differentials. Granule size and number may be different, and nuclear to cytoplasm ratios are often higher in cultured cells. Studies of chemotaxis and phagocytosis and oxidative killing have demonstrated that cells derived from serum-containing cultures are functional [34,66]. In the study by Lill et al. [34], serum free cultures produced cells with diminished phagocytic and bactericidal activity. The addition of 1% autologous plasma was able to produce fully functional cells. In another study [66], the comparison of G-CSF with GM-CSF in combination with IL-3 showed that GM-CSF stimulation results in oxygen radical production and chemotaxis equivalent to normal neutrophils, however phagocytic function was poor. In contrast, G-CSF stimulation resulted in cells that had all of the functions, but were not at the level of normal cells.

Studies of enzyme content indicate that the neutrophil phenotypes ($CD15^+CD11b^-$ and $CD15^+CD11b^+$) are positive for both MPO and NACE [15] (Fig. 5). As with other functions, MPO expression is augmented in cultures stimulated with G-CSF and GM-CSF [68]. $CD34^+$ cells from BM have been shown to be about 35% MPO positive, representing committed GM-progenitors [29], and negative for lactoferrin and lysozyme [68]. Studies of *in vitro* derived neutrophils showed that by day 14 of culture, all cells were positive for MPO, lactoferrin and lysozyme [68]. Interestingly, as with CD15 and CD11b expression, in neutrophils MPO was expressed first followed by lysozyme expression, while in monocytes the sequence was reversed. These results were extended by Lubbert et al. [30] who showed the augmentation of MPO and lysozyme expression with G-CSF.

Together, these studies demonstrate that neutrophil precursors can be produced *in vitro* and emphasize the importance of undefined serum components as well as G-CSF and GM-CSF in the terminal maturation of neutrophils. These neutrophil precursors also have significant proliferative capacity, suggesting that when re-infused they would give rise to large numbers of mature cells *in vivo*.

5. UTILITY OF SYSTEM

Neutropenia following high dose chemotherapy for the treatment of cancer is a major cause of infection, prolonged hospitalization and increased risk of mortality. Granulocytes harvested from allogeneic donors have been used to treat neutropenia [69], and recently mobilization of donors with G-

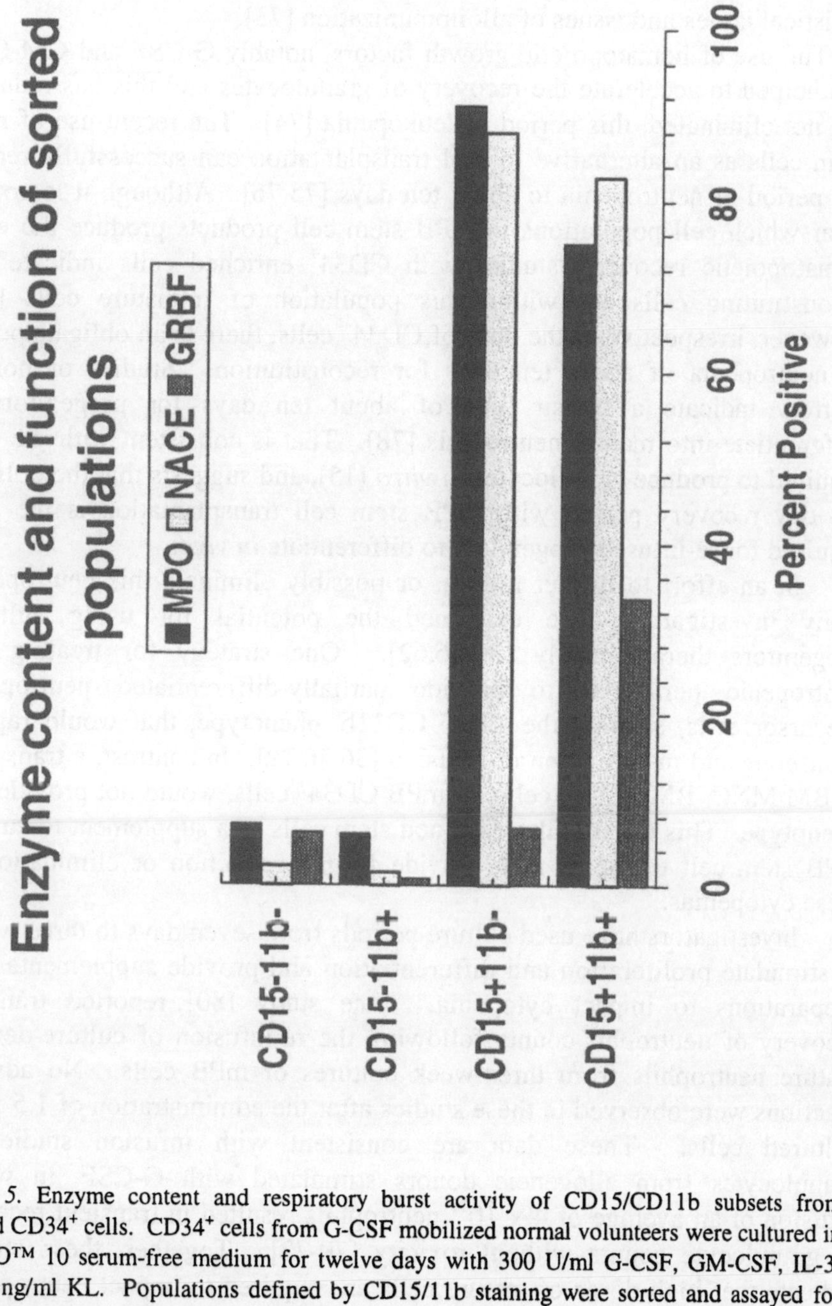

Figure 5. Enzyme content and respiratory burst activity of CD15/CD11b subsets from cultured CD34$^+$ cells. CD34$^+$ cells from G-CSF mobilized normal volunteers were cultured in X-VIVO™ 10 serum-free medium for twelve days with 300 U/ml G-CSF, GM-CSF, IL-3, and 10 ng/ml KL. Populations defined by CD15/11b staining were sorted and assayed for myeloperoxidase (MPO), specific esterase (naphthol ASD chloroacetate esterase, NACE) and granulocyte respiratory burst activity (GRBF).

CSF has been explored as a way to increase the number of granulocytes collected [70-72]. This approach, however, has not been widely used due to logistical issues and issues of alloimmunization [73].

The use of hematopoietic growth factors, notably G-CSF and GM-CSF, has helped to accelerate the recovery of granulocytes and this has reduced, but not eliminated, this period of leukopenia [74]. The recent use of mPB stem cells as an alternative to BM transplantation can successfully reduce the period of neutropenia to about ten days [75,76]. Although it is not yet clear which cell populations in mPB stem cell products produce the early hematopoietic recovery, studies with $CD34^+$ enriched cells indicate that reconstituting cells are within this population of immature cells [77]. However, irrespective of the dose of $CD34^+$ cells, there is an obligate period of neutropenia of about ten days for reconstitution. Studies of normal marrow indicate a transit time of about ten days for progenitors to differentiate into mature neutrophils [78]. That is consistent with the time required to produce granulocytes *in vitro* [15], and suggests that the obligate ten day recovery period with mPB stem cell transplantation is the time required for re-infused progenitors to differentiate *in vivo*.

In an effort to further reduce, or possibly eliminate this neutropenia, many investigators have examined the potential for using cultured progenitors therapeutically [36,46,62]. One strategy for treating this neutropenic period is to provide partially-differentiated neutrophilic precursor cells, such as the $CD15^+CD11b^-$ phenotype, that would rapidly proliferate and mature upon re-infusion [36,46,79]. In contrast, a transplant of BM MNC, BM $CD34^+$ cells, or mPB $CD34^+$ cells, would not provide this phenotype. This use of cultured blood stem cells as a supplement to current mPB stem cell therapies may provide further reduction or elimination of these cytopenias.

Investigators have used culture periods from seven days to three weeks to stimulate proliferation and differentiation and provide supplemental cell preparations to impact cytopenia. One study [80] reported transient recovery of neutrophil counts following the re-infusion of culture-derived mature neutrophils from three-week cultures of mPB cells. No adverse reactions were observed in these studies after the administration of 1.5×10^9 cultured cells. These data are consistent with infusion studies of granulocytes from allogeneic donors stimulated with G-CSF, in which infusion of an average of 4×10^{10} neutrophils resulted in transient recovery of granulocyte counts without toxicity [70-72]. Together, these studies indicate that high doses of mature cells are required to impact neutropenia, and that this can be done safely.

In another study, $CD34^+$ cells were expanded in PIXY321 (IL-3/GM-CSF) to produce neutrophil precursors [46]. In this study, patients with metastatic breast cancer underwent PB mobilization and $CD34^+$ cells from the second apheresis were enriched and cultured in serum-free medium and PIXY321. The cells were cultured in one-liter volumes in gas permeable bags for twelve days with the addition of medium and growth factor at day 7. A median 16.9-fold expansion of cell numbers was observed. These patients then underwent high dose chemotherapy and cultured cells were re-infused one day after the unmanipulated cryopreserved cells. Seventeen patients were re-infused with cultured cells at doses up to 1.6×10^8 /kg with no adverse reactions. A trend toward faster recovery and less neutropenic fever was observed with higher doses of cultured cells.

Studies have shown that mature neutrophils harvested by apheresis after G-CSF stimulation can effectively migrate to sites of inflammation and infection [81]. To study the fate of cultured neutrophil precursors, an aliquot of cultured cells radiolabeled with ^{111}Indium was re-infused and tracked [82]. Immediately upon reinfusion, localization to the lungs was observed. However, 24 and 48 hours after re-infusion, considerable activity was noted in the pelvis and vertebral bodies, indicating that *ex vivo* cultured neutrophil precursor cells can home to the BM.

6. CONCLUDING REMARKS

It is clear from the studies described here that proliferating and differentiating $CD34^+$ cells into neutrophils *in vitro* can be done easily, provided the appropriate factors, medium components and feeding schedules are chosen. Significant characterization of these cells has been done showing the similarities to and differences from normal cells. Limited clinical studies have been performed examining the potential therapeutic value of these cell phenotypes. A major issue is the complexity of hematopoietic cultures. Differentiation stages really represent a dynamic continuum rather than the discrete populations we work to define. Likewise, the number of factors that are shown to regulate differentiation increases yearly. Indeed, a review in 1996 by Broxmeyer [83] estimated the number of factors regulating hematopoiesis to be fifty-five. Because so many variables are involved in developing a system to produce therapeutic cells in this lineage, it becomes difficult to design clinical trials since it is unclear which maturation stage may be the most appropriate. To address this, some investigators have looked to computer modeling of neutrophil development to provide information as to what phenotypes may be useful [84]. This type of modeling is attractive because data from *in vitro* studies can be used to refine the assumptions that are used to predict the performance of cell populations *in vivo*.

Significant variability in expansion is also an important issue in utilizing these cell populations routinely in clinical practice. The source of this variability is not well understood. One source may be related to intrinsic properties of the CD34 population related to prior therapies or inherent variability between individuals. Another is the adverse effect of the *ex vivo* processing prior to culture related to fractionation, hydrodynamic damage, hypoxic conditions or physical damage from enzymatic or mechanical release of cells. Variability in the culturing process may also result from differential consumption of factors or nutrients, as well as density or environmental effects from non-homogeneous distribution of cells within the culture. When non-fractionated cells such as MNC are used, there can also be effects of accessory cells, depending on their composition. Finally, materials and surfaces can also affect the outcome, with progenitors preferring stromal contact for optimal proliferation.

Overall, the concept of using *ex vivo* expansion to produce therapeutic neutrophil precursors for treating neutropenia is attractive for many reasons, and is part of a larger field of *ex vivo* cell and tissue engineering. At this point, only a few clinical trials and theoretical modeling have suggested the value of the approach of culturing hematopoietic cells.

7. LITERATURE CITED

1. Warner HR and Athens JW. (1964) An analysis of granulocyte kinetics in blood and BM. Ann NY Acad Sci *113*:523-527.
2. Boggs DR. (1967) The kinetics of neutrophilic leukocytes in health and disease. Semin Hematol *4*:359-386.
3. Malech HL and Nauseef WM. (1997) Primary inherited defects in neutrophil function: etiology and treatment. Semin Hematol *34*:279-290.
4. Offner F, Schoch G, Fisher LD, Torok-Storb B, and Martin PJ. (1996) Mortality hazard functions as related to neutropenia at different times after marrow transplantation. Blood *88*:4058-4062.
5. Bradley TR and Metcalf D. (1966) The growth of mouse BM cells *in vitro*. J Exp Med Sci *44*:297-304.
6. Bender JG, Unverzagt KL, Walker DE, Lee W, Van Epps DE, Smith DH, Stewart CC, and To LB. (1991) Identification and comparison of CD34-positive cells and their subpopulations from normal PB and BM using multicolor flow cytometry. Blood *77*:2591-2596.
7. To LB, Haylock DN, Simmons PJ, and Juttner CA. (1997) The biology and clinical uses of blood stem cells. Blood *89*:2233-2258.
8. To LB, Haylock DN, Dowse T, Simmons PJ, Trimboli S, Ashman LK, and Juttner CA. (1994) A comparative study of the phenotype and proliferative capacity of PB (PB) $CD34^+$ cells mobilized by four different protocols and those of steady-phase PB and BM $CD34^+$ cells. Blood *84* :2930-2939.
9. Boyum A. (1968) Isolation of MNC and granulocytes from human blood. Scand J Clin Lab Invest *21*:77-81.

10. Bender JG and Van Epps DE. (1983) Analysis of the bimodal chemiluminescence pattern stimulated in human neutrophils by chemotactic factors. Infect Immun *41*:1062-1070.

11. Maloney CG, Kutchera WA, Albertine KH, McIntyre TM, Prescott SM, and Zimmerman GA. (1998) Inflammatory agonists induce cyclooxygenase type 2 expression by human neutrophils. J Immunol *160*:1402-1410.

12. Lopez AF, Williamson DJ, Gamble JR, Begley CG, Harian JM, Klebanoff SJ, Waltersdorph A, and Wong G. (1986) Recombinant human granulocyte-macrophage colony-stimulating factor stimulates *in vitro* mature human neutrophil and eosinophil function, surface receptor expression, and survival. J Clin Invest *78*:1220-1228.

13. Klebanoff SJ, Olszowski S, Van Voorhis WC, Ledbetter JA, Waltersdorph AM, and Schlechte KG. (1992) Effects of gamma interferon on human neutrophils: protection from deterioration on storage. Blood *80*:225-234.

14. Ling CJ, Owen WF, and Austen KF. (1990) Human fibroblasts maintain the viability and augment the functional response of human neutrophils in culture. J Clin Invest *85*:601-604.

15. Smith SL, Bender JG, Maples PB, Unverzagt K, Schilling M, Lum L, Williams S, and Van Epps DE. (1993) Expansion of neutrophil precursors and progenitors in suspension cultures of $CD34^+$ cells enriched from human BM. Experimental Hematology *21*:870-877.

16. Unverzagt KL, Bender JG, Loudovaris M, Martinson JA, Hazelton B, and Weaver C. (1997) Characterization of a culture-derived $CD15^+CD11b$- promyelocytic population from $CD34^+$ PB cells. J Leukoc Biol *62*:480-484.

17. Smith SL, Bender JG, Berger C, Lee WJ, Loudovaris M, Martinson JA, Opotowsky JD, Qiao X, Schneidkraut M, Sweeney P, Unverzagt KL, Van Epps DE, Williams DE, Williams SF, and Zimmerman TM. (1997) Neutrophil maturation of $CD34^+$ cells from PB and BM in serum-free culture medium with PIXY321 and granulocyte-colony stimulating factor (G-CSF). J Hematother *6*:323-334.

18. Siena S, Bregni M, Brando B, Belli N, Ravagnani F, Gandola L, Stern AC, Lansdorp PM, Bonadonna G, and Gianni AM. (1991) Flow cytometry for clinical estimation of circulating hematopoietic progenitors for autologous transplantation in cancer patients. Blood *77*:400-409.

19. Johnsen HE and Knudsen LM. (1996) Nordic flow cytometry standards for $CD34^+$ cell enumeration in blood and leukapheresis products: Report from the Second Nordic Workshop. J Hematother *5*:237-245.

20. Sutherland DR, Anderson L, Keeney M, Nayar R, and Chin-Yee I. (1996) The ISHAGE guidelines for $CD34^+$ cell determination by flow cytometry. J Hematother *5*:213-226.

21. Sims LC, Brecher ME, Gertis K, Jenkins A, Nickischer D, Schmitz JL, Sparks S, Wright P, and Bentley SA. (1997) Enumeration of CD34-positive stem cells: evaluation and comparison of three methods. J Hematother *6*:213-226.

22. Newburger, PE and Parmley RT. (1998). White blood cells, in R Hoffman, EJ Benz, SJ Shattil, B Furie, HJ Cohen and LE Silberstein, (eds.) *Hematology, Basic principles and practice*. Churchill Livingstone, Edinburgh,

23. Jandl JH. (1987). Blood, Textbook of Hematology. Little Brown and Company, Boston, p. 27.

24. Terstappen LW, Safford M, and Loken MR. (1990) Flow cytometric analysis of human BM. III. Neutrophil maturation. Leukemia *4*:657-663.

25. Olweus J, Lund Johansen F, and Terstappen LW. (1995) CD64/Fc gamma RI is a granulo-monocytic lineage marker on CD34$^+$ hematopoietic progenitor cells. Blood *85*:2402-2413.

26. Kerst JM, Slaper-Cortenbach ICM, von dem Borne AEG, van der Schoot CE, and van Oers RHJ. (1992) Combined measurement of growth and differentiation in suspension cultures of purified CD34 positive cells enables a detailed analysis of myelopoiesis. Exp Hematol *20*:1188-1193.

27. Terstappen LW, Buescher S, Nguyen M, and Reading C. (1992) Differentiation and maturation of growth factor expanded human hematopoietic progenitors assessed by multidimensional flow cytometry. Leukemia *6*:1001-1010.

28. Bainton DF. (1993) Neutrophilic leukocyte granules: from structure to function. Adv Exp Med Biol *336*:17-33.

29. Strobl H, Takimoto M, Majdic O, Fritsch G, Scheinecker C, Hocker P, and Knapp W. (1993) MPO expression in CD34$^+$ normal human hematopoietic cells. Blood *82*:2069-2078.

30. Lubbert M, Brugger W, Mertelsmann R, and Kanz L. (1996) Developmental regulation of myeloid gene expression and demethylation during *ex vivo* culture of PB progenitor cells. Blood *87*:447-455.

31. Nakamura K, Takahashi T, Sasaki Y, Tsuyuoka R, Okuno Y, Kurino M, Ohmori K, Iho S, and Nakao K. (1996) 1,25-dihydroxyvitamin D3 differentiates normal neutrophilic promyelocytes to monocytes/macrophages *in vitro*. Blood *87*:2693-2701.

32. Vowells SJ, Sekhsaria S, Malech HL, Shalit M, and Fleisher TA. (1995) Flow cytometric analysis of the granulocyte respiratory burst: a comparison study of fluorescent probes. J Immunol Methods *178*:89-97.

33. van Pelt LJ, van Zwieten R, Weening RS, Roos D, Verhoeven AJ, and Bolscher BG. (1996) Limitations on the use of dihydrorhodamine 123 for flow cytometric analysis of the neutrophil respiratory burst. J Immunol Methods *191*:187-196.

34. Lill MC, Lynch M, Fraser JK, Chung GY, Schiller G, Glaspy JA, Souza L, Baldwin GC, and Gasson JC. (1994) Production of functional myeloid cells from CD34-selected hematopoietic progenitor cells using a clinically relevant *ex vivo* expansion system. Stem Cells Dayt *12*:626-637.

35. Steinkamp JA, Wilson JS, Saunders GC, and Stewart CC. (1982) Phagocytosis: flow cytometric quantitation with fluorescent microspheres. Science *215*:64-66.

36. Haylock DN, To LB, Dowse TL, Juttner CA, and Simmons PJ. (1992) *Ex vivo* expansion and maturation of PB CD34$^+$ cells into the myeloid lineage. Blood *80*:1405-1412.

37. Bender JG, Smith SL, Williams SF, Zimmerman TM, Lee WJ, Loudovaris M, Schilling M, Unverzagt KL, and Van Epps DE. (1997) Large scale gas permeable bag system for the expansion of mobilised blood CD34$^+$ cells or CB cells. BM Transplant *19*:S159

38. Sandstrom CE, Bender JG, Papoutsakis ET, and Miller WM. (1995) Effects of CD34$^+$ cell selection and perfusion on *ex vivo* expansion of PB MNC. Blood *86*:958-970.

39. Koller MR, Bender JG, Papoutsakis ET, and Miller WM. (1992) Effects of synergistic cytokine combinations, low oxygen, and irradiated stroma on the expansion of human CB progenitors. Blood *80*:403-411.

40. Vogel W, Behringer D, Scheding S, Kanz L, and Brugger W. (1996) *Ex vivo* expansion of CD34$^+$ PB progenitor cells: Implications for the expansion of contaminating epithelial tumor cells. Blood *88*:2707-2713.

41. Muul LM, Director EP, Hyatt CL, and Rosenberg SA. (1986) Large scale production of human lymphokine activated killer cells for use in adoptive immunotherapy. J Immunol Methods *88*:265-275.

42. Muul LM, Nason Burchenal K, Carter CS, Cullis H, Slavin D, Hyatt C, Director EP, Leitman SF, Klein HG, and Rosenberg SA. (1987) Development of an automated closed system for generation of human lymphokine-activated killer (LAK) cells for use in adoptive immunotherapy. J Immunol Methods *101* :171-181.

43. Carter CS, Leitman SF, Cullis H, Muul LM, Nason Burchenal K, Rosenberg SA, and Klein HG. (1988) Technical aspects of lymphokine-activated killer cell production. J Clin Apheresis *4*:113-117.

44. Holyoake, T.L., M.J. Alcorn, L. Richmond, C. Pearson, E. Farrell, B. Kyle, D.J. Dunlop, E. Fitzsimons, W.B. Steward, I.B. Pragnell, and I.M. Franklin. 1995. A Phase 1 study to evaluate the safety of re-infusing CD34 cells expanded *ex vivo* as part or all of a PBPC transplant procedure. *Blood 86*:1161A.(Abstract)

45. LaIuppa JA, McAdams TA, Papoutsakis ET, and Miller WM. (1997) Culture materials affect *ex vivo* expansion of hematopoietic progenitor cells. J Biomed Mater Res *36*:347-359.

46. Williams SF, Lee WJ, Bender JG, Zimmerman T, Swinney P, Blake M, Carreon J, Schilling M, Smith S, Williams DE, Oldham F, and Van Epps DE. (1996) Selection and expansion of PB CD34$^+$ cells in autologous stem cell transplantation for breast cancer. Blood *87*:1687-1691.

47. Zimmerman TM, Bender JG, Lee WJ, Loudovaris M, Qiao X, Schilling M, Smith SL, Unverzagt KL, Van Epps DE, Blake M, and Williams DE. (1996) Large scale selection of CD34$^+$ PB progenitors and expansion of neutrophil precursors for clinical applications. J Hematotherapy *5*:247-254.

48. Martinson JA, Unverzagt K, Schaeffer A, Smith SL, Loudovaris M, Schneidkraut MJ, Bender JG, and Van Epps DE. (1998) Neutrophil precursor generation: effects of culture conditions. J Hematother 7:463-471.

49. Koller MR, Manchel I, Newsom BS, Palsson MA, and Palsson BO. (1995) Bioreactor expansion of human BM: Comparison of unprocessed, density-separated, and CD34 enriched cells. J Hematother *4*:159-169.

50. Zandstra PW, Eaves CJ, and Piret JM. (1994) Expansion of hematopoietic progenitor cell populations in stirred suspension bioreactors of normal human BM cells. Biotechnology *12*:909-914.

51. Sardonini CA and Wu YJ. (1993) Expansion and differentiation of human hematopoietic cells from static cultures through small-scale bioreactors. Biotechnol Prog *9*:131-137.

52. Koller MR, Bender JG, Miller WM, and Papoutsakis ET. (1993) Expansion of primitive human hematopoietic progenitors in a perfusion bioreactor system with IL-3, IL-6, and stem cell factor. Biotechnology N Y *11*:358-363.

53. Hillman GG, Wolf ML, Montecillo E, Younes E, Ali E, Pontes JE, and Haas GP. (1994) Expansion of activated lymphocytes obtained from renal carcinoma in an automated hollow fiber reactor. Cell Transplant *3*:263-271.

54. Koller MR, Palsson MA, Manchel I, and Palsson BO. (1995) Long-term culture-initiating cell expansion is dependent on frequent medium exchange combined with stromal and other accessory cell effects. Blood *86*:1784-1793.

55. McAdams TA, Sandstrom CE, Miller WM, Bender JG, and Papoutsakis ET. (1996) *Ex vivo* expansion of primitive hematopoietic cells for cellular therapies: An overview. Cytotechnology *18*:133-146.

56. Moore MA. (1995) Expansion of myeloid stem cells in culture. Semin Hematol *32*:183-200.
57. Emerson SG. (1996) *Ex vivo* expansion of hematopoietic precursors, progenitors and stem cells: The next generation of cellular therapeutics. Blood *87*:3082-3088.
58. Serke S, Sauberlich S, and Huhn D. (1991) A liquid culture method for the *in vitro* growth of hemopoietic progenitor cells from normal human adult PB allowing for analysis by multiparameter flow-cytometry. Eur J Haematol *46* :85-92.
59. Sato N, Sawada K, Koizumi K, Tarumi T, Ieko M, Yasukouchi T, Yamaguchi M, Takahashi TA, Sekiguchi S, and Koike T. (1993) *In vitro* expansion of human PB $CD34^+$ cells. Blood *82*:3600-3609.
60. Makino S, Haylock DN, Dowse T, Trimboli S, Niutta S, To LB, Juttner CA, and Simmons PJ. (1997) *Ex vivo* culture of PB $CD34^+$ cells: effects of hematopoietic growth factors on production of neutrophilic precursors. J Hematotherapy *6*:475-489.
61. Sandstrom CE, Collins PC, McAdams TA, Bender JG, Papoutsakis ET, and Miller WM. (1996) Comparison of whole serum-deprived media for the *ex vivo* expansion of hematopoietic progenitor cells from cord blood and mobilized peripheral blood mononuclear cells. J Hematotherapy *5*:461.
62. Brugger W, Heimfeld S, Berenson RJ, Mertelsmann R, and Kanz L. (1995) Reconstitution of hematopoiesis after high-dose chemotherapy by autologous progenitor cells generated *ex vivo*. New Eng J Med *333*:283-287.
63. Shapiro F, Yao TJ, Raptis G, Reich L, Norton L, and Moore MA. (1994) Optimization of conditions for *ex vivo* expansion of $CD34^+$ cells from patients with stage IV breast cancer. Blood *84*:3567-3574.
64. Loudovaris, M., S. Smith, M. Schilling, J. Martinson, W. Lee, J. Bender, D.E. Van Epps, and S. Williams. 1993. Human plasma as a supplement to media for the culture of human $CD34^+$ cells. *Blood 82:* (Abstract)
65. Scheding, S., H.J. Buhring, B. Ziegler, T. Bock, L. Kanz, and W. Brugger. 1995. Ex-vivo generation of myeloid post-progenitor cells from mobilized PB $CD34^+$ cells. *Blood 86*:912a.(Abstract)
66. Frenck RW, Ventura GJ, Krannig G, Felix EA, Zipf TF, VanWinkle WB, and Buescher ES. (1991) Studies of the function and structure of *in vitro* propagated human granulocytes. Pediatr Res *30*:135-140.
67. Ezekowitz RAB, Sieff CA, Dinauer MC, Nathan DG, Orkin SH, and Newburger PE. (1990) Restoration of phagocytic function by interferon-gamma in X-linked chronic granulomatous disease occurs at the level of a progenitor cell. Blood *76*:2443-2448.
68. Scheinecker C, Strobl H, Fritsch G, Csmarits B, Krieger O, Majdic O, and Knapp W. (1995) Granulomonocyte-associated lysosomal protein expression during *in vitro* expansion and differentiation of $CD34^+$ hematopoietic progenitor cells. Blood *86*:4115-4123.
69. Strauss RG. (1993) Therapeutic granulocyte transfusions in 1993. Blood *81*:1675-1678.
70. Bensinger WI, Price TH, Dale DC, Appelbaum FR, Clift R, Lilleby K, Williams B, Storb R, Thomas ED, and Buckner CD. (1993) The effects of daily recombinant human granulocyte colony stimulating factor administration on normal granulocyte donors undergoing leukapheresis. Blood *81*:1883-1888.
71. Caspar CB, Seger RA, Burger J, and Gmur J. (1993) Effective stimulation of donors for granulocyte transfusions with recombinant methionyl granulocyte colony-stimulating factor. Blood *81*:2866-2871.
72. Adkins D, Spitzer G, Johnston M, Velasquez W, Dunphy F, and Petruska P. (1997) Transfusions of granulocyte-colony-stimulating factor-mobilized granulocyte components

to allogeneic transplant recipients: analysis of kinetics and factors determining posttransfusion neutrophil and platelet counts. Transfusion *37*:737-748.
73. Stroncek DF, Leonard K, Eiber G, Malech HL, Gallin JI, and Leitman SF. (1996) Alloimmunization after granulocyte transfusions. Transfusion *36*:1009-1015.
74. Neidhart J, Mangalik A, Kohler W, Stidley C, Saiki J, Duncan P, Souza L, and Downing M. (1989) Granulocyte colony-stimulating factor stimulates recovery of granulocytes in patients receiving dose intensive chemotherapy without BM transplantation. J Clin Oncol *7*:1685-1692.
75. To LB, Roberts MM, Haylock DN, Dyson PG, Branford AL, Thorp D, Ho JQK, Dart GW, Horvath N, Davy MLJ, Olweny CLM, Abdi E, and Juttner CA. (1992) Comparison of haematological recovery times and supportive care requirements of autologous recovery phase PB stem cell transplants, autologous BM transplants and allogeneic BM transplants. BM Transplantation *9*:277-284.
76. Sheridan WP, Begley CG, Juttner CA, Szer J, To LB, Maher D, McGrath KM, Morstyn G, and Fox RM. (1992) Effect of PB progenitor cells mobilised by filgrastim (G-CSF) on platelet recovery after high-dose chemotherapy. Lancet *339* :640-644.
77. Shpall EJ, Jones RB, Bearman SI, Franklin WA, Archer PG, Curiel T, Bitter M, Claman HN , Stemmer SM, Purdy M, and et al. (1994) Transplantation of enriched CD34-positive autologous marrow into breast cancer patients following high-dose chemotherapy: influence of CD34-positive peripheral-blood progenitors and growth factors on engraftment. J Clin Oncology *12*:28-36.
78. Athens JW, Haab OP, and Raab SO. (1961) Leukokinetic studies IV. The total blood, circulating and marginal granulocyte pools and granulocyte turnover rate in normal subjects. J Clin Invest *40*:989-991.
79. Qiao X, Loudovaris M, Unverzagt K, Walker DE, Smith SL, Martinson J, Schilling M, Lee W, Williams SF, Van Epps DE, Cohen I, and Bender JG. (1996) Immunochemistry and flow cytometry evaluation of human megakaryocytes in fresh samples and cultures of $CD34^+$ cells. Cytometry *23*:250-259.
80. Gluck S, Chadderton T, Porter K, Dietz G, and Maruyama M. (1995) Characterization and transfusion of *in vitro* cultivated hematopoietic progenitor cells. Transfus Sci *16*:273-281.
81. Adkins D, Goodgold H, Hendershott L, Johnston M, Cravens D, and Spitzer G. (1997) Indium-labeled white blood cells apheresed from donors receiving G-CSF localize to sites of inflammation when infused into allogeneic BM transplant recipients. BM Transplant *19*:809-812.
82. Zimmerman, T.M., J.G. Bender, W.J. Lee, D.E. Van Epps, F.B. Oldham, and S.F. Williams. 1996. Infusion and localization of *ex vivo* expanded myeloid progenitors after high-dose chemotherapy in a patient with metastatic breast cancer. *Blood* *88*:605a.(Abstract)
83. Broxmeyer HE. (1996) Is interleukin 17, an inducible cytokine that stimulates production of other cytokines, merely a redundant player in a sea of other biomolecules? J Exp Med *183*:2411-2415.
84. Scheding, S., H. Franke, W. Brugger, L. Kanz, and S. Schmitz. 1995. How many post-progenitor cells have to be transplanted to completely abrogate neutropenia after high-dose chemotherapy and PB progenitor cell transplantation? Blood *86*:884A. (Abstract)

Chapter 9

Isolation and Culture of Eosinophils

Helene F Rosenberg
Laboratory of Host Defenses, National Institute of Allergy and Infectious Diseases, National Institutes of Health, Building 10, Room 11N104, 9000 Rockville Pike, Bethesda, MD 20892. Tel: 001-301-402-9131; Fax: 001-301-402-4369; E-mail: hr2k@nih.gov

1. INTRODUCTION

Eosinophils are leukocytes of the granulocyte series that are noted for their distinctive morphology and unusual staining properties. Although eosinophils are recognized for their detrimental contributions to the pathophysiology of asthma and other allergic disorders, the beneficial roles played by these cells remain poorly (if at all) understood. Although many texts still refer to eosinophils as providing host defense against helminthic parasites, recent findings have presented a dramatic challenge to this hypothesis [1-4]. Other roles for eosinophils currently under exploration include antigen presentation [5-7], wound healing [8, 9], tumoricidal activity [10-13] and other aspects of innate host defense [14-16].

The life cycle of the eosinophil has been described in detail [17]. Eosinophils develop in the bone marrow (BM) from pluripotent stem cells, and mature, non-dividing eosinophils are released into the circulation where they persist for several hours. Ultimately, eosinophils migrate into the tissues (primarily lung, gastrointestinal, and genitourinary), where they survive for several days. Blood and tissue eosinophilia results from increased production of mature eosinophils in BM, accompanied by a decreased rate of cell death, or apoptosis, in the periphery. Eosinophilia is typically associated with allergic disorders, parasitic infection, and specific idiopathic syndromes [18-24].

The mature peripheral blood (PB) eosinophil has a characteristic bi-lobed nucleus and large cytoplasmic granules that stain prominently with acidic dyes (Fig. 1). These granules contain secretory effectors, including eosinophil peroxidase (EPO), the ribonucleases eosinophil cationic protein (ECP) and eosinophil-derived neurotoxin (EDN), the cationic toxin, major basic protein (MBP), and various enzymes and cytokines (Table 1). Another secretory effector of eosinophils, the Charcot-Leyden crystal (CLC) protein, is associated with a distinct primary granule population in resting cells [25]. Eosinophils express Fc receptors for IgA, IgE and IgG, several seven-transmembrane chemokine receptors, adhesion molecules, and receptors for growth factors, complement components and lipid mediators (Table 2), all described in detail in recent reviews [18-23, 26-33].

This chapter is focused on methods for isolation and culture of human eosinophils. To this end, the isolation and culture techniques are described in significant detail, and references to their use in specific experimental settings are provided within the text.

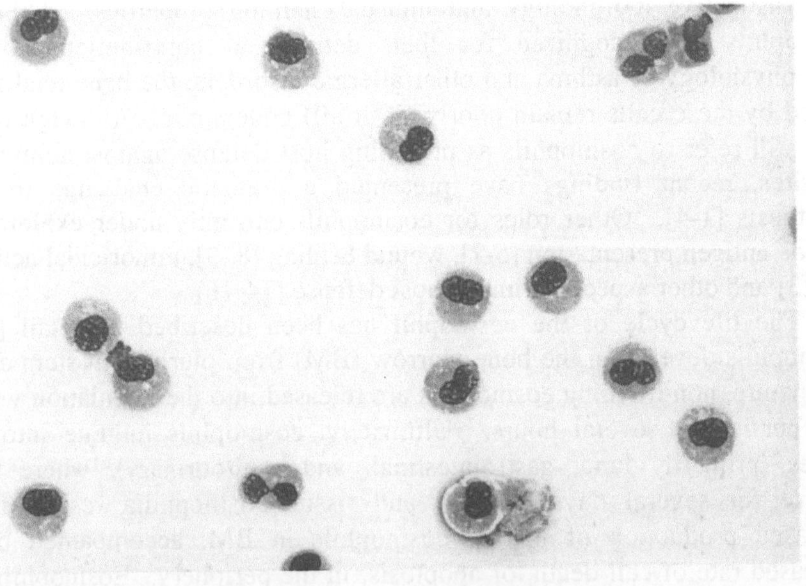

Figure 1. Normal PB eosinophils isolated by anti-CD16 negative selection and stained with Wright-Giemsa. Photograph courtesy of Dr Michael Centola (NIAMS, NIH).

Table 1. Secreted mediators from eosinophilic leukocytes

Major granule proteins	Eosinophil peroxidase (EPO)
	Major basic protein (MBP)
	Eosinophil cationic protein (ECP)
	Eosinophil-derived neurotoxin (EDN)
Lipid mediators	Platelet-activating factor (PAF)
	Leukotriene C4
	Prostaglandins
	Thromboxane
Cytokines	Interleukin-1 alpha (IL-1α)
	Interleukin-3 (IL-3)
	Interleukin-5 (IL-5)
	Granulocyte-macrophage colony-stimulating factor (GM-CSF)
	Interleukin-6 (IL-6)
	Interleukin-8 (IL-8)
	T-cell growth factor-alpha
	T-cell growth factor beta-1
	Tumor necrosis factor-alpha (TNF-α)
	Macrophage inflammatory peptide-1alpha (MIP-1α)
Enzymes and peptides	Arylsulfatase
	Histaminase
	Acid phosphatase

2. ISOLATION OF EOSINOPHILS AND EOSINOPHIL PROGENITORS

2.1 Normal Peripheral Blood Eosinophils

Until recently, isolation of PB eosinophils from normal or slightly hypereosinophilic donors was a painstaking and often frustrating task. These earlier methods, such as discontinuous Percoll [34] and metrizamide gradients [35], were based on the observation that the average density of a normal PB eosinophil is (very) slightly higher than that of a neutrophil. Thus, in a very finely prepared gradient, normal-density eosinophils would be expected to collect at a lower interface than their neutrophilic counterparts. These gradient methods have been largely supplanted by an immunomagnetic bead separation technique initially described by Miltenyi and colleagues [36, 37] (Fig. 2). This technique utilizes the differential expression of the leukocyte antigen, CD16, a cell surface protein expressed by neutrophils, but not by eosinophils, as the basis for separation.

Table 2. Major eosinophil cell surface receptors

Growth factors and cytokines	Interleukin-3 (IL-3)
	Granulocyte-macrophage colony-stimulating factor (GM-CSF)
	Interleukin-5 (IL-5)
	Interleukin-2 (IL-2)
	Interferon-gamma (IFN-γ)
	Tumor necrosis factor-alpha (TNF-α)
Chemokines	RANTES
	Macrophage inflammatory peptide-1 alpha (MIP-1α)
	Interleukin-8 (IL-8)
	Eotaxin
	Monocyte chemoattractant protein (MCP-2,3,4)
Adhesion	CD11a/CD18 (LFA-1)
	CD11b/CD18 (Mac-1, CR3)
	CD11c/CD18 (p150, 95)
	Very late antigen-4 (VLA-4)
	L-selectin
Lipid mediators	Platelet-activating factor (PAF)
	Leukotriene B4
Complement components	C5a
	C3b/C4b
	C1q
	CR1
Immunoglobulin Fc receptors	Fc alpha R (IgA)
	Fc gamma (IgG)
	Fc epsilon RI (IgE)
	Fc epsilon RII (IgE)

To begin freshly-drawn, heparinized PB is first separated into light density mononuclear cell (MNC) and high density granulocyte/erythrocyte fractions by low speed (350 *g*) centrifugation over Ficoll-Hypaque (d = 1.077; a more complete description of Ficoll-Hypaque separations can be found in reference 38). The granulocyte/erythrocyte fraction is washed, and the erythrocytes lysed by brief suspension in hypotonic saline (0.2%) or ammonium chloride lysis buffer (ACK lysis buffer, Biowhittaker, Walkersville, MD). The granulocytes are then washed and resuspended in PBE buffer (PBS + 0.5% BSA + 1 mM EDTA). Magnetic microparticle beads coupled to mouse monoclonal antibody (mAb) anti-CD16 are then added, and incubated with the cells at 4 °C. The cells are diluted and applied to a stainless steel wool-filled column placed within a strong magnetic field. The neutrophils, coupled to the magnetic beads, remain fixed to the magnetized resin, while the eosinophils flow through unimpeded. With this technique, we generally obtain 8-12 x 10^6 eosinophils from 60 ml of normal blood, with purities ranging from 95 to 98% [15]. Due to the simplicity of the technique and availability of reagents (Miltenyi Biotec, Sunnyvale CA or Dynal, Oslo, Norway), this level of purity has become the gold standard for functional studies with PB eosinophils. The PB eosinophils shown in Fig. 1 were prepared by CD16 negative selection.

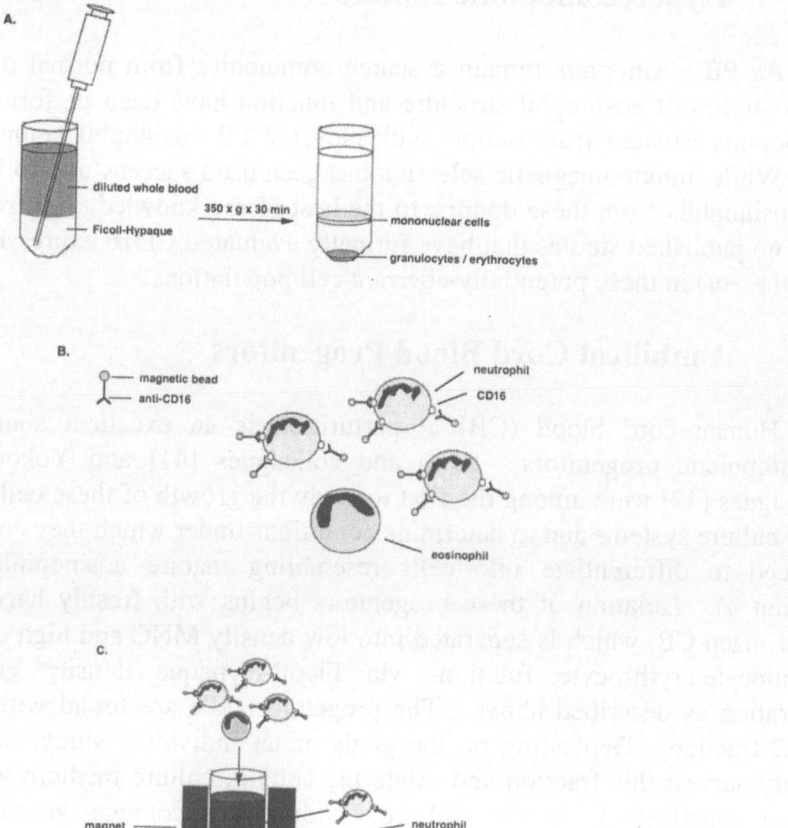

Figure 2. Isolation of PB eosinophils by Ficoll-Hypaque density gradient centrifugation followed by negative selection with anti-CD16 conjugated magnetic beads. (A) A Ficoll-Hypaque underlayer (d = 1.077) is placed beneath the heparinized PB suspension (diluted 1:1 in HBSS) and centrifuged at 350 *g* for 30 min at room temperature. The MNC layer is discarded, and the erythrocytes in the granulocyte-containing layer are lysed as described in the text. (B) The remaining granulocytes are washed and resuspended with anti-CD16 conjugated magnetic beads. Neutrophils, which express CD16, become attached to the beads. After labeling, the cells are introduced into a steel wool-containing column that has been placed within a magnetic field. The neutrophils, bound to magnetic beads, remain fixed to the steel wool resin, while the eosinophils pass through unimpeded. The eosinophils isolated in this fashion can then be maintained in culture (see text).

2.2 Hypereosinophilic Donors

As PB eosinophils remain a scarce commodity from normal donors, many studies of eosinophil structure and function have been performed on eosinophils isolated from donors with elevated PB eosinophil counts [39, 40]. While immunomagnetic selection has been used successfully to isolate PB eosinophils from these donors, to the best of my knowledge, there have been no published studies that have formally evaluated CD16 expression (or lack thereof) in these potentially-aberrant cell populations.

2.3 Umbilical Cord Blood Progenitors

Human cord blood (CB) at parturition is an excellent source of hematopoietic progenitors. Saito and colleagues [41] and Yokota and colleagues [42] were among the first to study the growth of these cells in *in vitro* culture systems and to determine conditions under which they could be induced to differentiate into cells resembling mature eosinophils (see Section 4). Isolation of these progenitors begins with freshly harvested, heparinized CB, which is separated into low density MNC and high density granulocyte/erythrocyte fractions via Ficoll-Hypaque density gradient separation as described above. The progenitor cells are found within the MNC fraction. Depending on the goals of an individual study, one can simply harvest this fraction and dilute the cells in culture medium without further purification, as the culture conditions encourage growth and differentiation specifically of progenitor cells. CB progenitors can be isolated by positive selection by either biotin-avidin-based (Cellpro, Bothell, WA) or immunomagnetic bead (Miltenyi) separation techniques, both of which are based on the expression of the leukocyte antigen CD34 by these cells [43, 44]. In the biotin-avidin based system, CB MNC are first incubated with biotin-linked murine anti-CD34 mAb. These antibody-bound cells bind tightly to a column-based avidin-linked cellulose resin. After extensive washing, the $CD34^+$ cells are removed by manually compressing the cellulose resin. The immunomagnetic bead separation proceeds in an analogous fashion. After incubation with anti-CD34-linked magnetic microparticle beads, the cells are passed over a steel wool-containing column within a magnetic field; the $CD34^+$ stem cells remain attached, and are washed through after the column is removed from the magnetic field (Fig. 3). This purified cell population can be cultured *in vitro*, and induced to differentiate into cells of the eosinophil lineage as described below.

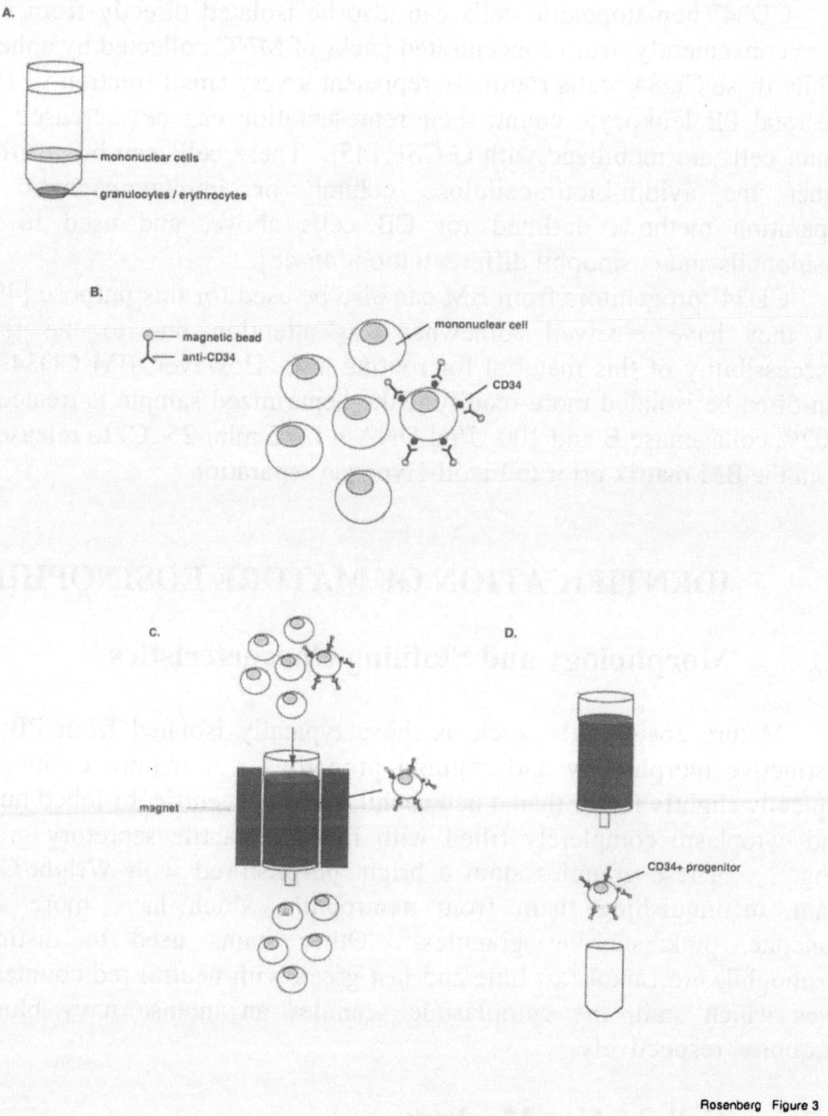

Figure 3. Isolation of CD34$^+$ cells by positive selection. (A) MNC are harvested after Ficoll Hypaque density centrifugation (see Fig. 2), and (B) resuspended with anti-CD34 conjugated magnetic beads. (C) After labeling, the cells are introduced into a steel wool-containing column placed within a magnetic field. The CD34$^+$ cells remain fixed to the column, while the other cells pass through. (D) The CD34$^+$ cells are then eluted after the column is separated from the magnet. They can then be maintained and differentiated in culture as described in the text.

2.4 Peripheral Blood and Bone Marrow Progenitors

$CD34^+$ hematopoietic cells can also be isolated directly from PB, or more conveniently, from concentrated packs of MNC collected by apheresis. While these $CD34^+$ cells routinely represent a very small fraction (0.1%) of the total PB leukocyte count, their representation can be increased if the donor cells are mobilized with G-CSF [45]. These cells can be purified by either the avidin-biotin-cellulose column or immunomagnetic bead separation methods outlined for CB cells above, and used to study eosinophils and eosinophil differentiation [46-48].

$CD34^+$ progenitors from BM can also be used for this purpose [49-51], but they have received somewhat less attention due to the relative inaccessibility of this material for routine use. However, BM $CD34^+$ cells can often be isolated more readily if the heparinized sample is treated with 0.02% collagenase B and 100 U/ml DNAse I (45 min/ 25 °C) to release cells from the BM matrix prior to Ficoll-Hypaque separation.

3. IDENTIFICATION OF MATURE EOSINOPHILS

3.1 Morphology and Staining Characteristics

Mature eosinophils, such as those typically isolated from PB, have distinctive morphology and staining properties. A mature eosinophil is typically slightly larger than a neutrophil, has an eccentric, bi-lobed nucleus, and cytoplasm completely filled with large, refractile secretory granules (Fig. 1). These granules stain a bright purplish-red with Wright-Giemsa stain, distinguishing them from neutrophils which have more subtle, punctate, pink staining granules. Other stains used to distinguish eosinophils are Luxol fast blue and fast green with neutral red counterstain, dyes which stain the cytoplasmic granules an intense navy blue and turquoise, respectively.

3.2 Cell Surface Markers

Eosinophils can be distinguished from neutrophils by the absence of the surface antigen CD16. Although fluorescence-activated cell sorting (FACS) technology is not routinely used for identification purposes, Gopinath and Nutman [52] have recently described a method using CD16-negativity and side scatter to identify eosinophils by FACS in mixed cell populations, and Sehmi and colleagues [53] presented a method based on the expression of the interleukin-5 receptor on both developing as well as mature eosinophils. Fattah and colleagues [54] have described an assay to delineate the activation state of isolated eosinophils based on differential expression of receptors for immobilized ICAM-1, VCAM-1 and IgG.

3.3 Specific Granule Proteins

There are several proteins expressed uniquely in mature and maturing eosinophils that can be used for positive identification of this lineage. EPO is one such protein. Found in the cytoplasmic granules, EPO can be detected by cytochemical staining for peroxide production, and can be distinguished from its neutrophil counterpart, myeloperoxidase (MPO), on the basis of its resistance to inhibition by cyanide anion [55]. MBP and CLC are unique to eosinophils and basophils, but can only be detected using specific antibodies that are not commercially available. The commercially-available mAb EG2 [56, 57] (Pharmacia Biotech) directed against the eosinophil-specific ECP has been used extensively to detect eosinophils in tissue sections. However, EG2 also detects EDN, an eosinophil granule protein with more widespread distribution. Although EG2 was initially described as a marker for activated eosinophils [56] and has been used extensively for this purpose [58-62], recent evidence suggests that this application may require more specific methods of staining and fixation than has been previously appreciated [63].

4. METHODS FOR EOSINOPHIL CULTURE

4.1 Maintaining Viability of Eosinophils Isolated from Peripheral Blood

Rothenberg and colleagues [64] were the first to demonstrate that the viability of isolated PB eosinophils cultured *in vitro* could be enhanced dramatically by co-culturing with endothelial cells, and that enhanced viability was dependent on soluble factors. Since then, three cytokines, interleukin-5 (IL-5), interleukin-3 (IL-3) and granulocyte-macrophage colony-stimulating factor (GM-CSF), have been identified as soluble factors promoting enhanced eosinophil viability *in vitro* [reviewed in 29]. While culture conditions differ slightly, protocols describe resuspension of isolated PB eosinophils at concentrations ranging from 0.25 to 2.0 x 10^6 cells/ml in RPMI 1640 medium with 10-20% fetal bovine serum (FBS), 2 mM glutamine, $^+$/- antibiotics, and $^+$/- 0.1 mM non-essential amino acids, followed by the addition of one or more of these cytokines. Yamaguchi and colleagues [65] initially tested a range of IL-5 concentrations for this enhancing capacity, and used 50 ng/ml in their subsequent studies [66]. Rothenberg and colleagues [67] performed a similar analysis of IL-3, and found that enhanced viability reached a plateau at 10 pM final concentration. GM-CSF appears to function somewhat less effectively than IL-5 or IL-3, but Owen and colleagues [68] have shown that GM-CSF (50 pM) can effectively promote enhanced viability of eosinophils co-cultured with murine 3T3 fibroblasts. Interestingly, all the aforementioned studies report conversion of normodense eosinophils to an activated hypodense state under these culture conditions, a state in which both toxicity and mediator release is enhanced. Recent results suggest that these cytokines function to prolong viability *in vitro* by preventing spontaneous apoptosis [69-74], a regulatory mechanism that has also been shown to operate at sites of eosinophil-mediated inflammation *in vivo* [75-78].

4.2 Differentiation and Culture of Eosinophils from BM and CB Progenitors

Interestingly, the cytokines that have been found to promote longevity of isolated PB eosinophils are also those known to induce differentiation of stem cells into maturing eosinophils. While various culture conditions have been reported, progenitors isolated from either BM or CB have been successfully differentiated into eosinophils when seeded at densities between 0.5-2.0 x 10^6 cells/ml in RPMI 1640 medium supplemented with 10% FBS, 2 mM L-glutamine, 50 μM 2-mercaptoethanol, with rhIL-3 and rhIL-5. Some have also included rhGM-CSF [79, 80] and/or 0.1 mM nonessential amino acids [81]. The use of recombinant human cytokines has largely supplanted the use of EL4 (mouse thymoma)-conditioned medium [50, 82], previously utilized as a source of (murine) eosinophilopoeitic cytokines. For example, using recombinant cytokines, Walsh and colleagues [83] reported that 88-90% of the cells evaluated at days 21-41 of culture have morphologic characteristics of mature or maturing eosinophils, as determined by May / Grunwald / Giemsa staining. Under similar conditions, Zardini and colleagues [84] found that ~90% of the cells stained positively for EPO on day 21 of culture. These results can be compared with those originally reported by Saito and colleagues [41] who determined that 81-91% of the total cells were morphologically eosinophils at day 21 of culture with IL-5 alone. Modifications to this basic protocol have been suggested. Hamann and colleagues [85] reported that wells coated with 100 μg/ml hyaluronic acid result in augmented proliferation and enhanced differentiation of CB progenitors into eosinophils, and Saito and colleagues [86] demonstrated that addition of platelet-activating factor (PAF; 10-100 pM) also enhanced differentiation toward eosinophils, an effect that may be mediated via PAF-induced production of IL-3.

4.3 Differentiation and Culture of Eosinophils from Peripheral Blood Progenitors

To date, there have been only three published studies using $CD34^+$ cells found in PB as a source of eosinophil progenitors [46-48]. In our study, $CD34^+$ PB cells isolated by the biotin-conjugated anti-CD34 method described above (CellPro) were resuspended at 0.5 x 10^6 cells/ml in IMDM supplemented with 20% FBS, c-kit ligand (KL; 50 ng/ml) and IL-6 (10 ng/ml) along with all three eosinophilopoeitic cytokines, IL-3 (20 ng/ml), GM-CSF (20 ng/ml) and IL-5 (10 ng/ml) [46]. Within three days, transcripts encoding the four major granule proteins as well as the eosinophil CLC protein were evident (Fig. 4). Similar results were obtained by Shalit and colleagues [47, 48], who reported that this method yielded 83% morphologically mature eosinophils by day 28. As these PB $CD34^+$ cells have not been studied as extensively in this context as have CB progenitors, it is unclear at the present time whether or not the populations of eosinophils produced are functionally equivalent to one another.

4.4 Quantitative Comparisons Between Natural and Cultured Eosinophils

Eosinophils derived in culture from progenitors (BM, CB or PB) are not necessarily the functional and structural equivalents of mature PB eosinophils. For example, Bach and colleagues [87] found that cultured eosinophils derived from CB progenitors were less dense than eosinophils isolated directly from PB, had reduced content of EPO and EDN, and responded more vigorously to activating agents C5a, C3a, fMLP and aggregated IgG. Walsh and colleagues [83] found increased expression of surface antigens LFA-1alpha, p15095 alpha, ICAM-1 and HLA-DR on cultured CB progenitor-derived eosinophils when compared to their mature PB counterparts. Similarly, we have found that immunoreactive EDN present in cultured eosinophils derived from PB $CD34^+$ cells was heterogeneous and of higher molecular weight than that identified in mature eosinophils, likely a result of aberrant glycosylation [88]. In all cases, it is unclear as to whether the structural and functional differences reflect the immaturity of the cultured eosinophil population, and/or activation due to the presence of stimulatory cytokines in the culture medium. For these reasons, caution must be taken when interpreting the results of functional studies performed with cultured cells, and efforts to compare results

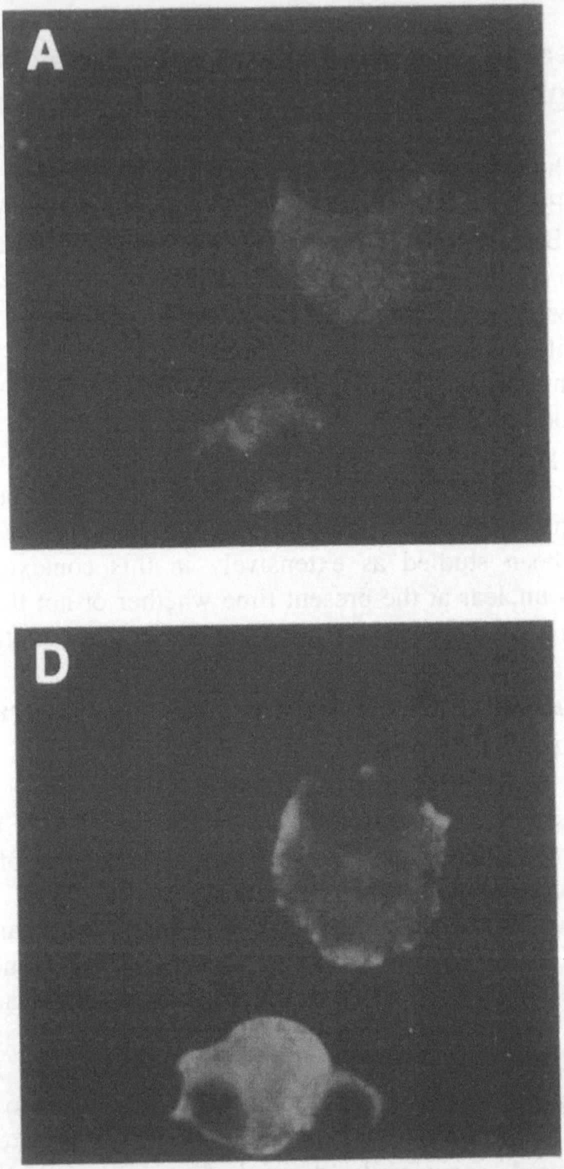

Figure 4. Eosinophilic promyelocytes derived from PB CD34$^+$ cells cultured for ten days with IL-3, IL-5 and GM-CSF as described [46]. The cells shown are double stained with monoclonal anti-eosinophil peroxidase (EPO) and polyclonal anti-Charcot-Leyden crystal protein (CLC), in the upper and lower panels, respectively. Of particular note is the apparent formation of the large cytoplasmic granules, clearly staining with anti-EPO in the upper panel. Reprinted with permission from reference [46].

obtained with cultured cells to those obtained with mature PB eosinophils should be made if at all possible.

5. APPLICATIONS

5.1 Characterization of Eosinophil-Mediated Functions *In Vitro*

The CD16-negative selection method for isolation of PB eosinophils represented a major advance for eosinophil-related studies. With purities ranging upwards of 95%, the confounding effects of contaminating neutrophils and/or MNC can be ruled out in all but the most sensitive assays. For example, this methodology has permitted us to examine the innate, ribonuclease-dependent antiviral activity of eosinophils *in vitro* [15, 16], and to set parameters with which to test these findings *in vivo*. This method has facilitated the identification of cell surface receptors [89-93] and their signal transduction mechanisms, [94-99] as well as novel mediators and cytokines [100-103] expressed in and secreted by PB eosinophils that might have otherwise gone unnoticed in the presence of even moderate contamination with neutrophils.

5.2 Apoptosis and the Control of Eosinophilic Inflammation

The ability to prolong the life of mature eosinophils *in vitro* has led to major advances in the study of eosinophil apoptosis, or programmed cell death. While IL-5 has been shown to be a primary force in sustaining eosinophils both *in vitro* and *in vivo* [69-74], the modulating roles of additional factors can now be tested. For example, several groups have found that eosinophil apoptosis is enhanced by antibody-mediated activation of the cell surface protein known as Fas receptor [93, 104, 105], a pro-apoptotic effect that effectively overcomes the anti-apoptotic activities of IL-5, IL-3 or GM-CSF [105]. Similar results have been obtained in studies utilizing a mAb directed against the cell surface antigen, CD69 [106]. Other studies have identified glucocorticoids [107-109], TGF-β1[110], and interferons alpha and gamma [111] as promoting apoptosis, and IL-13 as preventing apoptosis [99], all performed with PB eosinophils isolated and maintained *in vitro*.

5.3 Molecular Mechanisms of Differentiation and Transcriptional Control of Eosinophil-Specific Genes

While eosinophils can be clearly distinguished from related granulocyte lineages on both morphologic and biochemical bases, the molecular events that define and promote specific eosinophil differentiation remain obscure. While the promoters of several eosinophil genes have been identified and characterized [112-119], there are no transcriptional events or transcription factors known to operate uniquely in eosinophils. With the availability of these differentiating culture systems, the molecular events underlying commitment in and differentiation of the eosinophil lineage may be more amenable for study.

5.4 Identification of Novel Enhancers and Inhibitors of Eosinophil Differentiation

While IL-3, IL-5 and GM-CSF have been defined as the major factors promoting eosinophil differentiation, contributions from other modulators, both soluble and cellular, cannot yet be ruled out. The *in vitro* systems described will permit direct comparisons of the effects of novel mediators on eosinophil development and differentiation [120-122].

REFERENCES

1. Sher A, Coffman RL, Hieny S, and Cheever AW (1990) Ablation of eosinophil and IgE responses with anti-IL-5 or anti-IL-4 antibodies fails to affect immunity against *Schistosoma mansoni* in the mouse. *J Immunol* 145: 3911-3916.
2. Herndon FJ and Kayes SG (1992) Depletion of eosinophils by anti-IL-5 monoclonal antibody treatment of mice infected with *Trichinella spiralis* does not alter parasite burden or immunologic resistance to reinfection. *J Immunol* 149: 3642-3647.
3. Freeman GL, Tominaga A, Takatsu K, Secor WE, and Colley DG (1995) Elevated innate PB eosinophilia fails to autment irradiated cercarial vaccine-induced resistance to *Schistosoma mansoni* in IL-5 transgenic mice. *J Parasitol* 81: 1010-1011.
4. Kopf M, Brombacher F, Hodgkin PD, Ramsay AJ, Milbourne EA, Dai WJ, Ovington KS, Behm CA, Kohler G, Young IG, and Matthaei KI (1996) IL-5-deficient mice have a developmental defect in $CD5^+$ B-1 cells and lack eosinophilia but have normal antibody and cytotoxic T cell responses. *Immunity* 4: 15-24.
5. Lucey DR, Nicholson-Weller A, and Weller PF (1989) Mature human eosinophils have the capacity to express HLA-DR. *Proc Natl Acad Sci USA* 86: 1348-1351.
6. Weller PF, Rand TH, Barrett T, Elovic A, Wong DTW, and Finberg RW (1993) Accessory cell function of human eoinsophils: HLA-DR dependent MHC restricted antigen presentation and interleukin-1-alpha formation. *J Immunol* 150: 2554-2562.

7. Mawhorter SD, Kazura JW, and Boom WH (1994) Human eosinophils as antigen-presenting cells: relative efficiency for superantigen and antigen induced $CD4^+$ T-cell proliferation. *Immunology* 81: 584-591.
8. Wong DTW, Weller PF, Galli SJ, Rand TH, Elovic A, Chiang T, Chou MY, Gallagher GT, Matossian K, McBride J, and Todd R (1990) Human eosinophils express transforming growth factor alpha. *J Exp Med* 172: 673-681.
9. Todd R, Donoff BR, Chiang T, Chou MY, Elovic A, Gallagher GT, and Wong DTW. (1991) The eosinophil as a cellular source of transforming growth factor alpha in healing cutaneous wounds. *Am J Pathol* 138: 1307-1313.
10. Tepper RI, Pattengale PK, and Leder P (1989) Murine interleukin-4 displays potent antitumor activity in vivo. *Cell* 57: 503-512.
11. Tepper RI, Coffman RL, and Leder P (1992) An eosinophil-dependent mechanism for the antitumor effect of IL-4. *Science* 257: 548-551.
12. Golumbek PT, Lazenby AJ, Levitsky HI, Jaffee LM, Karasuyama H, Baker M, and Pardoll DM (1991) Treatment of established renal cancer by tumor cells engineered to secrete interleukin-4. Science 254: 713-716.
13. Tepper RI (1994) The eosinophil-mediated antitumor activity of interleukin-4. J Allergy Clin Immunol 94: 1225-1231.
14. Klebanoff SJ and Coombs RW (1996) Virucidal effect of stimulated eosinophils on human immunodeficiency virus type I. *AIDS Res Human Retrov* 12: 25-29.
15. Domachowske JB and Rosenberg HF (1997) Eosinophils inhibit retroviral transduction of human target cells by a ribonuclease-dependent mechanism. *J Leukoc Biol* 62: 363-368.
16. Domachowske JB, Dyer KD, Bonville CA, and Rosenberg HF (1998) Recombinant human eosinophil-derived neurotoxin / RNase 2 functions as an effective antiviral agent against respiratory syncytial virus. *J Infect Dis* 177:1458-1464.
17. Bainton DF (1992) Developmental biology of neutrophils and eosinophils. In: Gallin JI, Goldstein IM, and Snyderman R (eds) *Inflammation: basic principles and clinical correlates*, 2nd ed. Raven Press, New York.
18. Wardlaw AJ (1994) Eosinophils in the 1990s: new perspectives on their role in health and disease. *Postgrad Med* 70: 536-552.
19. Silberstein DS (1995) Eosinophil function in health and disease. *Crit Rev Oncol Hematol* 19: 47-77.
20. Wardlaw AJ, Moqbel R, and Kay AB (1995) Eosinophils: biology and role in disease. *Adv Immunol* 60: 151-266.
21. Weller PF (1991) The immunobiology of eosinophils. *N Engl J Med* 324: 1110-1118.
22. Gleich GJ, Adolphson CR and Leiferman KM (1993) The biology of the eosinophilic leukocyte. *Ann Rev Med* 44: 85-101.
23. Brito-Babapulle F (1997) Clonal eosinophilic disorders and the hypereosinophilic syndrome. *Blood Rev* 11: 129-145.
24. Weller PF and Bubley GJ (1994) The idiopathic hypereosinophilic syndrome. *Blood* 83: 2759-2779.
25. Dvorak AM, Letourneau L, Login GR, Weller PF, and Ackerman SJ. (1988) Ultrastructural localization of the Charcot-Leyden crystal protein (lysophospholipase) to a distinct crystalloid-free granule population in mature human eosinophils. *Blood* 72: 150-158.
26. Spry CJF (1988) Eosinophils. *A comprehensive review and guide to the scientific and medical literature.* Oxford University Press, Oxford UK.
27. Makino S and Fukuda T (1993) *Eosinophils. Biological and clinical aspects.* CRC Press, Boca Raton, FL

28. Capron M and Desreumaux P (1997) Immunobiology of eosinophils in allergy and inflammation. *Res Immunol* 148: 29-33.
29. Weller PF (1992) Cytokine regulation of eosinophil function. *Clinical Immunol Immunopathol* 62: S55-S59.
30. Weller PF (1994) Eosinophils: structure and functions. *Curr Opin Immunol* 6: 85-90.
31. Spry CJF, Kay AB, and Gleich GJ (1992) Eosinophils 1992. *Immunol Today* 13: 384-387.
32. Bignold LP (1995) The eosinophil leukocyte: controversies of recruitment and function. *Experientia* 51: 317-327.
33. Walsh GM (1997) Human eosinophils: their accumulation, activation and fate. *Br J Haematol* 97: 701-709.
34. Gartner I (1980) Separation of human eosinophils in density gradients of polyvinylpyrrolidone-coated silica gel (Percoll). *Immunology* 40: 133-136.
35. Vadas MA, David JR, Butterworth AE, Pisani NT, and Siongok TA (1979) A new method for the purification of human eosinophils and neutrophils and a comparison of the ability of these cells to damage schistosoma of *Schistosoma mansoni*. *J Immunol* 122: 1228-1336.
36. Hansel TT, de Vries JJ, Iff T, Rihs S, Wandzilak M, Betz, S, Blaser K, and Walker C (1991) An improved immunomagnetic procedure for isolation of highly purified human blood eosinophils. *J Immunol Methods* 145: 105-110.
37. Miltenyi S, Muller W, Weichel W, and Radbruch A (1990) High gradient magnetic cell separation with MACS. *Cytometry* 11: 231-240.
38. Metcalf JA, Gallin JI, Nauseef WM and Root RK (1986) *Laboratory Manual of Neutrophil Function*. Raven Press, New York.
39. Ishihara C, Ochiai K, Kagami M, Takashahi H, Matsuyama G, Yoshida S, Tomioka H, and Koya N (1997) Human peripheral eosinophils express functional interferon-gamma receptors (IFN-gammaR). *Clin Exp Immunol* 110: 524-529.
40. Simon UF, Tsao PW, Siminovitch KA, Mills GB, and Blaser K (1994) Functional platelet-activator factor receptors are expressed by monocytes and granulocytes but not by resting or activated T and B lymphocytes from normal individuals or patients with asthma. *J Immunol* 153: 364-377.
41. Saito H, Hatake K, Dvorak AM, Leiferman KM, Donnenberg A, Arai N, Ishizaka K, and Ishizaka T (1988) Selective differentiation and proliferation of hematopoietic cells induced by recombinant human interleukins. *Proc Natl Acad Sci USA* 85: 2288-2292.
42. Yokota T, Coffman RL, Hagiwara H, Rennick D, Takebe Y, Yokota K, Gemmel L, Shrader B,Yang G, Myerson P, Luh J, Hoy P, Pene J, Briere F, Banchereau J, De Vries J, Lee FD, Arai N, and Arai K. (1987) Isolation and characterization of lymphokine cDNA clones encoding mouse and human IgA-enhancing factor and eosinophil colony-stimulating factor activities: relationship to interleukin 5. *Proc Natl Acad Sci USA* 84: 7388-7392.
43. Civin CI, Strauss LC, Brovall C, Fackler MJ, Schwartz JF, and Shaper JH (1984) Antigenic analysis of hematopoiesis. III. A hematopoietic progenitor cell surface antigen defined by a monoclonal antibody raised against KG-1a cells. *J Immunol* 133: 157-165.
44. Civin CI, Trischmann RM, Fackler MJ, Bernstein ID, Buhring H-J, Campos L, Greaves MF, Kamoun M, Katz DR, Landsorp PM, Look AT, Seed B, Sutherland Dr, Tindle RW, and Kuchanska-Ziegler B (1989) Report on the CD34 cluster workshop. In: Knapp W, Dorken B, Gilks WR, Rieber EP, Schmidt RE, Stein H, and von dem Borne A (eds) *Leucocyte typing IV*. Oxford University Press, Oxford UK.
45. Sekhsaria S, Fleisher TA, Vowells S, Brown M, Miller J, Gordon I, Blaese RM, Dunbar CE, Leitman S, and Malech HL (1996) Granulocyte colony-stimulating factor recruitment

of CD34+ progenitors to PB: impaired mobilization in chronic granulomatous disease and adenosine deaminase-deficient severe combined immunodeficiency disease patients. *Blood* 88: 1104-1112.

46. Rosenberg HF, Dyer KD, and Li F (1996) Characterization of eosinophils generated in vitro from CD34+ PB stem cells. *Exp Hematol* 24: 888-893.
47. Shalit M, Sekhsaria S, Mawhorter S, Mahanti S, and Malech HL (1996) Early commitment to the eosinophil lineage by cultured human PB CD34+ cells: messenger RNA analysis. *J Allergy Clin Immunol* 98: 344-354.
48. Shalit M (1997) Growth and differentiation of eosinophils from human PB CD34+ cells. *Allerg Immunol (Paris)* 29: 7-10.
49. Clutterbuck EJ and Sanderson CJ (1988) Human eosinophil hematopoeisis studied in vitro by means of murine eosinophil differentiation factor (IL5): production of functionally active eosinophils from normal human BM. *Blood* 71: 646-651.
50. Butterfield JH and Weiler D (1986) Giant eosinophil colonies from cultures of BM cells. *In Vitro Cell Dev Biol* 22: 157-163.
51. Clutterbuck EJ and Sanderson CJ (1990) Regulation of human eosinophil precursor production by cytokines: a comparison of recombinant human interleukin-1 (rhIL-1), rhIl-3, rhIL-5, rhIL-6, and rhGM-CSF. *Blood* 75: 1774-1779.
52. Gopinath R and Nutman TB (1997) Identification of eosinophils in lysed whole-blood using side scatter and CD16 negativity. *Cytometry* 30: 313-316.
53. Sehmi R, Wood LJ, Watson R, Foley R, Hamid Q, O'Byrne PM, and Denburg JA (1997) Allergen-induced increases in IL-5 receptor alpha-subunit expression on BM derived CD34+ cells from asthmatic subjects. A novel marker of progenitor cell commitment towards eosinophilic differentiation. *J Clin Invest* 100: 2466-2475.
54. Fattah D, Page KR, Bezbaruah S, Priest RC, Horgan CM, and Solari R (1996) A rapid activation assay for human eosinophils based on adhesion to immobilized ICAM-1, VCAM-1 and IgG. *Cytokine* 8: 248-259.
55. Zucker-Franklin D and Grusky G (1976) The identification of eosinophil colonies in soft-agar cultures by differential staining for peroxidase. *J Histochem Cytochem* 24: 1270-1272.
56. Tai P-C, Spry CJF, Peterson C, Venge P, and Olsson I (1984) Monoclonal antibodies distinguish between storage and secreted forms of eosinophil cationic protein. *Nature* 309: 182-184.
57. Rosenberg HF and Tiffany HL (1994) Characterization of the proteins recognized by the activation specific monoclonal antibody, EG2. *J Leukoc Biol* 56: 502-506.
58. Tai P-C, Holt ME, Denny P, Gibbs AR, Williams BD and Spry CJF (1984) Deposition of eosinophil cationic protein in granulomas in allergic granulomatosis and vasculitis: the Churg-Strauss syndrome. *Br Med J* 298: 400-402.
59. Fredens K, Dybdahl H, Dahl R, and Baandrup U (1988) Extracellular deposits of the cationic proteins ECP and EPX in tissue infiltrations of eosinophils related to tissue damage. APMIS 96: 711-719.
60. Lundin A, Fredens K, Michaelsson G, and Venge P (1990) The eosinophil granulocyte in psoriasis. *Br J Dermatol* 122: 181-193.
61. Moqbel R, Barkans J, Bradley BL, Durham SR, and Kay AG (1992) Application of monoclonal antibodies against major basic protein (BMK-13) and eosinophil cationic protein (EG1 and EG2) for quantifying eosinophils in bronchial biopsies from atopic asthma. *Clin Exp Allergy* 22: 265-273.
62. Varney VA, Hamid QA, Gaga M, Ying S, Jacobson M, Frew AJ, Kay AB, and Durham SR (1993) Influence of grass pollen immunotherapy on cellular infiltration and cytokine

mRNA expression during allergen-induced late-phase cutaneous responses. *J Clin Invest* 92: 644-650.
63. Jahnsen FL, Brandtzaeg P, and Halstensen TS (1994) Monoclonal antibody EG2 does not provide reliable immunohistochemical discrimination between resting and activated eosinophils. *J Immunol Methods* 175:23-36.
64. Rothenberg ME, Owen WF Jr, Silberstein DS, Soberman RJ, Austen KF, and Stevens RL (1987) Eosinophils cocultured with endothelial cells have increased survival and functional properties. *Science* 237: 645-647.
65. Yamaguchi Y, Hayashi Y, Sugama Y, Miura Y, Kasahara T, Kitamura S, Torisu M, Mita S, Tominaga A, and Takatsu K (1988) Highly purified murine interleukin 5 (IL-5) stimulates eosinophil function and prolongs in vitro survival: IL-5 as an eosinophil chemotactic factor. *J Exp Med* 167: 1737-1742.
66. Yamaguchi Y, Suda T, Ohta S, Tominaga K, Miura Y, and Kasahara T (1991) Analysis of the survival of mature human eosinophils: interleukin-5 prevents apoptosis in mature human eosinophils. *Blood* 78: 2542-2547.
67. Rothenberg ME, Owen WF Jr, Silberstein DS, Woods J, Soberman RJ, Austen KF and Stevens RL (1988) Human eosinophils have prolonged survival, enhanced functional properties and become hypodense when exposed to human interleukin 3. *J Clin Invest* 81: 1986-1992.
68. Owen WF Jr, Rothenberg ME, Silberstein DS, Gasson JC, Stevens RL, Austen KF, and Soberman RJ (1987) Regulation of human eosinophil viability, density, and function by granulocyte/macrophage colony-stimulating factor in the presence of 3T3 fibroblasts. *J Exp Med* 166: 129-141.
69. Haslett C, Savill JS, Whyte MK, Stern M, Dransfield I, and Meagher LC. (1994) Granulocyte apoptosis and the control of inflammation. *Phils Trans R Soc London B Biol Sci* 345: 327-333.
70. Simon HU and Blaser K (1995) Inhibition of programmed eosinophil death: a key pathogenic event for eosinophilia? *Immunol Today* 16: 53-55.
71. Walsh GM (1997) Mechanisms of human eosinophil survival and apoptosis. *Clin Exp Allergy* 27: 482-487.
72. Simon HU (1997) Molecular mechanisms of defective eosinophil apoptosis in diseases associated with eosinophilia. *Int Arch Allergy Immunol* 113: 206-208.
73. Yousefi S, Blaser K, and Simon HU (1997) Activation of signaling pathways and prevention of apoptosis by cytokines in eosinophils. *Int Arch Allergy Immunol* 112: 9-12.
74. Simon HU, Yousefi S, and Blaser K. (1995) Tyrosine phosphorylation regulates activation and inhibition of apoptosis in human eosinophils and neutrophils. *Int Arch Allergy Immunol* 107: 338-339.
75. Simon HU, Yousefi S, Schranz C, Schapowal A, Bachert C, and Blaser K. (1997) Direct demonstration of delayed eosinophil apoptosis as a mechanism causing tissue eosinophilia. *J Immunol* 158: 3902-3908.
76. Tsuyuki S, Bertrand C, Erard F, Trifilieff A, Tsuyuki J, West M, Anderson GP, and Coyle AJ. (1995) Activation of the Fas receptor on lung eosinophils leads to apoptosis and the resolution of eosinophilic inflammation of the airways. *J Clin Invest* 96: 2924-2931.
77. Simon HU, Yousefi S, Dommann-Scherrer CC, Zimmermann DR, Bauer S, Barandun J, and Blaser K. (1996) Expansion of cytokine-producing CD4-CD8- T cells associated with abnormal Fas expression and hypereosinophilia. *J Exp Med* 183: 1071-1082
78. Coyle AJ, Tsuyuki S, Bertrand C, Huang S, Aguet M, Alkan SS, and Anderson GP. (1996) Mice lacking the IFN-gamma receptor have impaired ability to resolve a lung

eosinophilic inflammatory response associated with a prolonged capacity of T cells to exhibit a Th2 cytokine profile. *J Immunol* 156: 2680-2685.

79. Capron M, MOrita M, Woerly G, Lengrand F, Soussi Gounni A, Delaporte E, and Capron A (1997) Differentiation of eosinophils from CB cell precursors: kinetics of Fc Epsilon RI and Fc Epsilon RII expression. *Int Arch Allergy Immunol* 113: 48-50.
80. Gruart V, Truong MJ, Plumas J, Zandecki M, Kusnierz JP, Prin L, Vinatier D, Capron A and Capron M (1992) Decreased expression of eosinophil peroxidase and major basic protein messenger RNAs during eosinophil maturation. *Blood* 79: 2592-2597.
81. Boyce JA, Lam BK, Penrose JF, Friend DS, Parsons S, Owen WF, and Austen KF (1996) Expression of LTC4 synthase during the development of eosinophils in vitro from CB progenitors. Blood 99: 4338-4347.
82. Ten RM, Butterfield JH, Kita H, Weiler DA, Fischkoff S, Ishizaka T, Sanderson CJ and Gleich GJ (1991) Eosinophil differentiation of human CB MNCs and prolonged survival of mature eosinophils by murine EL-4 thymoma cell conditioned medium. *Cytokine* 3: 350-359.
83. Walsh GM, Hartnell A, Moqbel R, Cromwell O, Hagy L, Bradley B, Furitsu T, Ishizaka T, and Kay AB (1990) Receptor expression and functional status of cultured human eosinophils derived from umbilical CB MNCs. *Blood* 76: 105-111.
84. Zardini DM, Heuschling P, Gallois A, Bueb J-L, and Tschirhart EJ (1997) Human umbilical CB-derived eosinophils cultured in the presence of IL-3 and IL-5 respond to fMLP with [$Ca2^+$]i variation and superoxide production. *J Immunol Methods* 205: 1-9.
85. Hamann KJ, dowling TL, Neeley SP, Grant JA, and Leff AR (1995) Hyaluronic acid enhances cell proliferation during eosinopoiesis through the CD44 surface antigen. *J Immunol* 154: 4073-4080.
86. Saito H, Koshio T, Yanagihara Y, Akiyama K, and Shida T (1993) Platelet-activating-factor- induced augmentation of production of eosinophil-lineage cells in hematopoeitic precursor cells obtained from human umbilical CB. *Int Arch Allergy Immunol* 102: 375-382.
87. Bach MK, Brashler JR, Sanders ME, Furitsu T, and Ishizaka T (1990) How similar are in vitro differentiated CB derived eosinophils to PB eosinophils? *Int Arch Allergy Appl Immunol* 93: 323-329.
88. Tiffany HL, Li F, and Rosenberg HF (1995) Hyperglycosylation of eosinophil ribonucleases in a promyeloocytic leukemia cell line and in differentiated PB stem cells. *J Leukoc Biol* 58: 49-54.
89. Aldebert D, Lamkhioued B, Desaint C, Soussi Gounni AS, Goldman M, Capron A, Prin L, and Capron M (1996) Eosinophils express a functional receptor for interferon alpha; inhibitory role of interferon alpha on the release of mediators. *Blood* 87: 2354-2360.
90. Garcia-Zepeda EA, Rothenberg ME, Ownbey RT, Celestin J, Leder P and Luster AD (1996) Human eotaxin is a specific chemoattractant for eosinophil cells and provides a new mechanism to explain tissue eosinophilia. *Nature Medicine* 2: 449-456.
91. Garcia-Zepeda E, Combadiere C, Rothenberg M, Sarafi M, Lavigne F, Hamid Q, Murphy P, and Luster A (1996) Human monocyte chemoattractant protein (MCP-4) is a novel CC chemokine with activities on monocytes, eosinophils and basophils induced in allergic and nonallergic inflammation that signals throught the CC chemokine receptors CCR-2 and -3. *J Immunol* 157: 5613-5626.
92. Kishimoto S, Shimadzu W, Izumi T, Shimizu T, Fukuda T, Makino S, Sugiura T, and Waku K (1996) Regulation by IL5 of expression of functional platelet activating factor receptors on human eosinophils. *J Immunol* 157: 4126-4132.

93. Hebestreit H, Yousefi S, Balatti I, Weber M, Crameri R, Simon D, Hartung K, Schapowal A, Blaser K, and Simon HU (1996) Expression and function of the Fas receptor on human blood and tissue eosinophils. *Eur J Immunol* 26: 1775-1778.
94. Blom M, Tool ATJ, Mut FPJ, Knol EF, Roos D, and Verhoeven AJ (1995) Eosinophils isolated with two different methods show different characteristics of activation. *J Immunol Methods* 178: 183-193.
95. Kitayama J, Carr M, Roth S, Buccola J, Springer T (1997) Contrasting responses to multiple chemotactic stimuli in transendothelial migration. *J Immunol* 158: 2340-2349.
96. Neeley SP, Hamann KJ, White SR, Baranowske SL, Burch RA, and Leff AR (1993) Selective regulation of expression of surface adhesion molecules MAC-1, L-selectin, and VLA-4 on human eosinophils and neutrophils. *Am J Resp Cell Mol Biol* 8: 633-639.
97. Sung PKL, Yang L, Elices M, Jin G, Sriramarao P, and Broide DH (1997) Granulocyte-macrophage colony-stimulating factor regulates the functional adhesive state of very late antigen-4 expressed by eosinophils. *J Immunol* 158: 919-927.
98. Yousefi, Hoessli DC, Blaser K, Mills GB and Simon HU (1996) Requirement of Lyn and Syk tyrosine kinases for the prevention of apoptosis by cytokines in human eosinophils. *J Exp Med* 183: 1407-1414.
99. Luttman W, Knoechel B, Foerster M, Matthys H, Virchow JC Jr, and Kroegel C (1996) Activation of human eosinophils by IL-13. Induction of CD69 surface antigen, its relationship to messenger RNA expression, and promotion of cellular viability. *J Immunol* 157: 1678-1683.
100. Levi-Schaffer F, Lacy P, Servers NJ, Newman RM, North J, Gomperts B, Kay AB and Moqbel R (1995) Association of granulocyte-macrophage colony-stimulating factor with the crystalloid granules of human eosinophils. *Blood* 85: 2579-2586.
101. Nakajima H, Gleich GJ, and Kita H (1996) Constitutive production of IL-4 and IL-10 and stimulated production of IL-8 by normal PB eosinophils. *J Immunol* 156: 4859-4866.
102. Ying S, Meng Q, Taborda-Barata L, Corrigan C, Barkans J, and Kay A (1996) Human eosinophils express messenger RNA encoding RANTES and store and release biologically active RANTES protein. *Eur J Immunol* 26: 70-76.
103. Yousefi S, Hermann S, Weber M, Holzer C, Hartung K, Blaser K, and Simon HU (1995) IL-8 is expressed by human PB eosinophils. *J Immunol* 154: 5481-5490.
104. Druilhe A, Cai Z, Haile S, Chouaib S, and Pretolani M. (1996) Fas-mediated apoptosis in cultured human eosinophils. *Blood* 87: 2822-2830.
105. Matsumoto K, Schleimer RP, Saito H, Iikura Y, and Bochner BS (1995) Induction of apoptosis in human eosinophils by anti-Fas antibody treatment *in vitro*. *Blood* 86: 1437-1443.
106. Walsh GM, Williamson ML, Symon FA, Willars GB, Wardlaw AJ (1996) Ligation of CD69 induced apoptosis and cell death in human eosinophils cultured with granulocyte-macrophage colony-stimulating factor. *Blood* 87: 2815-2821.
107. Adachi T, Motojima S, Hirata A, Fukuda T, Kihara N, Kosaku A, Ohtake H, and Makino S (1996) Eosinophil apoptosis caused by theophylline, glucocorticoids, and macrolides after stimulation with IL-5. *J Allergy Clin Immunol* 98: S207-S215.
108. Woolley KL, Gibson PG, Carty K, Wilson AJ, Twaddell SH, and Woolley MJ (1996) Eosinophil apoptosis and resolution of airway inflammation in asthma. *Am J Respir Crit Care Med* 154: 237-243.
109. Meagher LC, Cousin JM, Seckel JR, Haslett C (1996) Opposing effects of glucocorticoids on the rate of apoptosis in neutrophilic and eosinophilic granulocytes. *J Immunol* 156: 4422-4428.

110. Atsuka J, Fujisawa T, Iguchi K, Terada A, Kamiya H, Sakurai M (1995) Inhibitory effect of transforming growth factor beta 1 on cytokine-enhanced eosinophil survival and degranulation. *Int Arch Allergy Immunol* 108 (suppl 1): 31-35.
111. Morita M, Lamkhioued B, Soussi Gounni A, Aldebert D, Delaporte E, Capron A, Capron M (1996) Induction by interferons of human eosinophil apoptosis and regulation by interleukin-3, granulocyte/macrophage colony stimulating factor and interleukin-5. *Eur Cytokine Netw* 7: 725-732.
112. Yamaguchi Y, Zhang DE, Sun Z, Albee EA, Nagata S, Tenen DG, and Ackerman SJ (1994) Functional characterization of the promoter for the gene encoding human eosinophil peroxidase. *J Biol Chem* 269: 19410-19419.
113. Yamaguchi Y, Tenen DG, and Ackerman SJ (1994) Transcriptional regulation of the human eosinophil peroxidase genes: characterization of a peroxidase promoter. *Int Arch Allergy Immunol* 104 (suppl); 30-31.
114. Gomolin HI, Yamaguchi Y, Paulpillai AV, Dvorak LA, Ackerman SJ, and Tenen DG (1993) Human eosinophil Charcot-Leyden crystal protein: cloning and characterization of a lysophospholipase gene promoter. *Blood* 82: 1868-1874.
115. Li MS, Sun L, Satoh T, Fisher LM, and Spry CJ (1995) Human eosinophil major basic protein, a mediator of allergic inflammation, is expressed by alternative splicing from two promoters. *Biochem J* 305: 921-927.
116. Sun Z, Yergeau DA, Tuypens T, Tavernier J, paul CC, Baumann MA, Tenen DG, and Ackerman SJ (1995) Identification and characterization of a functional promoter region in the human eosinophil IL-5 receptor alpha subunit gene. *J Biol Chem* 270: 1462-1471.
117. Sun Z, Yergeau DA, Wong IC, Tuypens T, Tavernier J, Paul CC, Baumann MA, Auron PE, Tenen DG, and Ackerman SJ (1996) Interleukin-5 receptor alpha subunit gene regulation in human eosinophil development: identification of a unique cis-element that acts like an enhancer in regulating activity of the IL-5R alpha promoter. *Curr Top Microbiol Immunol* 211: 173-187.
118. Tiffany HL, Handen JS, and Rosenberg HF (1996) Enhanced expression of the eosinophil-derived neurotoxin ribonuclease (RNS2) gene requires interaction between the promoter and intron. *J Biol Chem* 271: 12387-12393.
119. Handen JS and Rosenberg HF (1997) Intronic enhancer activity of the eosinophil-derived neurotoxin (RNS2) and eosinophil cationic protein (RNS3) genes is mediated by an NFAT-1 consensus binding sequence. *J Biol Chem* 272: 1665-1669.
120. Boyce JA, Friend D, Matsumoto R, Austen KF, and Owen WF (1995) Differentiation in vitro of hybrid eosinophil-basophil granulocytes: autocrine function of an eosinophil developmental intermediate. *J Exp Med* 182: 49-57.
121. Wang SR and Kao TH (1995) Eosinophil differentiation and hypodensity alteration activities in Dermatophagoides pteronyssinus-stimulated MNC culture supernatants derived from asthmatics. *Int Arch Allergy Immunol* 106: 297-301.
122. Ochiai K, Iwamoto I, Takahashi H, Yoshida S, Tomioka H, and Yoshida S (1995) Effect of IL-4 and interferon-gamma (IFN-gamma) on IL-3- and IL-5-induced eosinophil differentiation from human CB MNCs. *Clin Exp Immunol* 99: 124-128.

Chapter 10

Isolation and Culture of Mast Cells and Basophils

Peter Valent
Department of Internal Medicine I, Division of Hematology and Hemastaseology, The University of Vienna, Währinger Gürtel 18-20, A-1090 Vienna, Austria. Tel: 0043-1-404-006-085; Fax: 0043-1-404-6930; E-mail:

1. INTRODUCTION

Based on their unique dye-binding properties, the blood basophil (ba) and tissue mast cell (MC) were discovered by Paul Ehrlich in 1879 [1]. Both types of cells contain proteoglycan-rich granules staining purple after exposure to basic dyes. However, these granules not only express proteoglycans, but also an array of vasoactive and pro-inflammatory substances, like histamine, leukotrienes, or cytokines [2-6]. Some of these mediators (histamine) are expressed in both types of cells. Other substances are expressed in either MC or ba (Table 1). During an allergic reaction or other inflammatory event, MC and ba can release their granular mediators into the extracellular space. The allergen-dependent release is mediated by specific IgE molecules binding to high-affinity IgE receptors (FcεRI) expressed on MC and ba [7]. Allergen-binding to IgE is followed by IgER-cross-linking with consecutive signal transduction events culminating in degranulation [2,3,7]. The IgE-independent release of mediators from MC or ba involves cell surface receptors for cytokines, complement-degradation products, and other surface receptors [2,3,8] (Table 2). MC and ba differ in expression of surface membrane receptors and response to respective ligands [8]. Furthermore, depending on the tissue environment, maturation stage, and activation, MC may differentially respond to distinct agonists. Likewise, MC obtained from the juvenile foreskin, but not those derived from adult human lung (lung cancer patients), express C5aR (CD88) and release histamine in response to C5a [9-11].

Table 1. Expression of mediators in human mast cells and basophils

Mediator	Expression in	
	Mast cells	Basophils
Histamine	+++	++
PGD_2	++	+/-
Tryptase	+++	+/-
Chymase	+/-	-
tPA	+/-	-
TNF-α	+	-
IL-4	-(+)*	+
MCP-1	+	-
Heparin	+/-	-

+/-, mediator expressed in subset of cells or in small amounts. *IL-4 is produced in mouse MC lines.

For more than a century, the origin of human MC and ba remained an enigma. Then, it turned out that both cells belong to the hematopoietic cell family and develop from uncommitted hematopoïetic progenitor cells. MC and ba progenitor cells are detectable in the bone marrow (BM) and peripheral blood (PB) [12-15]. These cells are also detectable in fetal liver and cord blood (CB) [16-18]. The ba progenitor cells usually undergo differentiation and maturation in BM. By contrast, MC differentiation and maturation takes place in extramedullary organs [19]. The MC progenitor cells ($CD34^+$) supposedly leave the blood stream by transmigration through the endothelial cell layer and consecutive homing before differentiation and terminal maturation occurs [19,20]. The ba is a short-lived cell with a lifespan of several days. By contrast, the estimated lifespan of a human MC *in vivo* amounts to several months (and perhaps >1 year) [21]. The mature MC is located in extravascular sites exclusively, and is not detectable in PB. By contrast, mature ba are primarily found in the bloodstream. Under distinct pathologic conditions, however, ba can also translocate from the blood into tissue and appear at the site of disease [22].

The differentiation of MC and ba is regulated by distinct cytokines. Interleukin-3 (IL-3) is a major differentiation factor for human ba [13,14,16]. This cytokine induces development of ba from immature $CD34^+$ progenitor cells *in vitro* [14]. Furthermore, IL-3 regulates the functional properties of mature ba, including mediator secretion, adhesiveness, and migration [23-28]. The effects of IL-3 on human ba are mediated through high affinity IL-3-binding sites [27]. Other cytokines acting on human ba are granulocyte-macrophage colony-stimulating factor (GM-CSF), IL-5, nerve growth factor (NGF), and several chemokines such as IL-8 and monocyte chemoattractant protein-1 (MCP-1) [29-33].

Table 2. Expression of surface receptors on human mast cells and basophils

Receptor	CD	Expression on (function) Mast cells	Basophils
IgERI (FcεRI)	n.c.	+ (HR)	+ (HR)
KLR/*c-kit*	117	+ (HR, S)	-/+
C5aR	88	+/-* (HR, C)	+ (HR, C)
uPAR	87	+(C)	+ (?)
IL-3R	123 + 131	-	+ (HR, S)
GM-CSFR	116 + 131	-	+ (HR, S)
IL-5R	125 + 131	-	+ (HR, S)
IL-8R	128	-	+ (HR, C)

CD, cluster of differentiation; HR, histamine release; S, survival; C, chemotaxis; n.c. not yet clustered; *mast cells obtained from juvenile foreskin express C5aR.

The differentiation of human MC is regulated by the ligand of the *c-kit* proto-oncogene product, *c-kit* ligand (KL). KL induces differentiation of human MC from their CD34$^+$ progenitor cells [14,15,17,18,34,35]. Furthermore, KL promotes the functional properties of MC, including mediator secretion, adhesion, and chemotaxis [36-39]. Other cytokines have little or no effect on mature tissue MC. The effects of KL on MC are mediated through the *c-kit* product, the tyrosine kinase receptor for KL.

Immature multipotent CD34$^+$ progenitor cells express both *c-kit* and IL-3 receptor [40,41]. However, during differentiation into MC or ba, receptor expression changes. Thus, mature ba express IL-3 receptors, but do not express large amounts of *c-kit*, whereas MC express large amounts of *c-kit*, but do not express IL-3 receptors [8,20,27,42]. Other surface structures are also differentially expressed on human MC and ba (see Table 2). For example, ba express substantial amounts of CD11, CD17, CD18, CD25, or CD31 [3]. The vitronectin receptor CD51/CD61 is detectable on tissue MC, but not on ba [3]. With regard to granular mediators, MC, but not ba, store huge amounts of tryptase and heparin [2]. During differentiation of MC and ba from their progenitor cells, diverse mediators and antigens are synthesized. The production of histamine and tryptase in MC progenitors is induced by KL [15,17,18]. However, other MC antigens like chymase or IgER require the stimulatory activity of additional cytokines. Thus, IL-4 promotes expression of IgER and chymase in KL-induced MC progenitor cells [43,44].

Based on expression of tryptase and chymase, two types of mature tissue MC have been defined [45]. The MC_{TC} type contains large amounts of both enzymes in its granules [45]. In contrast, the MC_T type of MC is stained by anti-tryptase, but not anti-chymase antibody. Large amounts of MC_{TC} are detectable in skin, whereas the MC_T type is found in the connective tissue of many visceral organs [45].

MC and ba have been implicated in various inflammatory conditions, including allergic reactions, infectious diseases, and chronic inflammation [2-7,46]. Interestingly, however, very little is known about possible physiologic functions of MC and ba. In the case of MC, several lines of evidence point to the production of repair molecules regulating tissue homeostasis. Such repair molecules include heparin, proteolytic enzymes, and pro-fibrinolytic antigens [2,45,47]. One explanation for the apparent lack of information about MC and ba is their relatively small number in blood and tissues, and the difficulty in isolating sufficient numbers of cells from various sources. Another problem is that few standard techniques for the isolation of MC and ba have been presented to date. Moreover, these techniques are sometimes "tricky," and often require special considerations and technology. The aim of this chapter is to provide guidelines for the isolation and culture of human MC and ba.

2. TISSUE PROCUREMENT AND PROCESSING

2.1 Progenitor Cells

MC and ba progenitor cells are detectable in fetal liver, CB, adult BM, and adult PB [13-18]. The numbers of progenitor cells vary between sources, and correlate with the numbers of CD34$^+$ cells [14,34,35]. In contrast to other sources, relatively low numbers of progenitors are present in adult PB. BM is a rich source of ba progenitor cells [13]. In patients receiving granulocyte colony-stimulating factor (G-CSF) or GM-CSF for stem cell mobilization, elevated numbers of circulating CD34$^+$ cells are found in the blood. However, many of these circulating progenitors show precommitment to neutrophils or monocytes and therefore do not represent an optimal tool for studying the *in vitro* differentiation of human MC or ba. Also, the progenitors should not be obtained from individuals who suffer from infectious diseases, overt allergy, or chronic inflammation. Some leukemias, especially chronic myeloid leukemia (CML), have high levels of circulating progenitors and ba. From these patients, the culture of ba from progenitors may yield high numbers of ba with a longer lifespan, as compared to normal ba. However, these cells are often quite immature and contaminated (or even outnumbered) by myeloblasts and promyelocytes. In patients with mastocytosis, CD34$^+$ progenitor cells may give rise to neoplastic MC [48].

No special consideration is required when MC or ba progenitor cells are prepared from PB or BM samples. As anticoagulants, heparin or EDTA can be used. The time from sampling of specimens until start of tissue processing should not exceed six hours. Room temperature is recommended for collection and initial processing of cells. MC or ba progenitor cells can be enriched in a first step by density gradient centrifugation (using Ficoll or a similar reagent), and are recovered together with mononuclear cells (MNC). The PB MNC fractions are composed of lymphocytes, monocytes, ba and CD34$^+$ progenitor cells. In the case of BM, multiple stages of cell maturation (myeloid cells) are also present. After isolation, MNC should be washed in medium containing EDTA. In each case, the composition of cells, morphology, and cell viability should be analyzed by Giemsa (or similar) staining. The cytospin preparation should be mounted and recovered for proper documentation. Cell viability can be checked by a trypan blue exclusion test. Usually, the cell viability of MNC is >90%. If the viability is <80%, no culture experiments should be started. If the MNC fraction is subjected to further enrichment of progenitors, the cells should be kept at 4 °C. Freeze-thawing of MNC or CD34$^+$ cells can be performed, but is not recommended for cells that are used in long-term culture experiments. Prior to purification of CD34$^+$ progenitor cells, it is useful to confirm the presence (percentage) of CD34$^+$ cells by monoclonal antibody (mAb) and light microscopy or by flow cytometry. For the processing and isolation of fetal liver cells, a different protocol is applied. Usually, these progenitor cells are recovered by mincing the fetal liver specimens, followed by filtration through an appropriate membrane, and gradient centrifugation [44].

2.2 Peripheral Blood Basophils

Blood ba are short-lived granulocytic cells which exhibit densities similar to those of PB monocytes and lymphocytes. Therefore, these cells are enriched in the MNC fraction of blood. PB ba can be obtained from healthy individuals or patients suffering from disease. Blood samples are recovered in syringes containing anticoagulant. In the case of untreated CML, large numbers of leukemic ba (viable and long-lived) can be obtained [49]. CML patients treated with interferon-alpha, hydroxyurea, or other cytoreductive drugs may have normal (or even reduced) numbers of ba. In patients with atopic or severe allergic disease, or chronic inflammatory diseases, ba often tend to undergo spontaneous degranulation during sampling or isolation. CB samples also contain sufficient numbers of ba. In contrast to other sources, CB ba express lower levels of receptor-bound IgE on their surface. In each case, MNC should be washed after isolation and checked for morphology, composition of cells (presence of ba), and cell viability (see above). If ba are subjected to further purification, MNC should be kept at 4 °C.

2.3 Tissue Mast Cells

MC can be enriched from a number of different organs including lung, skin, uterus, intestine or heart. Depending on the amount of tissue sample, sufficient numbers of cells can be obtained from lung, uterus, intestine and skin. The most common sources that have been used in our laboratory are lung and foreskin. Before tissue is procured, one should exclude recent chemotherapy, infectious diseases affecting the organ (tuberculosis and others), and organ necrosis. In the case of lung, intestine and heart, it is of great importance to put the tissue into Ca/Mg-free Tyrode's buffer immediately after resection [50,51]. In the case of uterus, MC are preserved quite well in the tissue for several hours (reason unknown), so the tissue can be transported into the laboratory before buffer is added. Foreskin or skin obtained from other sites should be placed into buffer as soon as possible. The first critical step in the laboratory is to remove all blood cells and all other non-structural cells from the tissue [50]. For this purpose, the tissue is cut into small pieces and washed extensively in Tyrode´s buffer. In our laboratory, the washing procedure is performed at room temperature using 50 ml plastic tubes. Usually, 10-30 washing steps should be adequate. The washing is best performed by use of an electric shaker. The washing procedure comes to an end when the buffer remains clear after heavy shaking. In case of severe contamination with blood cells (seen with lung samples), the washed specimens should be kept in Tyrode´s buffer overnight (room temperature), and again washed on the next day.

2.4 Isolation of $CD34^+$ Progenitor Cells

The isolation and purification of $CD34^+$ progenitor cells or subsets of $CD34^+$ cells has been described extensively [52-55]. Most of these techniques utilize mAb against CD34 and flow cytometry. With BM or CB, these progenitors can be directly sorted from MNC samples (although a pre-enrichment step is helpful). However, in the case of PB obtained from normal adult individuals, a pre-enrichment step using magnetic beads or an elutriator (to remove lymphocytes and/or monocytes) is required. Despite such pre-enrichment, the numbers of $CD34^+$ cells in normal adult PB are very low, such that excessive amounts of blood (0.5 l) are required as the starting material in order to obtain sufficient cell numbers.

2.5 Isolation of Basophils from CML Blood

Depending on the stage of disease and other factors, the percentage of ba in PB smears in patients with CML may range from a few to 40%, or even more. In the MNC fraction, the percentage of ba is further increased compared to the differential count. Often, these MNC fractions already contain 20-30% ba, and sometimes even exceed 50%. A further enrichment can be obtained by elutriation or gradient centrifugation. However, the purity reached without antigen-based isolation usually does not exceed 90%.

The purification of CML ba using antigen-specific mAb can be performed by either positive selection or negative selection techniques. Positive selection is usually performed by flow cytometry, although other techniques may also be applicable. The mAb that have successfully been used to purify CML ba include anti-IgE [56-59], Bsp-1 [60], anti-IL-3Rα/CD123 [unpublished], and anti-lactosylceramide/CDw17 [61]. CDw17 is recommended for isolation of mature ba (immature ba lack CDw17). All of the above antigens are also expressed on other (non-ba) cells, albeit in lower amounts. This means that the ba have to be defined and separated as brightly positive (ba^{++}) cells in flow cytometry procedures. No other special consideration is required for the sorting procedure itself, although some of the ba may undergo degranulation during cell sorting. It is therefore recommended that the sorting procedure be performed using sterile conditions and the cells kept at continuous low temperature. In most cases, the sorting procedure has to be repeated to obtain ultrapure cell fractions (>99%). Purification of CML ba by negative selection has been performed using a cocktail of mAb and complement [49]. The cocktail is essentially composed of mAb against CD3, CD7, CD14, CD15, CD20, CD24, CD57, and HLA-DR (each approximately 25 μg per 10^8 cells). In our laboratory, complement lysis has been performed using rabbit complement (1-3 ml per 10^8 cells) [49]. Complement lysis should be performed twice in order to remove all residual cells [51]. However, the lysis protocol is expensive and the resulting purity usually does not exceed 95%. By contrast, the positive selection protocols (sorting) usually result in a purity of >97%. Such purity allows for investigation of mRNA expression by RT-PCR [61]. An advantage of the negative selection protocol (over positive selection techniques) is that the surface receptors on isolated ba are not occupied by mAb. When using the negative selection technique, one should reconfirm the correct (complement lysis-mediating) isotype of the mAb.

2.6 Isolation of Basophils from Normal Blood

These conditions and techniques are similar to those for CML ba. However, the numbers of ba are much lower in most cases. This requires special considerations. The donors must donate a significant amount of blood (0.3-0.5 l) [61]. Also, prior to positive or negative selection of cells from MNC, a pre-enrichment step is required [34,61]. This step can be performed by elutriation or negative selection by beads [34,61]. The cell fractions that are used for sorting should contain at least 10% ba. Recently, a multi-step selection procedure for normal human ba has been established in our laboratory [61]. This protocol includes i) Ficoll separation of blood, ii) elutriation of MNC (to remove lymphocytes), iii) magnetic cell separation of $CD14^+$ (monocytes) and $CD14^-$ (ba-containing) cell fractions, and iv) sorting of $CD14^-$ cells for $CDw17^{++}$ cells (ba). Using this protocol, we have been able to reproducibly purify normal ba to near homogeneity (>98%) [61]. Also, the purified ba were found to be viable, and can be analyzed for mRNA expression by RT-PCR [61].

2.7 Isolation and Purification of Mast Cells from Tissues

A number of protocols suitable for the isolation of human MC from various tissues including lung, skin, heart, uterus, foreskin or intestine, have been described [9,10,50,51,61-65]. These protocols use various combinations of enzymes or single-enzyme digestion to disperse the primary tissue specimens. The most useful enzyme appears to be collagenase type II at a concentration of 2-4 mg per ml (approximately 1-5 mg per g net tissue weight). Other enzymes used for the isolation of MC are pronases, chymotryptic enzymes, hyaluronidase, DNAse, or papain. However, we recommend the use of collagenase as a single enzyme for all tissues except heart (see below). Such uniform procedure for most MC types allows for comparative analysis and helps to keep control experiments (excluding enzyme effects) to a minimum. Thus, in contrast to proteases or other enzymes, collagenase appears to degrade few cellular substrates. After digestion, dispersed cells are recovered by nytex cloth or similar material. Then the cells should be washed twice in medium or buffer and examined for the percentage of MC by Giemsa staining. Cell viability is checked by trypan blue exclusion. In the case of heart MC, it is important to remove myocytes as soon as possible during the isolation procedure. This is because after lysis, myocyte-derived substances critically affect the viability of MC. We usually remove myocyte cell ghosts during an initial purification step by using pronase-E, hyaluronidase, and DNAse [64].

As with ba, MC can be purified to near homogeneity by either positive or negative selection techniques. In most cases, MC are purified from lung or uterus specimens because they contain sufficient cell numbers. Less frequently, MC are purified from foreskin or other sources (low MC numbers). Today, MC are mostly purified by positive selection using mAb against *c-kit* (CD117) [61,66]. If there are large numbers of MC in the primary cell suspensions, the final purity usually exceeds 97% (after resorting). A pre-enrichment step (elutriation or magnetic separation) may also be helpful prior to sorting with *c-kit* [61]. The *c-kit*-sorted MC are functionally active and can be kept in culture over a longer time period [61]. Moreover, these ultrapure *c-kit*$^+$ MC can be examined for expression of mRNA by RT-PCR [61]. Negative selection of MC has also been described [51]. This isolation technique is performed using a mixture of mAb to remove non-MC lineage cells [51]. The general considerations and technical details are similar to those for ba (see above). However, since MC express only a limited number of myeloid cell surface antigens [8], it is relatively easy to select a useful cocktail of mAb. We have used mAb against CD2, CD3, CD5, CD11b, CD14, CD15, CD20, CD24, CD57, and HLA-DR to enrich human lung MC [51].

3. CULTURE TECHNIQUES

3.1 Culture of Basophils from Their Progenitor Cells

Human ba can be cultured and grown from their progenitor cells derived from various sources. Unfractionated MNC or purified CD34$^+$ progenitor cells (all sources) may be used as the starting cell material. In the case of unfractionated precursor cells (MNC), BM should be used [13]. PB MNC contain few ba progenitor cells. We recommend the culture of ba progenitor cells in liquid cell suspension, although these cells also grow and differentiate in semi-solid media [12]. The culture period (time of differentiation of ba from their progenitors) usually takes 2-3 weeks. The differentiation-inducing cytokines are added on day zero. We usually add rhIL-3 (100 U/ml) to 0.5 x 10^6 BM MNC or 10^4 CD34$^+$ cells in 1 ml assay volume [13]. The starting suspension contains RPMI 1640 medium (or similar culture medium), 10% fetal bovine serum (FBS, pretested for growth of progenitors in the investigator's laboratory), antibiotics, and 100 U rhIL-3 per ml. Cultures are maintained in a humidified 37 °C atmosphere containing 5% CO_2. After 14 days (standard time point) or later (days 21 or 28), cells are harvested. The percentage of ba in IL-3-triggered BM cells after 14 days amounts to aproximately 10-40%. Other cells frequently found in such cultures are eosinophils and monocytes [13]. A higher percentage of ba (compared with IL-3 alone) is obtained when cultures are maintained in a combination of IL-3 (100 U/ml) and transforming growth factor-β TGF-β (1-10 ng/ml) [67]. By contrast, addition of GM-CSF or IL-5 does not result in enhanced ba growth. Using CD34$^+$ cells as the starting material, culture conditions should be essentially the same as those for MNC. The differentiation of ba from CD34$^+$ cells may yield higher numbers of ba compared to BM MNC. In contrast, the amounts of ba in cultures of PB MNC are rather low. As a measure of ba lineage differentiation, several methods have been described. The total number of metachromatic cells per ml can be calculated based on the total number of cultured cells and the percentage of metachromatic cells (evaluated by Giemsa staining). As a non-subjective parameter, the amount of total histamine in the cultures can be measured. The calculated amount of histamine per cultured ba should be determined as a control (average: 0.1-1.0 pg per cell) [13].

3.2 Culture of Mature Isolated Basophils (Normal And CML-Derived)

The culture of mature ba (obtained from blood) requires special considerations. In the case of CML ba, a lifespan of several days can be expected. Nevertheless, CML ba rapidly lose their functional properties and viability unless exposed to IL-3 (IL-5 and GM-CSF can also be used). Likewise, in the absence of IL-3, cultured CML ba do not release histamine when exposed to anti-IgE. When CML ba are maintained in IL-3 (1-10 U/ml), however, these cells are responsive to IgE-dependent stimuli. Human ba obtained from healthy individuals should be kept in culture with IL-3 unless they are used immediately for further experiments. Using IL-3 (1-10 U/ml) as a culture adjunct, these cells can be kept alive for several days [61]. In the case of CML, the ba can even be kept alive in IL-3 for several weeks [49]. Both CML and normal ba are maintained in RPMI 1640 medium (or similar culture medium) with 10% (pre-tested) FBS and antibiotics.

3.3 Culture of Mast Cells from Their Progenitors

The differentiation of MC from uncommitted hematopoietic $CD34^+$ progenitor cells is a prolonged process when compared to the differentiation time of granulocytes. In particular, while granulocytes (including ba) differentiate and mature within 2-3 weeks, MC differentiation takes several months. Therefore, the *in vitro* differentiation of MC is best analyzed in a long-term culture system (1-3 months). The progenitor cells are maintained in culture medium with 10% (pre-tested) FBS, rhKL (starting concentration 10-100 ng/ml) and antibiotics in a fully humidified atmosphere of 5% CO_2 at 37 °C [15,17,18]. Fresh medium and growth factors are added every two weeks (1-10 ng/ml of rhKL). Addition of IL-6 to the culture may yield higher numbers of MC [68,69]. The differentiation and growth of chymase-containing MC requires addition of IL-4 (10-20 ng/ml) to the culture medium [44]. IL-4 also promotes expression of IgE receptor on cultured human MC [44]. As a measure of MC differentiation, several methods have been described. The number of metachromatic cells per ml can be calculated based on the total number of cultured cells and the percentage of metachromatic cells (evaluated by tryptase staining of cultured cells by immunohistochemistry on cytospin slides). A non-subjective parameter is the amount of total tryptase (and total histamine) in cell cultures (measured as tryptase per ml cell suspension) [15]. The calculated amount of tryptase per cultured MC should be determined as control (average 0.1-5.0 pg per cell).

3.4 Culture of Isolated Mature Tissue Mast Cells

In contrast to ba, mature MC are long-lived cells. Under appropriate culture conditions, mature human MC can be maintained for several weeks without significant loss of cell viability. Usually, MC are kept in complete medium with 10% FBS, antibiotics, and rhKL (1 ng/ml). A stromal cell layer can also be used instead of KL. In the absence of KL or accessory cells (ultrapure MC preparations), the viability of cells may decrease after several days in culture due to apoptosis.

4. UTILITY OF SYSTEMS

The apparent clinical significance of MC and ba deserves a more thorough investigation of these cells. Therefore, the isolation and culture of human MC and ba is of particular interest. It is also noteworthy that various animal models (murine) have been established in order to investigate the biology of these cells, such as the important models of MC deficient animals. Despite their usefulness, such models have several disadvantages. First, human MC differ from murine (and other animal-derived) MC in several biological aspects [8,42]. Second, many animal species (including mouse) have low numbers of ba, preventing extensive analysis of such cells. During the last few years, many human MC and ba models have been established. These models have been found to be suitable and applicable to various research fields.

4.1 Basic Science and Cytokine Research

One example of the utility of MC and ba technologies is cytokine research. It has been described that MC are a potent source of TNF-α, and ba a potent source of IL-4. These observations may be of pathophysiologic as well as clinical significance. Other cytokines and chemokines are also produced by MC and/or ba. Another aspect relates to the clinical use of growth factors. For example, the growth factors G-CSF, GM-CSF, IL-3 and KL have major effects on progenitor cells, but may also influence the functional properties of mature hematopoietic effector cells [23-27,36-39]. In the case of ba and MC, such effects may lead to adverse clinical symptoms. Therefore, the pre-clinical phase of investigation of novel factors should include screening for possible activating effects on ba and/or MC. Likewise, KL administration is known to induce local flushing and a reversible MC hyperplasia in patients. Allergy-like symptoms have also been reported in patients receiving IL-3 or GM-CSF.

4.2 Clinical Immunology and Allergy Research

Another important aspect of MC and ba is their role in allergic and other inflammatory (immunologic) reactions. Thus, both cells can release a number of pro-inflammatory mediators in response to IgE-dependent or other immunologic stimuli. The conditions of mediator release and regulation of the release of MC and ba is the subject of ongoing research. Likewise, mediator release from human MC and ba can be used as a tool for screening novel anti-allergic drugs. MC or ba can also be isolated from patients with inflammatory or allergic diseases in order to investigate their functional properties. Such analysis can lead to a better understanding of mechanisms of allergic and other inflammatory disorders. The allergen-dependent release of histamine (or other mediators) from ba and MC can also be used as a sensitive and reliable *in vitro* test assay for allergy typing using recombinant allergens. The advantage over cutaneous tests is that patients are not exposed to allergenic (and potentially sensitizing) material. Allergy research is one of the most important applications of MC and ba technologies.

4.3 Pharmacology and Toxicology

A number of drugs reportedly exert effects on human MC or ba. Such effects are of considerable clinical interest. Likewise, anti-allergic drugs may show a deactivating effect on MC or ba. On the other hand, pharmacologists often screen for toxic effects of certain drugs or other agents using MC or ba. Likewise, high concentrations of Ag^{2+} can induce liberation of histamine from MC and ba [70]. Another example is the effect of morphine on human cutaneous MC [9,10]. Since histamine is a well-known toxic agent, the knowledge about histamine releasing effects of potentially noxious substances may be of importance.

5. CONCLUDING REMARKS

Several technical details and considerations have to be borne in mind when planning the isolation and culture of human MC and ba. In the case of primary (mature) cells, it may be difficult to obtain sufficient amounts of pure cells. In the case of immature progenitor cells, knowledge about optimal growth factors and differentiation time in culture is helpful. The present article provides a guide for isolating and culturing human MC and ba.

LITERATURE CITED

1. Ehrlich P (1879) Beiträge zur Kenntnis der granulierten Bindegewebszellen und der eosinophilen Leukozyten. *Arch Anat Physiol* 3:166-169.
2. Schwartz LB, Huff TF (1991) Mast cells. In: RG Crystal and JB West (eds.), *The Lung*, New York, NY Raven Press, pp. 601-616.
3. Valent P, Bettelheim P (1990) The Human Basophil. *Crit Rev Hematol Oncol* 10:327-352.
4. Johnston SL, Holgate ST (1990) Cellular and chemical mediators - their roles in allergic diseases. *Curr Opin Immunol* 2:513-524.
5. Serafin WE, Austen KF (1987) Mediators of immediate hypersensitivity reactions, *N Engl J Med* 317:30-34.
6. Lewis AR, Austen KF (1981) Mediation of local homeostasis and inflammation by leukotrienes and other mast cell-dependent compounds. *Nature* 293:103-108.
7. Ishizaka T, Ishizaka K (1984) Activation of mast cells for mediator release through IgE receptors. *Prog Allergy* 34:188-235.
8. Valent P, Bettelheim P (1992) Cell surface structures on human basophils and mast cells: Biochemical and functional characterization. *Adv Immunol* 52:333-421.
9. Benyon RC, Lowman MA, Church MK (1987) Human skin mast cells: Their dispersion, purification, and secretory characterization. *J Immunol* 138:861-867.
10. Lawrence ID, Warner JA, Cohan VL, Hubbard WC, Kagey-Sobotka A, Lichtenstein LM (1987) Purification and characterization of human skin mast cells. *J Immunol* 139:3062-3069.
11. Füreder W, Agis H, Willheim M, Bankl HC, Maier U, Kishi K, Müller MR, Czerwenka K, Radaszkiewicz T, Butterfield JH, Klappacher G, Sperr WR, Oppermann M, Lechner K, Valent P (1995) Differential expression of complement receptors on human basophils and mast cells: Evidence for mast cell heterogeneity and C5aR/CD88 expression on skin mast cells. *J Immunol* 155:3152-3160.
12. Denburg JA, Messner H, Lim B, Jamal N, Talizyn S, Bienenstock J (1985) Clonal origin of human basophil/mast cells from circulating multipotent hemopoietic progenitors. *Exp Hematol* 13:185-188.
13. Valent P, Schmidt G, Besemer J, Mayer P, Liehl E, Hinterberger W, Lechner K, Maurer D, Bettelheim P (1989) Interleukin-3 is a differentiation factor for human basophils. *Blood* 73:1763-1769.
14. Kirshenbaum AS, Goff JP, Kessler SW, Mican JM, Zsebo KM, Metcalfe DD (1992) Effects of IL-3 and stem cell factor on the appearance of human basophils and mast cells from $CD34^+$ pluripotent progenitor cells. *J Immunol* 148:772-777.

15. Valent P, Spanblöchl E, Sperr WR, Sillaber C, Agis H, Strobl H, Zsebo KM, Geissler K, Bettelheim P, Lechner K (1992) Induction of differentiation of human mast cells from BM and PB mononuclear cells by recombinant human stem cell factor (KL) / kit ligand (KL) in long term culture. *Blood* 80:2237-2245.
16. Saito H, Hatake K, Dvorak AM, Leiferman KM, Donnenberg AD, Arai N, Ishizaka K, Ishizaka T (1988) Selective differentiation and proliferation of hematopoietic cells induced by recombinant human interleukins. *Proc Natl Acad Sci (USA)* 85:2288-2292.
17. Irani AM, Nilsson G, Miettinen U, Craig SS, Ashman LK, Ishizaka T, Zsebo KM, Schwartz LB (1992) Recombinant human stem cell factor stimulates differentiation of human mast cells from dispersed fetal liver cells. *Blood* 80:3009-3016.
18. Mitsui H, Furitsu T, Dvorak AM, Irani AMA, Schwartz LB, Inagaki N, Takei M, Ishizaka K, Zsebo KM, Gillis S, Ishizaka T (1993) Development of human mast cells from umbilical CB cells by recombinant human and murine stem cell factor, *c-kit* ligand. *Proc Natl Acad Sci (USA)* 90:735-740.
19. Galli SJ (1990) Biology of disease: New insights into ´the riddle of the mast cells´: microenvironmental regulation of mast cell development and phenotypic heterogeneity. *Lab Invest* 62:5-33.
20. Valent P (1994) The riddle of the mast cell: kit(CD117)-ligand as missing link ? *Immunol Today* 15:111-114.
21. Födinger M, Fritsch G, Winkler K, Emminger W, Mitterbauer G, Gadner H, Valent P, Mannhalter Ch (1994) Origin of human mast cells: Development from transplanted hemopoietic stem cells after allogeneic BM transplantation. *Blood* 84:2954-2459.
22. Charlesworth EN, Hood AF, Soter NA, Kagey-Sobotka A, Norman PS, Lichtenstein LM (1989) Cutaneous late-phase response to allergen. Mediator release and inflammatory cell infiltration. *J Clin Invest* 83:1519-1526.
23. Kurimoto Y, deWeck AL, Dahinden CA (1989) Interleukin 3-dependent mediator release in basophils triggered by C5a. *J Exp Med* 170:467-469.
24. Schleimer RP, Derse CP, Friedmann B, Gillis S, Plaut M, Lichtenstein LM, MacGlashan DW (1989) Regulation of basophil mediator release by cytokines. I. Interactions with antiinflammatory steroids. *J Immunol* 143:1310-1317.
25. Bochner BS, McKelvey AA, Sterbinsky SA, Hildreth JEK, Derse CP, Klunk DA, Lichtenstein LM, Schleimer RP (1990) Interleukin-3 augments adhesiveness for endothelium and CD11b expression in human basophils but not neutrophils. *J Immunol* 145:1832-1837.
26. Hirai K, Morita YM, Misaki Y, Ohta K, Takaishi T, Suzuki S, Motoyoshi K, Miyamoto T (1988) Modulation of human basophil histamine release by hemopoietic growth factors. *J Immunol* 141:3958-3964.
27. Valent P, Besemer J, Muhm M, Majdic O, Lechner K, Bettelheim P (1989) Interleukin 3 activates human blood basophils via high affinity binding sites. *Proc Natl Acad Sci (USA)* 86:5542-5547.
28. Yamaguchi M, Hirai K, Takaishi T, Ohta K, Morita Y, Suzuki S, Ito K (1992) Hemopoietic growth factors induce human basophil migration in vitro. *Clin Exp Allergy* 22:379-383.
29. Bishoff SC, Brunner T, deWeck AL, Dahinden CA (1990) Interleukin-5 modifies histamine release and leukotriene generation by human basophils in response to diverse agonists. *J Exp Med* 172:1577-1582.
30. Dahinden CA, Kurimoto Y, de Weck AL, Lindley I, Dewald B, Baggiolini M (1989) The neutrophil- activating peptide NAF/NAP-1 induces histamine and leukotriene release by interleukin 3 primed basophils. *J Exp Med* 170:1787-1792.

31. Krieger M, Brunner T, Bischoff SC, Tscharner V, Walz A, Moser B, Baggiolini M, Dahinden CA (1992) Activation of human basophils through the IL-8 receptor. *J Immunol* 149:2662-2667.
32. Kuna P, Reddigari SR, Schall TJ, Rucinski D, Sadick M, Kaplan AP (1993) Characterization of human basophil response to cytokines, growth factors, and histamine releasing factors of the intercrine / chemokine family. *J Immunol* 150:1932-1943.
33. Bischoff SC, Krieger M, Brunner T, Dahinden CA (1992) Monocyte chemotactic protein 1 is a potent activator of human basophils. *J Exp Med* 175:1271-1275.
34. Agis H, Willheim M, Sperr WR, Wilfing A, Krömer E, Kabrna E, Spanblöchl E, Strobl H, Geissler K, Spittler A, Boltz-Nitulescu G, Majdic O, Lechner K, Valent P (1993) Monocytes do not make mast cells when cultured in the presence of KL: Characterization of the circulating mast cell progenitor as a *c-kit*$^+$, CD34$^+$, Ly$^-$, CD14$^-$, CD17$^-$, colony forming cell. *J Immunol* 151:4221-4227.
35. Kirshenbaum AS, Kessler SW, Goff JP, Metcalfe DD (1991) Demonstration of the origin of human mast cells from CD34+ BM progenitor cells. *J Immunol* 146:1410-1415.
36. Bischoff SC, Dahinden CA (1992) *c-kit* Ligand: a unique potentiator of mediator release from human lung mast cells. *J Exp Med* 175:237-244.
37. Columbo M, Morowitz EM, Botana LM, McGlashan DW, Bochner BS, Gillis S, Zsebo KM, Galli SJ, Lichtenstein LM (1992) The human recombinant *c-kit* receptor ligand, rhKL, induces mediator release from human cutaneous mast cells and enhances IgE-dependent mediator release from both skin mast cells and PB basophils. *J Immunol* 149:599-608.
38. Sperr WR, Czerwenka K, Mundigler G, Müller MR, Semper H, Klappacher G, Glogar HD, Lechner K, Valent P (1993) Specific activation of human mast cells by the ligand for *c-kit*: comparison between lung, uterus and heart mast cells. *Int Arch Allergy Appl Immunol* 102:170-175.
39. Nilsson G, Butterfield JH, Nilsson K, Siegbahn A (1994) Stem cell factor is a chemotactic factor for human mast cells. *J Immunol* 153:3717-3724.
40. Ashman LK, Cambareri AC, To LB, Levinski RJ, Juttner CA (1991) Expression of *c-kit* antigen (*c-kit* proto-oncogene product) by normal human BM cells. *Blood* 78:30-37.
41. Sato N, Caux C, Kitamura T, Watanabe Y, Arai K, Banchereau J, Miyajima A (1993) Expression and factor-dependent modulation of the interleukin-3 receptor subunits on human hematopoietic cells. *Blood* 82:752-761.
42. Valent P, Besemer J, Sillaber Ch, Maurer D, Butterfield JH, Kishi K, Lechner K, Bettelheim P (1990) Failure to detect interleukin-3 binding sites on human mast cells. *J Immunol* 145:3432-3438.
43. Toru H, Ra C, Nonoyama S, Suzuki K, Yata J, Nakahata T (1996) Induction of the high affinity IgE receptor (Fc epsilon RI) on human mast cells by IL-4. *Int Immunol* 8:1367-1373.
44. Xia HZ, Du Z, Craig S, Klisch G, Noben-Trauth N, Kochan JP, Huff TH, Irani AM, Schwartz LB (1997) Effect of recombinant human IL-4 on tryptase, chymase, and Fc epsilon receptor type I expression in recombinant human stem cell factor-dependent fetal liver-derived human mast cells. *J Immunol* 159:2911-2921.
45. Irani AA, Schechter NM, Craig SS, DeBlois G, Schwartz LB (1986) Two human mast cell subsets with distinct neutral protease composition. *Proc Natl Acad Sci (USA)* 83:4464-4468.
46. Galli SJ (1993) New Concepts about the mast cell. *N Engl J Med* 328:257-265.
47. Valent P, Sillaber C, Baghestanian M, Bankl HC, Kiener HP, Lechner K, Binder BR (1998) What have mast cells to do with edema formation, the consecutive repair, and fibrinolysis ? *Int Arch Allergy Immunol* 115:2-8.

48. Rottem M, Okada T, Goff JP, Metcalfe DD (1994) Mast cells cultured from PB of normal donors and patients with mastocytosis originate from a $CD34^{+}/FceRI^{-}$ cell population. *Blood* 84:2489-2496.
49. Stain C, Stockinger H, Scharf M, Jäger U, Gössinger H, Lechner K, Bettelheim P (1987) Human blood basophils display a unique phenotype including activation-linked membrane structures. *Blood* 70:1872-1879.
50. Schulman ES, Mac Glashan DW, Peters SP, Schleimer RP, Newball HH, Lichtenstein LM (1981) Lung mast cells: purification and characterisation. *J Immunol* 129:2662-2667.
51. Valent P, Ashman LK, Hinterberger W, Eckersberger F, Majdic O, Lechner K, Bettelheim P (1989) Mast cell typing: demonstration of a distinct hemopoietic cell type and evidence for immunophenotypic relationship to mononuclear phagocytes. *Blood* 73:1778-1785.
52. Sowala H, Wunder E, Henon P (1990) Purification and characterization of the CD34+ hematopoietic precursor cell population by flow cytometry. *BM Transplant* 5:9-10.
53. Cardoso AA, Li ML, Batard P, Sansilvestri P, Hatzfeld A, Levesque JP, Lebkowski JS, Hatzfeld J (1993) Human umbilical CB CD34+ cell purification with high yield of early progenitors. *J Hematother* 2:275-279.
54. Charbord P, Newton I, Voillat L, Schaal JP, Herve P (1996) The purification of CD34+ cells from human CB: comparison of separation techniques and cytokine requirements for optimal growth of clonogenic progenitors. *Br J Haematol* 94:449-454.
55. deWynter EA, Coutinho LH, Pei X, Marsh JC, Hows J, Luft T, Testa NG (1995) Comparison of purity and enrichment of CD34+ cells from BM, umbilical cord and PB (primed for apheresis) using five separation systems. Stem Cells Dayt. 13:524-532.
56. Weil GJ, Leiserson WM, Chused TM (1983) Isolation of human basophils by flow microfluorometry. *J Immunol Methods* 58:359-363.
57. Graziano FM (1988) Basophils in allergic and inflammatory reactions. *Methods Enzymol* 162:522-538.
58. Pendy LM, Arra SJ, Anselmino LM, Thomas LL (1986) A flow cytometric procedure for the isolation of antigenically responsive basophils. *J Immunol Methods* 91:59-63.
59. Degenhart HJ, van der Maarel JW, Jongkind H, Raatgeep HC, Nejens HJ (1988) Properties of human basophils isolated by fluorescence-activated cell sorting. *J Allergy Clin Immunol* 82:455-461.
60. Bodger MP, Newton LA (1987) The purification of human basophils: Their immunophenotype and cytochemistry. *Br J Haematol* 67:281-284.
61. Willheim M, Agis H, Sperr WR, Köller M, Bankl HC, Kiener H, Fritsch G, Füreder W, Spittler A, Graninger W, Scheiner O, Gadner H, Lechner K, Boltz-Nitulescu G, Valent P (1995) Purification of human basophils and mast cells by multistep separation technique and mAb to CDw17 and CD117/*c-kit*. *J Immunol Methods* 182:115-129.
62. Massey WA, Guo CB, Dvorak AM, Hubbard WC, Bhagavan BS, Cohan VL, Warner WA, Kagey-Sobotka A, Lichtenstein LM (1991) Human uterine mast cells. Isolation, purification, characterization, and pharmacology. *J Immunol* 147:1621-1627.
63. Patella V, Marino I, Lamparter B, Arbustini E, Adt M, Marone G (1995) Human heart mast cells. Isolation, purification, ultrastructure, and immunologic characterization. *J Immunol* 154:2855-2865.
64. Sperr WR, Bankl HC, Mundigler G, Klappacher G, Großschmidt K, Agis H, Simon P, Laufer P, Imhof M, Radaszkiewicz Th, Glogar D, Lechner K, Valent P (1994) The human cardiac mast cell: Localization, isolation, phenotype and functional characterization. *Blood* 84:3876-3884.
65. Fox CC, Dvorak AM, Peters SP, Kagey-Sobotka A, Lichtenstein LM (1985) Isolation and characterization of human intestinal mucosal mast cells. *J Immunol* 135:483-491.

66. Okayama Y, Hunt TC, Kassel O, Ashman LK, Church MK (1994) Assessment of the anti-*c-kit* monoclonal antibody YB5.B8 in affinity magnetic enrichment of human lung mast cells. *J Immunol Methods* 169:153-161.
67. Sillaber C, Geissler K, Scherrer R, Kaltenbrunner R, Betelheim P, Lechner K, Valent P (1992) Type ß transforming growth factors promote interleukin-3 (IL-3)-dependent differentiation of human basophils but inhibit IL-3-dependent differentiation of human eosinophils. *Blood* 80:634-641.
68. Yanagida M, Fukamachi H, Ohgami K, Kuwaki T, Ishii H, Uzumaki H, Amano K, Tokiwa T, Mitsui H, Saito H, Iikura Y, Ishizaka T, Nakahata T (1995) Effects of T-helper 2-type cytokines, interleukin-3 (IL-3), IL-4, IL-5, and IL-6 on the survival of cultured human mast cells. *Blood* 86:3705-3714.
69. Saito H, Ebisawa M, Tachimoto H, Shichijo M, Fukagawa K, Matsumoto K, Iikura Y, Awaji T, Tsujimoto G, Yanagida M, Uzumaki H, Takahashi G, Tsuji K, Nakahata T (1996) Selective growth of human mast cells induced by Steel factor, IL-6, and prostaglandin E2 from CB mononuclear cells. *J Immunol* 157:343-350.
70. Schedle A, Samorapoompichit P, Rausch-Fan X, Franz W, Füreder W, Sperr WR, Sperr W, Ellinger A, Slavicek R, Boltz-Nitulescu G, Valent P (1995) Response of L-929 fibroblasts, human gingiva fibroblasts and human tissue mast cells to various metal cations. *J Dental Research* 74:1513-1520.

Chapter 11

Purification and Culture of Erythroid Progenitor Cells

Chun-Hua Dai, Amittha Wickrema and Sanford B Krantz
Department of Medicine/Hematology, Vanderbilt University Medical School, 547 Medical Research Building II, Nashville, TN 37232-6305. Tel: 001-615-327-5350; Fax: 001-615-321-6237; E-mail: krantz.sanford_b@nashville.va.gov

1. INTRODUCTION

The development of primary erythroid progenitor culture systems during the last decade has provided a means of studying biochemical and molecular events during growth factor mediated differentiation. Traditionally, hematopoiesis has been studied *in vitro* in semi-solid cultures where early progenitors are allowed to develop into discrete colonies which can be subsequently plucked for further analysis. With respect to the erythroid lineage, colonies are easily recognizable by their distinct color due to the presence of hemoglobin. Although this method was adequate for studies involving late erythroid progenitors, it did not provide adequate numbers of more immature progenitor cells to carry out molecular and cellular biology experiments during the proliferative phase of the differentiation program.

Human erythroid progenitor cells consist of immature blasts that are in various stages of maturation. Burst-forming units-erythroid (BFU-E) are early primitive erythroid progenitor cells that have a very high proliferative capacity and give rise to large colonies, generally consisting of more than 500 cells, when cultured *in vitro* [1]. Colony-forming units-erythroid (CFU-E) are derived from BFU-E and give rise to smaller colonies (8-49 cells) because of a lower proliferative capacity [2]. CFU-E further develop into more mature erythroid colony-forming cells which usually produce even smaller colonies containing 2-7 erythroblasts. All of the erythroid progenitor cells that form single colonies of 2-500 hemoglobinized cells have been collectively called erythroid colony-forming cells (ECFC) [3], and all of these progenitor cells eventually mature to late erythroblasts. While human fetal liver and bone marrow (BM) contain a high frequency of BFU-E and CFU-E, peripheral blood (PB) has mainly BFU-E, most of which are in a primitive stage of maturation, with only occasional rare CFU-E [1].

Purification of human BFU-E can be achieved from fetal liver and BM as well as PB [4-7]. However, it is more difficult to obtain fetal liver and human BM aspirates on a routine basis, and both tissues have a marked heterogeneity of developing erythroid progenitor cells. In contrast, PB is a readily available resource for purification of BFU-E, which are relatively free of mature erythroid progenitor cells, and basically in the same phase of development [8,9,10]. From these BFU-E, a large number of highly purified CFU-E can be generated *in vitro*, which are also in one selective phase of development.

Several primary erythroid cell culture systems have been developed where immature BFU-E have been purified from PB by utilizing negative selection methods to eliminate non-erythroid hematopoietic cells, thereby enriching for BFU-E [8,9]. These methods take advantage of the observation that normal PB has very few late CFU-E, unlike BM, and therefore, it is possible to obtain a relatively homogeneous population of early erythroid progenitors after purification. These early erythroid progenitors (BFU-E) can then be cultured with growth factors that favor survival and expansion of erythroid progenitors to give rise to a large population of early CFU-E, which can then be utilized for biochemical and molecular studies during the terminal differentiation program. Such a system was initially devised by Sawada et al. [8] where BFU-E and BFU-E-derived CFU-E were purified starting from a unit of whole blood. This system employed a series of monoclonal antibody (mAb)-mediated negative selection steps to deplete the majority of T-lymphocytes and other myeloid cells to obtain an enriched population of BFU-E. The BFU-E were then cultured for 6-7 days in semi-solid or liquid suspension cultures, and a further purification was performed before carrying out the second phase CFU-E cultures. This purification system was later modified by adding positive CD34 selection and enhanced negative selection to produce a very high purity of initial BFU-E [9].

Two other human erythroid progenitor purification systems have also been developed by Fibach et al. [11] and Peschle's group [12,13], in which PB has been utilized as the tissue source to enrich for erythroid progenitors. The system developed by Fibach et al. relied on T lymphocyte mAb followed by complement fixation to eliminate T-lymphocytes, adherent cell depletion to eliminate myeloid cells, and then the culturing of these cells during erythropoietin (Epo)-independent and dependent phases of erythroid development. The system developed by Peschle's group utilized a large panel of mAb and selected for multipotential hematopoietic progenitors that could be developed by different sets of growth factors to give rise to cells of the erythroid or granulocyte lineage. The methodology utilized relied on magnetic beads in addition to numerous de-bulking steps to reduce the lymphocytes and late myeloid cells. A further method described below is a modification from the original method developed by Sawada et al. [9], and incorporates a new magnetic microbead technology, as well as eliminating some of the time consuming steps during the selection process. Here we describe some of these methods for the purification of human PB BFU-E and generation of highly purified CFU-E in cell culture.

2. PURIFICATION OF BFU-E

Three methods for the purification of BFU-E are described here. They are Method I, through which highly purified day 1 BFU-E are obtained; Method II, through which lower purity but high yield day 1 BFU-E are obtained; and Method III, utilizing MACS magnetic microbeads to obtain enriched BFU-E with a simpler procedure and in a shorter time period, for subsequent culture to yield high purity and high yield CFU-E. A brief overall summary of these procedures is provided in Figure 1.

2.1 High Purity Method of BFU-E Purification (Method I) [9,10]

2.1.1 Materials and reagents

If not provided sterile, all reagents must be sterile filtered prior to use, and all manipulations must be performed in a laminar flow hood in order to maintain sterility throughout the protocol.

1. Heparin (Elkins-Sinn, Inc.)
2. Ficoll-Hypaque (FH:1.077 g/ml; Pharmacia, Piscataway, NY)
3. Dulbecco's phosphate-buffered saline (PBS; Sigma, St. Louis, MO), pH 7.3, containing sodium citrate (D-PBS; Sigma)
4. Bovine serum albumin (BSA; Intergen Co., Purchase, NY)
5. Alpha minimum essential medium (α-MEM; Sigma)
6. Deionized BSA (D-BSA; cf reference 3)

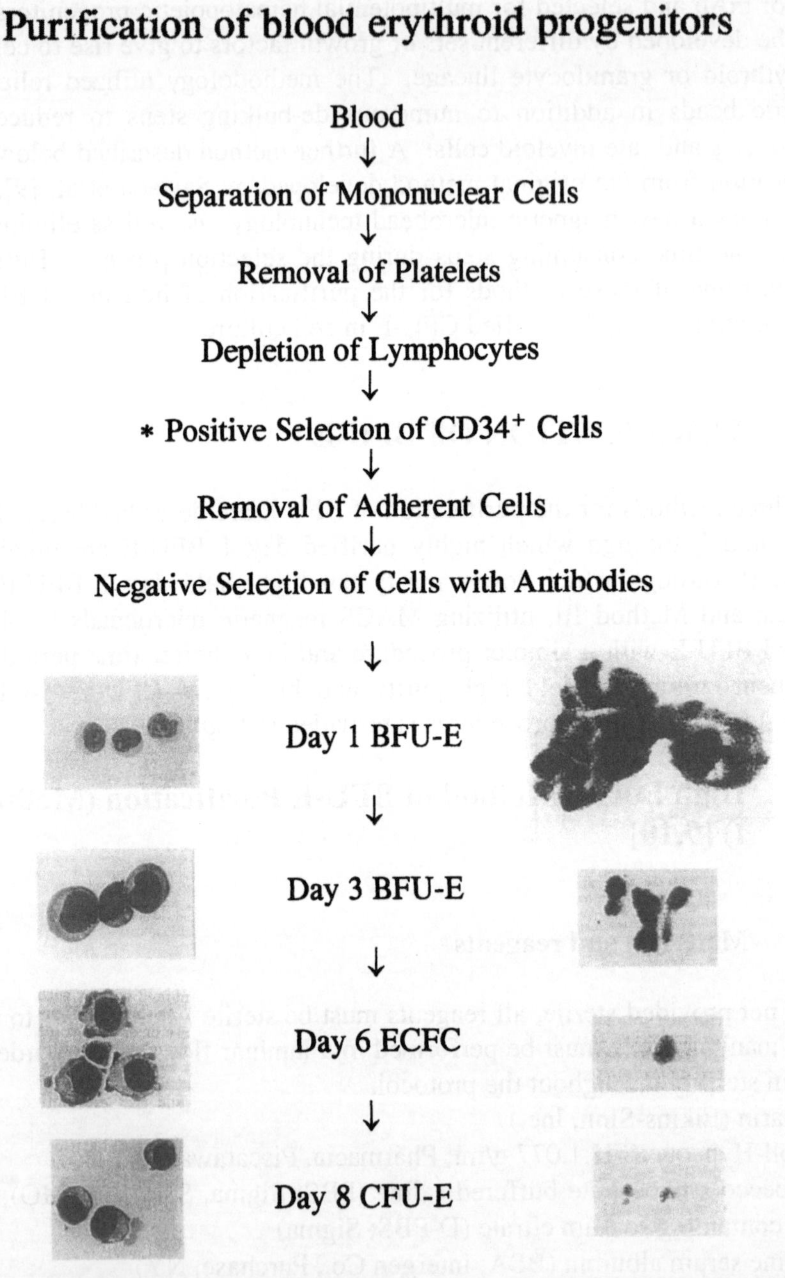

Figure 1. Purification of human blood BFU-E by Method I. Light microscopy cell morphology of the erythroid progenitors for each day is shown on the left, and colonies derived from these single progenitor cells are shown on the right. For Method II, CD34 positive selection is omitted and a lesser number of antibodies are used in negative selection.

7. Sheep red blood cells (SRBC) (Cappel, Organon Teknika Corp., West Chester, PA)
8. Neuraminidase (GIBCO Laboratories, Grand Island, NY
9. IgG fraction of rabbit anti-SRBC (Cappel)
10. Iscove's modified Dulbecco's medium (IMDM; Sigma)
11. CD34, CD15, CD33, CD45R, CD16, CD56, CD7, CD20, CD2, and CD11b antibodies (Becton Dickinson, San Jose, CA)
12. Immunomagnetic microspheres (Dynabeads M-450 coated with goat antimouse IgG, and goat antimouse IgM, Dynal Inc., Great Neck, NY)
13. Magnetic Plastic Concentrator (Dynal)
14. Disodium ethylenediamine tetraacetate (EDTA, Fisher Scientific Co., Fair Lawn, NJ)
15. Fetal bovine serum (FBS; Hyclone Laboratories, Logan, UT)
16. Recombinant human interleukin-3 (rhIL-3; Genetics Institute, Cambridge, MA)
17. Recombinant human erythropoietin (rhEpo; Amgen Inc, Thousand Oaks, CA)
18. Recombinant human c-kit ligand (KL; Amgen Inc., Thousand Oaks, CA)
19. Insulin (Calbiochem, Behring Diagnostics, LaJolla, CA)
20. 2-mercaptoethanol (2-ME; Eastman-Kodak Co., Rochester, NY)
21. Penicillin
22. Streptomycin
23. Trypan blue 0.4% solution (National Aniline, New York, NY; or Sigma)
24. Chymopapain (Chymocel-R; Baxter Biotech, Duarte, CA)

2.1.2 Detailed protocol

1. The initial steps in the purification scheme are designed to eliminate mature erythrocytes and platelets and obtain a mononuclear cell (MNC) fraction of cells that can be further enriched for BFU-E.
2. PB is obtained from normal adults with informed consent approved by the Committee for the Protection of Human Subjects. Four hundred ml blood are collected in sodium heparin at a final concentration of 20 U/ml.
3. Thirty-five ml blood are overlayed on 15 ml FH in 50 ml tubes and centrifuged at 400 *g* for 25 min at room temperature (RT, 24 °C). The interface MNC are collected, mixed with an equal volume of D-PBS, and centrifuged at 600 *g* for 15 min at RT. The cells are washed again with D-PBS and resuspended in 20 ml of D-PBS (FH cells).

4. The FH cell suspension is overlaid on 30 ml of 10% BSA in D-PBS and then centrifuged at 180 *g* for 10 min at RT to remove platelets. The cell pellet is resuspended in 20 ml of D-PBS, and this procedure is repeated. After removal of platelets, the cells are washed once with α-MEM and once with α-MEM containing 0.1% D-BSA, before suspension in α-MEM at a concentration of 1×10^7/ml (FH, PL^- cells).
5. This step is designed to remove lymphocytes. Only 10% of the MNC remain after this step. Ten ml or more of SRBC are washed three times with PBS and suspended in PBS at a concentration (vol/vol) of 10%. The SRBC suspension is divided into two groups and incubated separately with either neuraminidase at 25 U/ml or the IgG fraction of rabbit anti-SRBC at 46 μg/ml for 30 min at 37°C. After incubation, the SRBC are washed three times with PBS and are resuspended in α-MEM at a concentration of 2% for storage in ice until use.
6. The FH, PL^- cells are mixed with one-half volume of neuraminidase-treated SRBC and rabbit IgG-bound SRBC and are incubated for 10 min in a 37 °C water bath. After centrifugation at 130 *g* for 10 min at RT, the cells are incubated for 60 min in an ice bath. The cell pellet is dispersed and cells are centrifuged over FH at 400 *g* for 15 min at RT. The interface MNC (ERF^-, FcR^-) are then collected, washed twice, and resuspended in IMDM containing 36 mM NaCl instead of $NaHCO_3$, with the pH adjusted to 7.3 using NaOH. This IMDM is used for positive selection and panning (IMDM-P) and has 1% D-BSA with a final cell concentration of 2×10^7/ml.
7. The CD34 antigen is a marker of primitive progenitor cells. Through this step, most of the remaining contaminant cells are removed. Since the cell number is very small at the end of this step, BSA is always added to the medium to protect the cells.
8. The 1 ml ERF^-, FcR^- cell suspension is mixed with 50 μl of dialyzed, purified anti-CD34 mAb at a concentration of 10 μg/ml, and the cell suspension is incubated at 4 °C using an end-over-end rotator. After a 30-45 min incubation, the cell suspension is transferred to a 15 ml conical tube and 1.5 ml of 10% D-BSA is gently underlaid beneath the cells, and then the tube is centrifuged at 400 *g* for 5 min at 4 °C. The supernatant is removed and the cell pellet is washed one more time with IMDM-P containing 0.25% D-BSA (washing medium).
9. During the above cell incubation period, immunomagnetic microspheres are washed five times with washing medium using the Magnetic Plastic Concentrator and kept in ice until use.
10. The cell pellet is resuspended in IMDM-P containing 1% D-BSA at a concentration of 3×10^7/ml in a 12 x 75 cm polystyrene tube, and is incubated with the immunomagnetic microspheres using a microsphere:cell ratio of 3:1, at 4 °C on an end-over-end rotator. After

one hour of incubation, the cell-bound microspheres and free microspheres are attached to the tube wall by the magnet, and the free cells that did not bind microspheres are removed by washing the tube three times with 3 ml of washing medium at 4 °C and once with 3 ml of IMDM-P at 4 °C.

11. The cells with bound microspheres, plus the free microspheres, are gently pipetted for 30 sec in 1.5 ml of IMDM-P containing 130 U/ml of chymopapain and 0.02 M EDTA, prewarmed for 10 min at 37 °C, and are then incubated for an additional 30 sec at RT. The tube is then attached to the magnet for 1 min at RT. The cells released from the microspheres are collected and designated as $CD34^+$ cell fraction 1, and this fraction is immediately mixed with 1.5 ml of heat-inactivated FBS to dilute the chymopapain activity. This fraction is stored in case more cells are later needed, and the separation procedure is sequentially repeated three times to collect $CD34^+$ fractions 2-4, which are then pooled before underlaying 1.5 ml of 10% BSA and centrifuging the cells. These $CD34^+$ cells are then resuspended in 0.2 ml IMDM containing 0.1% BSA.
12. IMDM with 20% FBS, 10% pooled heat-inactivated human type AB serum, 1% D-BSA, 50 U/ml rhIL-3, 2 U/ml rhEpo, 10 µg/ml of insulin, 5 x 10^{-5} M 2-ME, penicillin at 500 U/ml, and streptomycin at 40 µg/ml are combined.
13. $CD34^+$ cells are incubated overnight in 5 ml of serum-containing medium in a 25 cm^2 tissue culture polystyrene flask in a 5% CO_2 / 95% air atmosphere incubator.
14. Nonadherent day 1 cells (AD^-) are collected in a 15 ml tube. Then, 5 ml of washing medium is added to the flask and this is pooled with the tube contents and centrifuged at 400 *g* for 5 min at 4 °C.
15. The day 1 AD^- cells are suspended in 0.8 ml IMDM-P containing 1% BSA in a 2 ml tube, and the mouse mAb listed in Table 1 are added to coat granulocytes, monocytes, natural killer cells, T- and B-lymphocytes, and colony-forming units-granulocyte-macrophage (CFU-GM) at the final concentrations shown. The cells are incubated for 30 min at 4 °C on an end-over-end rotator, and are then transferred to a 15 ml tube for washing procedures. After washing, the cells are resuspended in 0.8 ml IMDM-P containing 1% D-BSA and are then mixed with two classes of immunomagnetic microspheres, one coated with goat anti-mouse IgG with a microsphere:cell ratio of 80:1, and another coated with goat anti-mouse IgM with a microsphere:cell ratio of 40:1. The total volume of the mixture is adjusted to 2 ml and is then incubated on the end-over-end rotator at 4 °C. After 60 min, the cells that do not bind microspheres are collected by magnetic separation (AB^-).

16. The AB^- cells are resuspended in 2 ml of washing medium and are overlaid on FH in a 12 x 75 mm polystyrene tube. After centrifugation at 600 *g* for 5 min at RT, the interface MNC are collected in a 15 ml tube and washed with washing medium. These highly purified BFU-E (AB^-, FH) are then resuspended in IMDM containing 0.3% D-BSA to be kept at 4 °C until used.

Table 1. Mouse monoclonal antibodies used for negative selection

Antibody	Immunoglobulin Subclass	Concentration (μg/ml)	Specificity
CD15/anti-My	IgM	0.8	Pan-granulocytes, monocytes
CD33/anti-My	IgG_2	0.05	Early myeloid cells, CFU-GM
CD45R/anti-My11	IgG	0.2	T and B lymphocytes, CFU-GM
CD16/anti-My23	IgG	0.2	Mature granulocytes, monocytes
CD56/anti-My31	IgG	0.2	Natural killer cells
CD7/anti-Leu9	IgG_2	0.12	T-lymphocytes
CD20/anti-Leu16	IgG_1	0.2	B-lymphocytes
CD2/anti-OKT*11	IgG_2	0.05	T-lymphocytes
CD11b/anti-OKM*1	IgG_2	0.2	Monocytes, granulocytes, natural killer cells

All antibodies were purchased from Becton Dickinson (San Jose, CA). Table modified from reference 8 with permission.

2.1.3 Outcome of method I

The cell concentrations of the FH, PI^-, ERF^-, and FcR^- fractions are measured on a Coulter counter, while a hemocytometer is used for counting the $CD34^+$cells, AD^- cells, AB^- cells and AB^- FH cells. The viability of the cells is measured by dye exclusion using 0.2% trypan blue. The purity and yield of BFU-E from each step are shown in Table 2. Through this procedure, human blood BFU-E (day 1 cells) are purified from 0.04% to 57 ±14%, a 1400-fold purification with a 13% yield. Because these cells are primitive, they are able to generate an erythroid population of highly purified mature BFU-E at different stages (day 2-6 cells). The purity of day 4 cells was 57 ±10%, with a range of 44-72 %, while the purity of day 6 cells was 61 ±9%, with a range of 47-83%. The purity of CFU-E at day 8 was 80 ±18% with a range of 63-100%. These cell fractions are very good for studying early erythropoiesis and analyzing the molecular events that control hematopoietic progenitor cell development. However, the number of erythroid progenitor cells obtained with this method is limited. The average number of day 1 cells (AB^- FH) is 0.5-1 x 10^5, and day 8 CFU-E is 4-8 x 10^6. When more cells are needed for studies such as protein, DNA and RNA

Table 2. Purification of Human BFU-E from PB[1]

Fraction	Cell no.	Total cells Recovery (%)	Viability (%)	BFU-E Purity (%)	BFU-E Purification factor	BFU-E Yield (%)	Erythroid mix Purity (%)	Nonerythroid Purity (%)
	X 10^{-5}							
FH,PI$^-$	5320 ± 1300	100	95 ± 2	0.04 ± 0.02		100	0.01 ± 0.00	0.01 ± 0.00
ERF$^-$,FcR$^-$	500 ± 200	9.3 ± 3.9	97 ± 2	0.35 ± 0.17	9	81	0.05 ± 0.03	0.03 ± 0.01
CD34$^+$	10.0 ± 6.60	0.190 ± 0.125	99 ± 1	8.20 ± 5.14	205	39	1.36 ± 1.03	1.16 ± 1.32
AD$^-$	6.70 ± 4.60	0.125 ± 0.086	88 ± 5	14.30 ± 11.60	358	45	2.11 ± 2.12	1.10 ± 0.77
AB$^-$	1.00 ± 0.30	0.018 ± 0.006	90 ± 8	44.90 ± 21.00	1,121	22	7.45 ± 5.12	4.72 ± 2.64
AB FH	0.50 ± 0.15	0.009 ± 0.003	99 ± 1	56.60 ± 14.00	1,414	13	8.30 ± 3.28	6.04 ± 3.19

[1]Human BFU-E were purified from 400 ml of blood using density centrifugation (FH,PI$^-$), rosetting with a mixture of neuraminidase-treated and IgG-coated sheep erythrocytes (EFC,FcR$^-$), positive panning with anti-CD34 mAb (CD34$^+$), adherent cell depletion (AD$^-$), negative panning with mAb listed in Table I (AB$^-$), and density centrifugation (AB$^-$FH). The cells from each fraction were cultured in triplicate in 0.5 ml fibrin clots at a range of concentrations from 5 x 10^1 to 2 x 10^5 cells/ml with 2 U/ml of rhEpo and 50 U/ml of rhIL-3. Mean values for seven experiments ±SD are shown. Table reproduced from reference 8 with permission.

analysis, an alternate Method II is used to obtain a greater number of CFU-E.

2.2 High Yield Method of BFU-E Purification (Method II) [8,14]

In order to generate human CFU-E in large numbers from BFU-E, the BFU-E are first separated at a higher yield, but lower degree of purity from the blood.

2.2.1 Materials and reagents

1. Giant cell tumor (GCT)-conditioned medium (GCT-CM; Gibco)
2. Affinity-purified goat anti-mouse IgG (Boehringer Mannheim Biochemicals)
3. Anti-CD11b, CD2, CD45R and CD16 mAb (Becton Dickinson)

2.2.2 Detailed protocol

1. The first six steps, including MNC separation, platelet deletion, depletion of T-cells by SRBC rosetting (but without the use of neuraminidase treatment) are the same as described in Section 2.1.2. If erythrocytes are very hypochromic, as with blood from patients with polycythemia vera, some may still remain at the cell interface after this step, and these can be lysed by adding 5 ml of water at RT and pipetting several times to disperse the cells into the water before adding 45 ml of α-MEM and centrifuging the cells at 400 *g* for 5 min. If erythrocytes remain, this can be repeated, and also can be repeated after one of the next steps.
2. The ERF^-, FcR^- cells are counted and incubated overnight at 37 °C in a 5% CO_2 / 95% air incubator, in 10 ml IMDM with 5×10^7 cells containing 20% FBS, and 4% (vol/vol) GCT-CM. The nonadherent day 1 cells are then removed with gentle washes and centrifuged at 400 *g* for 5 min at 4 °C.
3. The nonadherent day 1 cells are suspended in 0.8 ml of cold IMDM and four mAb are added as follows: CD11b (0.25 μg per 1×10^7 cells), CD2 (0.06 μg per 1×10^7 cells), CD45R (1.15 μg per 1×10^7 cells), and CD16 (0.5 μg per 1×10^7 cells). These mAb bind selectively to granulocytes, monocytes, lymphocytes, and CFU-GM, respectively. The cell suspension is gently rocked on a platform in an ice bath for 60 min and then washed twice with α-MEM.

4. The cells are then resuspended in 5 ml IMDM-P containing 5% FBS per 1.5 x 10^7 cells and incubated at 4 °C for 90 min in 100 mm plastic tissue culture dishes that have been previously coated with affinity-purified goat anti-mouse IgG. The antibody-negative nonadherent cells are then removed and centrifuged.
5. The cells are then cultured in dishes with 5 ml IMDM containing 20% FBS at 37 °C for 30-60 min to remove additional adherent cells. The cells are collected, centrifuged, and resuspended in IMDM containing 0.1 % D-BSA to be maintained in an ice bath until further use (AB^-).

2.2.3 Outcome of method II

Through this procedure, a relatively large number of BFU-E can be obtained, but the purity of BFU-E is lower than from Method I. From 400 ml blood, the average number of total AB^- cells was 9 ±6 x 10^6. Human blood BFU-E are purified from 0.017% (FH cells) to 0.368% (AB^- cells), a 22-fold purification with a 43% yield. A high cell viability of 88-98% is maintained throughout the 24 hour period of purification. Culture in methylcellulose for seven days provides 1.04 ±0.95 x 10^7 cells of which 70 ±18% are ECFC. An average number of 5-10 x 10^7 cells can be obtained in liquid medium with similar purity.

2.3 Purification of BFU-E Using Magnetic Microbeads (Method III)

2.3.1 Materials and reagents

1. Vario MACS magnetic device (Miltenyi Biotec Inc., Auburn, CA)
2. CS or BS separation columns (Miltenyi Biotec Inc.)
3. 3-way stopcocks (Miltenyi Biotec Inc.)
4. Anti-CD3, CD11b, CD15 or CD16, CD45RA mAb microbeads (Miltenyi Biotec Inc.)
5. Priming buffer (PBS, 6% BSA)
6. Red cell lysis buffer (0.16 M ammonium chloride, 10 mM potassium bicarbonate, 5 mM EDTA)
7. Sample buffer (PBS, pH 7.3, 5 mM EDTA, 0.5% BSA); should be degassed before use
8. Cell collection tubes (15 ml polypropylene tubes)

2.3.2 Detailed protocol

1. In this method, the initial steps are similar to the previous methods, but SRBC rosetting is omitted.
2. After the FH procedure, if erythrocytes are still present, each cell pellet is suspended in 10 ml of erythrocyte lysis buffer at 37 °C and the cells are incubated for 5 min before bringing the volume to 50 ml with D-PBS. The cells are then centrifuged at 400 *g* for 7 min. In addition, this method limits adherent cell depletion to 1 hour at 4 °C to complete the entire procedure in one work day. At the end, the cells are centrifuged for 10 min at 400 *g* and 4 °C, and the pelleted cells are resuspended in 1 ml of cold sample buffer for negative selection.
3. This purification protocol is different from the previously published protocols since it takes advantage of the latest cell selection technology that is presently available by utilizing paramagnetic mAb-coated microbeads to negatively select contaminant cells. Although there are now multiple commercial sources to purchase such mAb, here we describe a method utilizing mAb-coated microbeads and a magnetic device purchased from Miltenyi Biotec, Inc. This protocol requires that the user have access to the Vario MACS magnetic device which is rather expensive. After the initial investment, the device can be utilized to purify not only early erythroid progenitors, but also other populations of hematopoietic cells since there is a wide range of mAb-coated microbeads available for both negative and positive selection. An advantage of the system is that, unlike other magnetic selection techniques, it employs microbeads that are extremely small (50 nm), and as a result, these beads do not need to be physically separated from the cells prior to culture (in positively selected cells). The attached microbeads do not interfere with cell viability and are bio-degradable. In addition, the small size of the beads makes them more efficient in the selection process.
4. During the negative selection process, four mAb-microbeads are utilized to deplete T-cells, B-cells, granulocytes, monocytes and NK cells. At this stage of purification, the largest contaminating cells are T-cells, and a greater amount of anti-T-cell mAb-coated microbeads are needed during the mAb binding step. The volumes of microbeads suggested in this protocol are not the optimum amounts required to eliminate all unwanted contaminant lymphoid and myeloid cells, although the amounts indicated produce a high enough initial BFU-E purity to attain 80-90% CFU-E by day 7 of *in vitro* culture. The initial culture conditions do not allow an environment that is conducive for long-term survival of many of the non-erythroid cells and, therefore, there is no need for excessive use of the mAb-coated microbeads if the intended

purpose is to obtain highly purified CFU-E for experimental studies. The amounts indicated below make the protocol more economical since mAb-coated microbeads are the most expensive reagent in this protocol. If 350-400 ml of blood is utilized, 2-3 x 10^8 cells are generally present at the end of the adherent cell depletion step. The size CS column is ideal under these conditions. If the protocol is carried out starting with 150-250 ml of blood, then the smaller BS size column can be utilized. It is suggested that two rounds of selection be performed as indicated in step 6 for maximum efficiency. Degassing of sample buffer also greatly improves the separation efficiency. At the end of the purification, the expected yield is between 10-20 x 10^6 cells. After seven days of culture in liquid suspension medium with growth factors, the expected yield is between 30-50 x 10^6 cells. The purity of day seven cells, as assessed by flow cytometry for transferrin receptor (CD71) and glycophorin A positivity, was 85 ±6%. Both CS and BS columns can be reused 5-10 times for subsequent purifications by cleaning them with a mild detergent, followed by flushing 90% ethanol through the column before allowing them to vacuum dry overnight. The columns can then be autoclaved and stored in a sterile environment until further use.

5. Eighty µl of CD3 mAb-coated microbeads and 60 µl each of CD15, CD11b and CD45RA mAb-coated microbeads are added to the FH, erythrocyte depleted cell pellet, and the cells are incubated for 1 h at 4 °C. The negative selection column (CS) is assembled and primed. This is performed by placing the column in the Vario MACS magnet, attaching the syringe filled with priming buffer via the three-way stopcock, and back flushing the priming buffer into the column. Full instructions for assembly and separation are provided with the Vario MACS system. The buffer is then allowed to drip through the column. The column is refilled a second time with the priming buffer and the three-way stopcock is closed until after application of cells. The cell collection tubes are primed by coating the tubes with the priming buffer.
6. Cells are applied to the column slowly while the column is firmly placed in the magnet, and the tube is rinsed with 1-2 ml of degassed sample buffer. The cells are allowed to pass through the column into a 15 ml sterile tube and are saved on ice, and then the column is washed with 20-30 ml of cold sample buffer and the wash is saved. The selected cells are pooled with the wash in a 50 ml tube and the cells are then centrifuged at 400 *g* for 7 min.
7. The cell pellet is resuspended in 1-2 ml of sample buffer, and this is reloaded a second time to repeat step 5. Although this can be omitted, it provides an additional step to ensure that most of the magnetically-labeled cells are eliminated during the second round of selection. The flow through cell preparation contains the BFU-E, and these cells are

centrifuged one more time at 400 *g* for 7 min before the sample buffer is aspirated to obtain the final cell pellet.

3. *IN VITRO* GENERATION OF CFU-E FROM PURIFIED BFU-E

3.1 Methylcellulose Cell Culture for Generation of CFU-E [8]

This method is used for CFU-E generation from BFU-E prepared by Method II. The methycellulose medium is a highly viscous fluid that maintains the separation of discrete colonies.

3.1.1 Materials and reagents

1. Methylcellulose (Fisher Scientific Co., Fair Lawn, NJ)

3.1.2 Detailed protocol

1. Day 1 AB^- cells are cultured at a concentration of 2×10^5 cells/ml in a mixture containing 30% FBS (vol/vol), 1% D-BSA, 10^{-4} M 2-ME, penicillin at 500 U/ml, streptomycin at 40 μg/ml, insulin at 10 μg/ml, rhEpo at 2 U/ml, IL-3 at 100 U/ml, and 0.9% methylcellulose in IMDM. The cells are plated in flat-bottomed 12-well tissue culture plates at 1 ml/well and are incubated at 37 °C in a high humidity 5% CO_2 / 95% air incubator.
2. On day 9 of incubation (day 8 cells), 2 ml of α-MEM is dispersed into each well to reduce the viscosity, and the medium is gently mixed using a 5 ml pipette. The mixture is collected in 50 ml tubes, and the wells are washed again with 2 ml α-MEM which is added to the 50 ml tubes. The cells are then centrifuged at 600 *g* for 10 min, at 4 °C, and are washed one more time (MC cells).
3. MC cells are resuspended in 2 ml α-MEM and layered over 4 ml of 10% BSA in a 15 ml tube. After centrifugation at 800 *g* for 5 min at 4 °C, the cells are collected in 10 ml α-MEM and are centrifuged. After the cells are resuspended in 5 ml IMDM containing 20% FBS, they are incubated in a flask for 1 h at 37 °C, in a 5% CO_2/95% air incubator. The nonadherent cells are collected with three gentle washes, centrifuged, and resuspended in 2 ml IMDM. If necessary, this step may be repeated at 4 °C instead of 37 °C (MC, AD^-).

4. The MC, AD⁻ cells are overlaid on the top of 2 ml of FH in a 12 x 75 mm polystyrene tube. After centrifugation at 600 *g* for 15 min at RT, the interface MNC (MC, FH) are collected, washed two times, and resuspended in IMDM with 0.3% D-BSA to be kept on ice until use.

3.1.3 Outcome of method

Through this procedure, an average of 1-2 x 10^7 cells can be obtained with enrichment to 70 ±18%, a 77% yield, and a high viability of 94 ±2%. CFU-E can be further enhanced at this stage if desired by a repeat of the Method II negative selection process. The advantage of this method is that discrete colony growth can be visualized for enumeration or plucking. The disadvantage is that a relatively large number of CFU-E are lost during the elaborate collection and washing procedure.

3.2 CFU-E Generation from BFU-E in Liquid Suspension Culture

This method applies to BFU-E purified by all three methods. The BFU-E at a concentration of 1-5 x 10^4 (Method I), or 0.3 x 10^6 (Method II), are cultured in a 1 ml mixture containing 20% FBS, 5% pooled human AB serum, 1% D-BSA, 5 x 10^{-5} M 2-ME, penicillin at 500 U/ml, streptomycin at 40 µg/ml, 10 µg/ml insulin, rhEpo at 2U/ml, rhIL-3 at 50 U/ml, and IMDM. KL at 50 ng/ml may be added to the medium, and this produces marked amplification of the cell number. Cells are incubated at 37 °C in a 5% CO_2/95% air incubator, and can be collected at successive days, as desired. When the cell concentration rises above 1 x 10^6 cells per ml, 0.5-1.0 ml of new medium can be added, or the cells can be centrifuged in tubes and replated with fresh medium. This enhances cell replication and the number of viable cells. The advantage of the liquid culture method is that it is simpler and a larger number of CFU-E can be obtained while maintaining a high degree of purity. At day 8, 5-10 x 10^7 cells can be obtained, and the purity of cells is similar to those produced with the methylcellulose method, while the yield is nearly 100%. In our laboratory, the liquid suspension culture method has virtually replaced the methylcellulose method.

The data indicate that day 1 cells derived by Method I consist predominantly of early, immature BFU-E from which very large colonies, with more than 2,000 cells, are derived [9]. These BFU-E gradually differentiate into more mature erythroid progenitors, and their proliferative capacity diminishes as the period in liquid culture increases. Day four cells consist of approximately 33% intermediate BFU-E (500-2000 cells/colony) and 59% late BFU-E (50-500 cells/colony), while day six cells consist of

50% late BFU-E and 50% CFU-E (8-49 cells/colony). By day eight, most cells are CFU-E [10].

4. ERYTHROID PROGENITOR CELL ASSAYS

4.1 Semi-Solid Culture of BFU-E and CFU-E

4.1.1 Materials and reagents

1. Aminocaproic acid (Elkins-Sinn, Inc.)
2. Fibrinogen (Fibrinogen Kabi Grade L; Kabi Diagnostica, Stockholm, Sweden)
3. Thrombin (Parke-Davis, Morris Plains, NJ)
4. F-12 [HAM] (Sigma)
5. Crystalline Bovine Serum Albumin (Sigma)
6. Transferrin (Sigma)
7. Lipids including oleic acid, L-α phosphatidylcholine, cholesterol (Sigma)

4.1.2 BFU-E plasma clot culture in serum-containing medium [9,10,14]

BFU-E at a concentration of 100 cells/ml (Method I) or 2 x 10^5 cells/ml (Method II), are cultured in triplicate in 24-well flat-bottomed tissue culture plates, with a 0.5 ml mixture containing 20% FBS, 10% pooled human AB serum, 1% D-BSA, 5 x 10^{-5} M 2-ME, penicillin at 500 U/ml, streptomycin at 40 μg/ml, 10 μg/ml insulin, rhEpo at 2U/ml, rhIL-3 at 50 U/ml, IMDM, 1.5 mM aminocaproic acid, 2 mg/ml human fibrinogen, and 0.2 U/ml thrombin. KL at 50 ng/ml may be added to the medium and will markedly increase the size of the colonies, but is not necessary for BFU-E growth in serum-containing medium [15].

After 15 days of incubation at 37 °C in a 5% CO_2/95% air incubator, the clots are taken out and fixed and stained with 3,3'-dimethoxybenzidine-hematoxylin [16]. Enumeration of BFU-E colonies is performed with a microscope, and colonies that have 50 or more hemoglobinized cells are counted as BFU-E [1].

4.1.3 BFU-E plasma clot culture in serum-free medium [17]

Serum-free medium consists of 50% IMDM/50% F-12 [HAM], 10 mg/ml deionized, delipidated, and dialyzed crystalline bovine serum albumin (C-BSA-3D), 300 μg/ml of iron-saturated transferrin, lipid suspension (5.6 μg/ml oleic acid, 8.0 μg/ml L-α-phosphatidylcholine, 7.8 μg/ml cholesterol), 5 x 10^{-5} M 2-ME, insulin at 10 μg/ml, rhEpo at 2U/ml, rhIL-3 at 100 U/ml, KL at 50 ng/ml, penicillin at 500 U/ml, streptomycin at 40 μg/ml, 1.5 mM aminocaproic acid, fibrinogen at 2 mg/ml and thrombin at 0.2 U/ml. The culture conditions and enumeration criteria are the same as with serum-containing medium. This serum-free medium is useful since it deletes many hormones, cytokines and growth factors, and is a reliable system that completely supports BFU-E development with an equivalent plating efficiency and somewhat enhanced proliferative capacity compared to serum-containing medium.

4.1.4 CFU-E plasma clot culture in serum-containing medium [18]

Day 7 or day 8 CFU-E are cultured at a concentration of 1,000 cells/ml in clots with 15% FBS, 15% human pooled AB serum, 0.5% HSA or 1% BSA, 1.5 mM aminocaproic acid, 1 U/ml of rhEpo, penicillin at 500 U/ml, streptomycin at 40 μg/ml, 2 mg/ml fibrinogen, 0.2 U/ml bovine thrombin and 50% IMDM/50% F-12 [HAM]. Aliquots of 0.2 or 0.25 ml of cell suspension are plated in 48-well flat-bottomed tissue culture plates.

The clots are incubated at 37 °C in a 5% CO_2/95% air incubator for seven days, and are then fixed and stained with 3,3' dimethoxybenzidine-hematoxylin. CFU-E are defined as cells that give rise to single colonies of 8-49 hemoglobinized cells [2,19], whereas ECFC give rise to colonies of 2 or more cells [8].

4.1.5 CFU-E plasma clot culture in serum-free medium [20]

This medium includes C-BSA-3D (10 mg/ml), iron-saturated transferrin (150 μg/ml), lipid suspension (5.6 μg/ml oleic acid, 8.0 μg/ml L-α-phosphatidylcholine, 7.8 μg/ml cholesterol), insulin at 10 μg/ml, rhEpo at 1U/ml, penicillin at 500 U/ml, streptomycin at 40 μg/ml, 1.5 mM aminocaproic acid, fibrinogen at 2 mg/ml, thrombin at 0.2 U/ml and 50% IMDM/50% F-12 [HAM]. Culture conditions and enumeration criteria are the same as for serum-containing culture medium.

4.2 Determination of Cell Purity in Liquid Culture

Cell purity has generally been determined by plating the cells in plasma clots and determining the number of erythroid colonies that are produced [8,9]. However, in liquid culture, cell purity may be determined by flow cytometry (Fig. 2) [14], or by staining individual cells with 3,3' dimethoxybenzidine-hematoxylin during erythroid differentiation [10,15]. Flow cytometric analysis can be performed beginning on day five for detecting the percentage of high density transferrin receptors (CD71) present on erythroid cells [14], and on successive days for analysis of high density CD71 and glycophorin A surface markers which are extremely specific for the erythroid lineage. Fig. 2 shows such an analysis where CFU-E development was followed all the way until terminal differentiation. The benzidine staining method can be used at and after day ten. The cells are cytocentrifuged onto glass sides, stained, and examined individually by light microscopy.

4.3 Morphology of BFU-E and CFU-E

Day 1 highly purified BFU-E [9] are very immature blast cells by light microscopy. The majority of these cells have a slightly oval cytoplasm, occasional pseudopods, high nuclear/cytoplasmic ratio, very fine nuclear chromatin, slightly basophilic cytoplasm, and multiple, large, well-defined nucleoli (see Fig. 1). On electron microscopy, the cell population is composed of a single blast cell type. The nucleus contains small amounts of clumped heterochromatin and a prominent nucleolus. The cytoplasm is abundant and contains a mixture of distributed single ribosomes and polyribosomes. Surface pinocytosis is occasionally noted. No recognizable erythroid precursors are encountered, but occasional small lymphocytes and mononuclear phagocytes are present.

CFU-E [8], by light microscopy, are very immature cells characterized by a fine nuclear chromatin, a well-defined large nucleolus, high nuclear/cytoplasm ratio, a perinuclear clear zone and basophilic cytoplasm with pseudopods. Although these cells vary in size and often have a vesicular, expanded cytoplasm that distorts the nucleus, they consistently retain these common features(see Fig. 1). On electron microscopy, the cells are primitive blasts with a high nuclear/cytoplasmic ratio, dispersed nuclear chromatin, and a prominent nucleolus. The cytoplasm contains clumps of mitochondria, granules are absent, and the cell surface has pinocytotic vesicles. When day 1 highly purified BFU-E are incubated in liquid culture for various times, erythroid progenitor cells at different stages of maturation can be obtained for morphologic study, size distribution, and proliferative capacity, as indicated by culture in plasma clots. As the incubation time in liquid culture is increased, the cells continuously develop a more mature morphology and have a diminished proliferative capacity, with a decreased number of erythroblasts in each colony (Fig. 3).

5. UTILITY OF PURIFICATION PROCEDURES

5.1 Transitional Change of Growth Factor Requirements and Receptors

Transitional change of growth factor requirements and receptors has been studied using the procedures described above. Erythroid progenitors at different stages have been shown to exhibit different growth factor requirements for their development as described below.

5.1.1 Erythropoietin

Delayed addition of rhEpo to cell cultures has shown that, in the presence of IL-3, more than 90% of immature, day 1 BFU-E had full erythroid burst formation after an absence of rhEpo for 48 hours and 50% after an absence of rhEpo for 120 hours, thus demonstrating that early primitive BFU-E are not Epo dependent [9]. The degree of Epo-dependency increased as the erythroid progenitors differentiated and matured. Fifty percent of day seven cells will undergo apoptosis after absence of Epo for 20 hours, and more than 90% will die after absence of Epo for 48 hours [21]. Specific binding of ^{125}I-rhEpo was detected by autoradiography in 19 ±7% of the initial day 1 BFU-E with a low grain distribution, and in all of the erythroid progenitors with a high grain distribution after incubation in liquid culture for 72 hours [9]. These data demonstrate that primitive BFU-E have a much lower number of Epo receptors than CFU-E, and develop an increased concentration of Epo receptors in association with their maturation and loss of proliferative capacity.

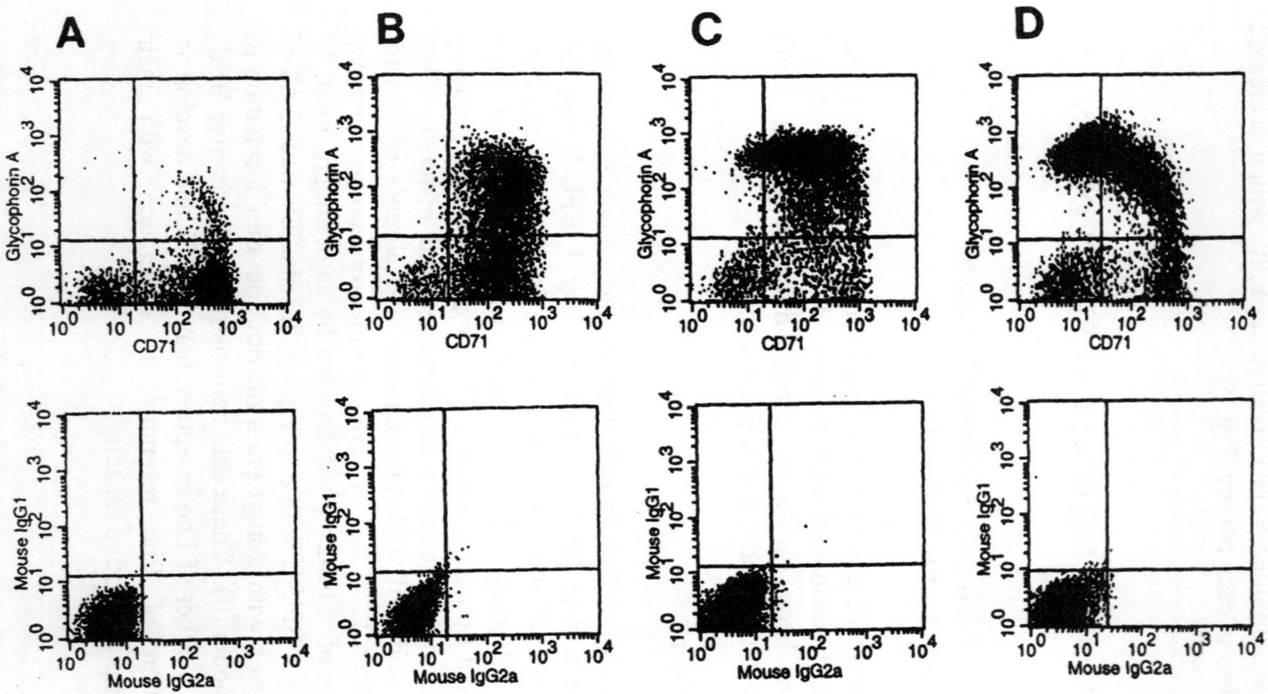

Figure 2. Phenotypic characterization during proliferation and differentiation of purified human erythroid progenitor cells. Human CFU-E were analyzed for expression of CD71 and glycophorin A by flow cytometry on days 7 (A), 9 (B), 11 (C) and 14 (D) of culture by staining samples of cells with fluorescein isothiocyanate- and phycoerythrin-conjugated antibodies to CD71 and glycophorin A, respectively. The bottom panel shows the isotype control to determine the background fluorescence. The top panel shows the appearance of the erythroid marker protein glycophorin A during later stages of erythroid development, and the decrease of high intensity CD71 transferrin receptors during terminal differentiation.

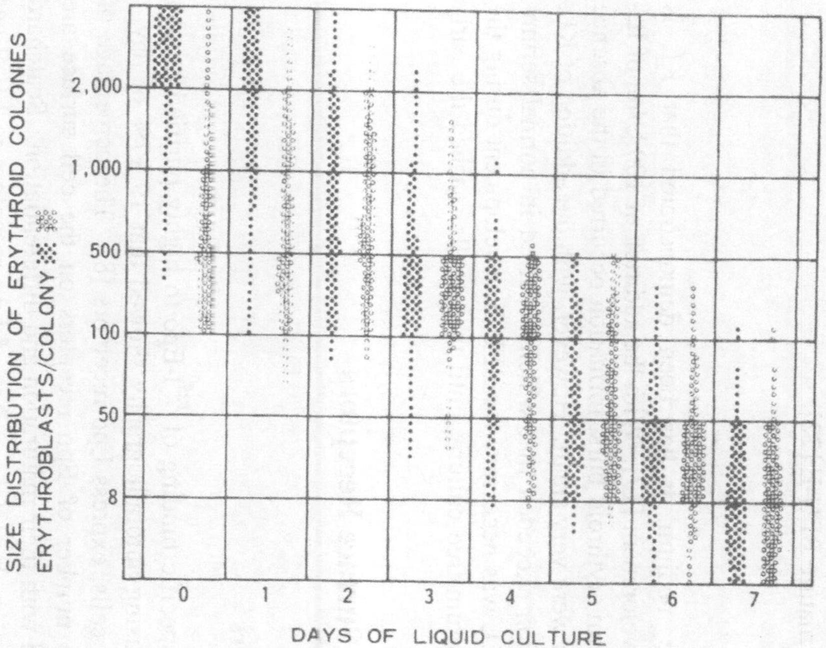

Figure 3. Size distribution of erythroid colonies. Purified cells were incubated in liquid culture for 0-7 days at 37 °C with 2 U/ml of rhEpo and 50 U/ml of rhIL-3. At the indicated days, cells were collected, washed twice, and plated into fibrin clots with 2 U/ml of rhEpo and 50 U/ml of rhIL-3 (closed dots), or 2 U/ml of rhEpo alone (open dots). After incubation of the plated cells until day 15, the clots were fixed and stained with benzidine-hematoxylin. The number of hemoglobinized erythroblasts was counted in each of 100 colonies in each of three experiments, and the mean data of three experiments are represented. Figure reproduced from reference 22 with permission.

5.1.2 Interleukin-3

In contrast with the initial lack of an Epo requirement, early BFU-E need IL-3 for their maintenance and growth. Omission of IL-3 for 24 hours of culture in serum-free medium led to a 45% decline in BFU-E colonies, and omission for 48 hours produced a 65% decline, indicating a marked dependence on IL-3 during the first 48 hours of BFU-E culture [17,22]. However, these cells had a graded loss of their dependency on IL-3 as the incubation time in liquid culture increased. A complete loss of dependency was observed after three days of liquid culture, at a time when 85% of the erythroid progenitors gave rise to colonies of more than 100 erythroblasts that were equivalent to mature BFU-E [22].

5.1.3 *c-kit* ligand

Using serum-free medium, it has been demonstrated that KL is essential for BFU-E development [17]. With the addition of 100 U/ml of IL-3, only 12% of maximum erythroid burst formation occurred in the absence of KL, and these bursts were very small. However, with the addition of KL, erythroid burst formation exceeded that which occurred in normal serum-containing medium. KL was necessary for BFU-E development during the initial seven days of serum-free culture, until these cells reached the early CFU-E stage.

5.2 Studies of Surface Receptors

5.2.1 Epo receptors

The analysis of specific binding of ^{125}I-Epo to highly purified human erythroid progenitors using autoradiography showed that 19 ±7% of day 1 cells, and all of day 8 cells, express Epo receptors [8]. The percentage of positive cells and the number of Epo receptors on the cell surface are continuously increased with their maturation and differentiation. Scatchard analysis demonstrated that the total number of Epo receptors is 1,050 per ECFC, and that two classes of Epo receptors are present on ECFC. One-fifth of them have a K_d of 0.1 nM, while the remainder have a K_d of 0.57 nM [23].

5.2.2 KL receptors

Binding experiments using radioiodinated rhKL were performed to characterize KL receptors on erythroid progenitor cells [24]. Approximately 69-82% of day 1 cells, and almost 100% of day 8 cells, have KL receptors. A single class of KL receptor, 23,000 per ECFC, was demonstrated with a high-binding affinity (K_d =17 pM). Saturation occurred at 0.5 nM (10ng/ml), which produces a nearly maximum biological effect, and one-half of the radiolabeled KL was internalized by the cells after 30 min at 37 °C. No significant differences in the receptor numbers, dissociation constants, or internalization rates were found between normal and polycythemia vera patients [24].

5.2.3 Insulin-like growth factor (IGF-1) receptors

Studies in serum-free medium have shown that CFU-E require insulin or IGF-1 and Epo for their development [20]. Binding studies with radioiodinated IGF-1 revealed that two classes of IGF-1 receptors were present on the CFU-E cell surface: a low-affinity class of 549 receptors per CFU-E with a K_d 0.44 nM, and a high-affinity class of 341 receptors with a K_d of 0.04 nM [25].

5.2.4 Interferon-gamma (IFN-γ) receptors

Binding experiments with ^{125}I-IFN-γ have shown that a high affinity, single class of IFN-γ receptors is present on human day 6 ECFC, which are mainly CFU-E. The number of receptors per ECFC is 2,070, with a K_d of 0.02 nM [26].

5.3 Studies on Apoptosis

Apoptosis, or programmed cell death, plays a major role in normal development, tissue homeostasis, plus defense against oncogenesis or viral invasion, and when dysregulated, produces many diseases.

Highly purified ECFC have been used to study the control of apoptosis during their maturation and differentiation [3,10,21]. The methods for delineating apoptosis have included morphology [3,21], DNA fragmentation as determined by agarose gel electrophoresis [3], flow cytometry analysis with *in situ* end-labeling of DNA breaks with digoxigenin-11-dUTP and anti-digoxigenin mAb conjugated with fluorescein [10], and staining with fluorescein-conjugated annexin.

Fifty percent of ECFC undergo apoptosis after incubation at 37 °C in serum-free medium without Epo for as little as 6 hours [3]. Concomitant incubation with Epo greatly reduced DNA breakdown to 23% of ECFC. KL and IGF-1 each reduced the amount of DNA breakdown to 38% and 46%, respectively, and when added together, to 24% [3]. Inhibition of heme synthesis also induces apoptosis in human ECFC [21].

When highly purified human blood BFU-E were used to study the effects of IFN-γ, it was demonstrated that IFN-γ inhibited erythroid colony formation, cell proliferation, and differentiation of day 3-6 mature BFU-E in a dose-dependent manner, while also inducing erythroblast apoptosis [10]. This reduction in the number of erythroid colonies and induction of apoptosis could be overcome by KL. Further investigations demonstrated that IFN-γ induced Fas expression in ECFC, but not Fas ligand (FasL) which was constitutively present [14]. IFN-γ has now been shown to activate and/or upregulate some caspases, such as caspase-8/FLICE, caspase-1/ICE, and caspase-3/CPP32/-APOPAIN [27]. This work has shown that the Fas/FasL system is an important pathway for inducing apoptosis in highly purified ECFC, and these observations may partly explain the mechanism by which IFN-γ inhibits erythropoiesis and contributes to the anemia of chronic disease [8, 9].

5.4 Studies on Signal Transduction and Protein Phosphorylation

Investigations on the effects of Epo and KL on protein phosphorylation in day 8 normal and polycythemia vera (PV) cells have shown that Epo and KL rapidly induced tyrosine phosphorylation of the Epo and KL receptors [18]. No additional phosphorylated proteins were evident in PV cells when compared to normal cells. Culture of normal erythroid progenitors with orthovanadate, an inhibitor of protein tyrosine phosphatases, resulted in an increased number of colonies and enhanced protein tyrosine phosphorylation. However, little enhancement was evident with PV cells. Further investigations have shown that PV erythroid progenitors have a two- to four-fold higher protein tyrosine phosphatase (PTP) activity than that observed in normal cells [30]. While this might be responsible for the increased erythropoiesis in PV patients, it may also be only an effect of the PV disease. Additional studies with highly purified human ECFC have most recently shown that KL and Epo have a synergistic effect on the activation of MAP kinase, and that the synergism is mediated by signaling through PI-3 kinase [31]. This provides a mechanism for the synergistic action of Epo and KL on erythropoiesis [14]. Epo, but not KL, led to activation of STAT-5, and KL had no enhancing effect on the Epo-dependent activation of

STAT-5 [31]. IFN-γ activated STAT-1 but not STAT-5 (C-H Dai, unpublished data).

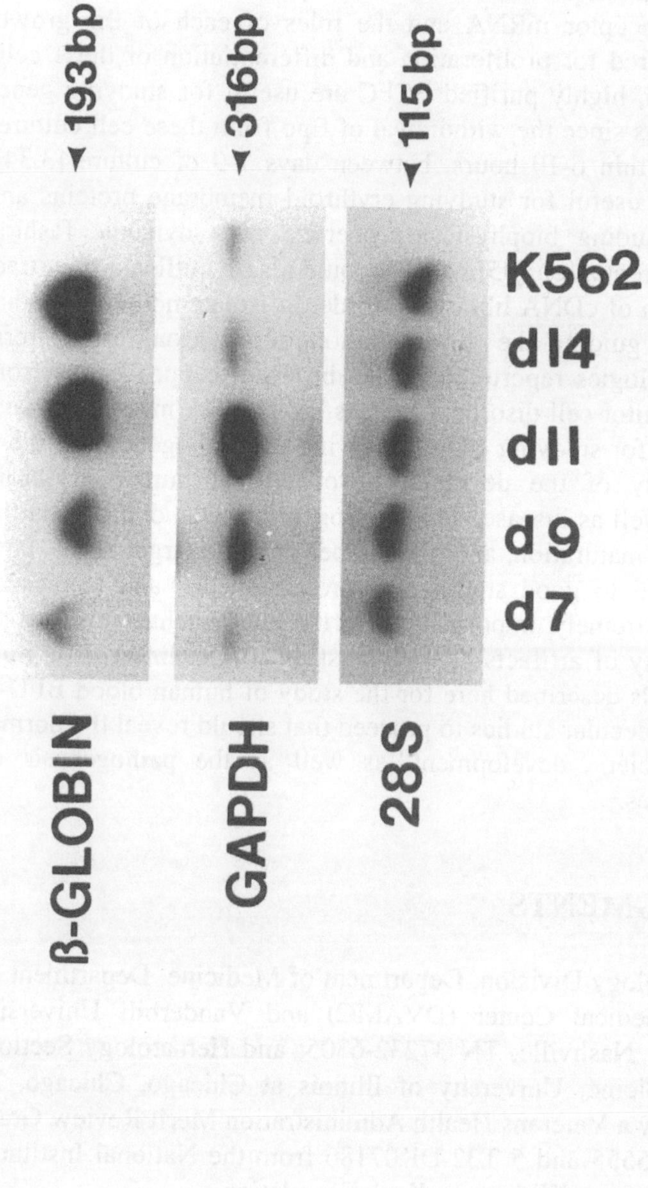

Figure 4. Expression of β-globin, glyceraldehyde phosphate dehydrogenase (GAPDH) and 28S ribosomal RNA genes during differentiation of human erythroid progenitors. RNAse protection assays were performed with 20 μg of total RNA isolated from purified erythroid progenitors between days 7-14 of culture. All three RNA probes were utilized in the same hybridization reaction in order to determine the relative expression levels. RNA from an uninduced K562 erythroleukemia cell line shows no expression of the β globin gene.

6. CONCLUSIONS

Purified cells can be utilized for studies involving growth factor regulated development of erythroid progenitor cells. For example, Fig. 4 demonstrates the expression pattern of β-globin and glyceraldehyde phosphate dehydrogenase (GAPDH) mRNA during differentiation of highly purified human erythroid progenitors. Studies have been performed to delineate the expression of Epo receptor mRNA and the roles of each of the growth factors that are required for proliferation and differentiation of these cells [32, 33]. In addition, highly purified ECFC are useful for studying genes that regulate apoptosis since the withdrawal of Epo from these cell cultures induces apoptosis within 6-10 hours, between days 7-9 of culture [3,34]. Also, these cells are useful for studying erythroid membrane proteins and their structures, including biophysical properties in a dynamic fashion during terminal differentiation [35]. ECFC could also be utilized to extract RNA for construction of cDNA libraries in order to isolate novel genes that may play a role in guiding the differentiation of erythroid progenitors. Finally, the methodologies reported here can be used to purify cells from hematopoietic progenitor cell disorders such as PV [18,24], myelodysplasia, and erythroleukemia for studying genes involving the pathogenesis of these diseases. The study of the development of normal human erythroid progenitor cells, as well as diseased human progenitor cells, depends on the purity, uniformity of maturation, and the number of these target cells. Pure cells are just as vital to good studies as pure chemicals, and studies of primary cells are extremely important to verify real mechanisms and to remove the possibility of artifacts due to transformed cell lines. It is now clear that the methods described here for the study of human blood BFU-E and CFU-E allow molecular studies to proceed that should reveal the normal control of erythropoietic development, as well as the pathogenesis of erythropoietic diseases.

ACKNOWLEDGMENTS

From the Hematology Division, Department of Medicine, Department of Veterans Affairs Medical Center (DVAMC) and Vanderbilt University School of Medicine, Nashville, TN 37232-6305; and Hematology Section, Department of Medicine, University of Illinois at Chicago, Chicago, IL 60607. Supported by a Veterans Health Administration Merit Review Grant and by grants DK-15555 and 5 T32-DK07186 from the National Institutes of Health and a grant from William B. Keck Foundation.

REFERENCES

1. Clarke BJ and Housman D (1977) Characterization of an erythroid precursor cell of high proliferative capacity in normal human PB. *Proc. Natl. Acad. Sci. USA* 74:1105-1109.
2. Eaves AC and Eaves CJ (1984) Erythropoiesis in culture. *Clin. Haematol.* 13:371-391.
3. Muta K and Krantz S (1993) Apoptosis of human erythroid colony-forming cells is decreased by stem cell factor and insulin-like growth factor-1 as well as erythropoietin. *J. Cell. Physiol.* 156:264-271.
4. Emerson SG, Sieff CA, Wang A, Wong W, Clark SC, and Nathan DG (1985) Purification of fetal hematopoietic progenitors and characterization of recombinant multipotential colony-stimulating activity. *J. Clin. Invest.* 76:1286-1290.
5. Sieff CA, Emerson SG, Mufson A, Gesner TG, and Nathan DG (1986) Dependence of highly enriched human BM progenitors on hemopoietic growth factors and their response to recombinant erythropoietin. *J. Clin. Invest.* 77:74-81.
6. Tepperman AD, Curtis JE, and McMulloch EA (1974) Erythropoietic colonies in cultures of human marrow. *Blood* 44:659-669.
7. Hasson MW, Lutton JD, Levere RD, Rieder RF, and Cederquist LL (1979) *In vitro* culture of erythroid colonies from human fetal liver and umbilical cord blood. *Br. J. Haematol.* 41:477-484.
8. Sawada K, Krantz SB, Kans JS, Dessypris EN, Sawyer S, Glick AD and Civin CI (1987) Purification of human erythroid colony-forming units and demonstration of specific binding of erythropoietin. *J. Clin. Invest.* 80:357-366.
9. Sawada K, Krantz SB, Dai CH, Koury ST, Horn ST, Glick AD and Civin CI (1990) Purification of human blood burst-forming units-erythroid and demonstration of the evolution of erythropoietin receptors. *J. Cell. Physiol.* 142:219-230.
10. Dai CH, Krantz SB, Kollar K, and Price JO. (1995) Stem cell factor can overcome inhibition of highly purified human burst-forming units-erythroid by interferon γ. *J. Cell. Physiol.* 165:323-332.
11. Fibach E, Manor D, Oppenheim A, and Rachmilewitz EA. (1989) Proliferation and maturation of human erythroid progenitors in liquid culture. *Blood* 73:100-103.
12. Gabbianelli M, Sargiacomo M, Pelosi E, Testa U, Isacchi G, and Peschle C. (1990) "Pure" human hematopoietic progenitors: Permissive action of basic fibroblast growth factor. *Science* 249:1561-1564.
13. Giampaola A, Sterpetti P, Bulgarini D, Samoggia P, Pelosi E, Valtieri M, and Peschle C. (1994) Key functional role and lineage-specific expression of selected HOXB genes in purified hematopoietic progenitor differentiation. *Blood* 84:3637-3647.
14. Dai CH, Price JO, Brunner T, and Krantz SB (1998) Fas ligand is present in human erythroid colony-forming cells and interacts with Fas induced by interferon γ to produce erythroid cell apoptosis. *Blood*, 91:1235-1242.
15. Muta K, Krantz SB, Bondurant MC, and Dai CH (1995) Stem cell factor retards differentiation of normal human erythroid progenitor cells while stimulating proliferation. *Blood* 86:572-580.
16. McLeod DL, Shreeve MM, and Axelrad AA (1974) Improved plasma culture system for production of erythrocytic colonies *in vitro*. Quantitative assay method for CFU-E. *Blood* 44:517-534.
17. Dai CH, Krantz SB, and Zsebo KM (1991) Human burst-forming units-erythroid need direct interaction with stem cell factor for further development. *Blood* 78:2493-2497.
18. Dai CH, Krantz SB, and Sawyer ST (1997) Polycythemia vera. V. Enhanced proliferation and phosphorylation due to vanadate are diminished in polycythemia vera erythroid

progenitors cells: a possible defect of phosphatase activity in polycythemia vera. *Blood* 89:3574-3581.

19. Ogawa M, MacEachern MD, and Avila L (1977) Human marrow erythropoiesis in culture: II. Heterogeneity in the morphology, time course of colony formation, and sedimentation velocities of the colony-forming cells. *Am. J. Hematol.* 3:29-36.
20. Sawada K, Krantz SB, Dessypris EN, Koury ST, and Sawyer ST (1989) Human colony-forming units-erythroid do not require accessory cells, but do require direct interaction with insulin-like growth factor I and/or insulin for erythroid development. *J. Clin. Invest.* 83:1701-1709.
21. Muta K and Krantz SB (1995) Inhibition of heme synthesis induces apoptosis in human erythroid progenitor cells. *J. Cell. Physiol.* 163:38-50.
22. Sawada K, Krantz SB, Dai CH, Sato N, Ieko M, Sakurama S, Yasukouchi T, and Nakagawa S (1991) Transitional change of colony stimulating factor requirements for erythroid progenitors. *J. Cell. Physiol.* 149:1-8.
23. Sawada K, Krantz SB, Sawyer ST, and Civin CI (1988) Quantitation of specific binding of erythropoietin to human erythroid colony-forming cells. *J. Cell. Physiol.* 137:337-345.
24. Dai CH, Krantz SB, Koury ST, and Kollar K (1994) Polycythemia vera. IV. Specific binding of stem cell factor to normal and polycythemia vera highly purified erythroid progenitor cells. *Br. J. Haematol.* 88:497-505.
25. Cotton EW, Means RT, Jr., Cline SM, and Krantz SB (1991) Quantitation of insulin-like growth factor-I binding to highly purified human erythroid colony-forming units. *Exp. Hematol.* 19:278-281.
26. Taniguchi S, Dai CH, and Krantz SB (1997) Specific binding of interferon-γ to high affinity receptors on human erythroid colony-forming cells. *Exp Hematol.* 25:193-198.
27. Dai CH and Krantz SB (1999) Interferon γ induces upregulation and activation of caspases 1, 3 and 8 to produce apoptosis in human erythroid progenitor cells. Blood 93: in press.
28. Krantz SB. (1994) Pathogenesis and treatment of the anemia of chronic disease. *Am. J. Med. Sci.* 307:353-359.
29. Means RT Jr and Krantz SB. (1992) Progress in understanding the pathogenesis of the anemia of chronic disease. *Blood* 80:1639-1647.
30. Sui X, Krantz SB, and Zhao ZZ (1997) Identification of increased protein tyrosine phosphatase activity in polycythemia vera erythroid progenitor cells. *Blood* 90:651-657.
31. Sui X, Krantz SB, You M, and Zhao Z. (1997) Insight into the synergistic effects of erythropoietin and stem cell factor on erythropoiesis. *Blood* 90(Suppl 1):309a.
32. Muta K, Krantz SB, Bondurant MC, and Wickrema A. (1994) Distinct roles of erythropoietin, insulin-like growth factor I, and stem cell factor in the development of erythroid progenitor cells. *J. Clin. Invest.* 94:34-43.
33. Wickrema A, Krantz SB, Winkelmann JC, and Bondurant MC. (1992) Differentiation and erythropoietin receptor gene expression in human erythroid progenitor cells. *Blood* 80:1940-1949.
34. Jacobs-Helber SM, Wickrema A, Birrer MJ, and Sawyer ST. (1998) AP1 regulation of proliferation and initiation of apoptosis in erythropoietin-dependent erythroid cells. *Mol. Cell. Biol.*,18:3699-3707.
35. Wickrema A, Koury ST, Dai CH, and Krantz SB (1994) Changes in cytoskeletal proteins and their mRNAs during maturation of human erythroid progenitor cells. *J. Cell. Physiol.* 160:417-426.

Chapter 12

In Vitro Development of Megakaryocytes and Platelets

[1]Marcus O Muench and [2]Jae-Hung Shieh
[1]Fetal Treatment Center Research Laboratory, University of California at San Francisco, 513 Parnassus Avenue, Room HSW-1601, San Francisco, CA 94143-0570 and [2]Memorial Sloan-Kettering Cancer Institute. Tel: 001-415-476-0449; Fax: 001-415-476-2314; E-mail: muench@itsa.ucsf.edu

1. INTRODUCTION

Megakaryocytes are a rare cell type representing <0.4% of the adult bone marrow (BM), where they serve as factories for the production of platelets [1]. Despite the paucity of megakaryocytes in BM, these cells are responsible for maintenance of the large pool of circulating blood platelets. Normal platelet numbers range from 150,000-450,000 per mm^3 of blood with an average lifespan of ten days [2]. Platelets are responsible for aiding in the process of blood clotting and subsequent wound healing, which occurs at a site of injury. Platelets are not nucleated cells, but rather the enucleated products of the cellular membrane and cytoplasm of megakaryocytes. Platelets are consequently smaller in size than any leukocytes or red blood cells.

Megakaryocytes are the progeny of hematopoietic stem cells and can be distinguished from the other blood cell lineages by several unique morphological features (Fig. 1). As their name suggests, megakaryocytes are multinucleated cells, typically with an 8N-16N DNA content [1, 3]. The nuclear content of megakaryocytes is variable, with up to 128N megakaryocytes having been generated *in vitro* [4, 5]. The multinucleated nature of megakaryocytes makes these cells large, ranging up to 65 µm in diameter [3, 6].

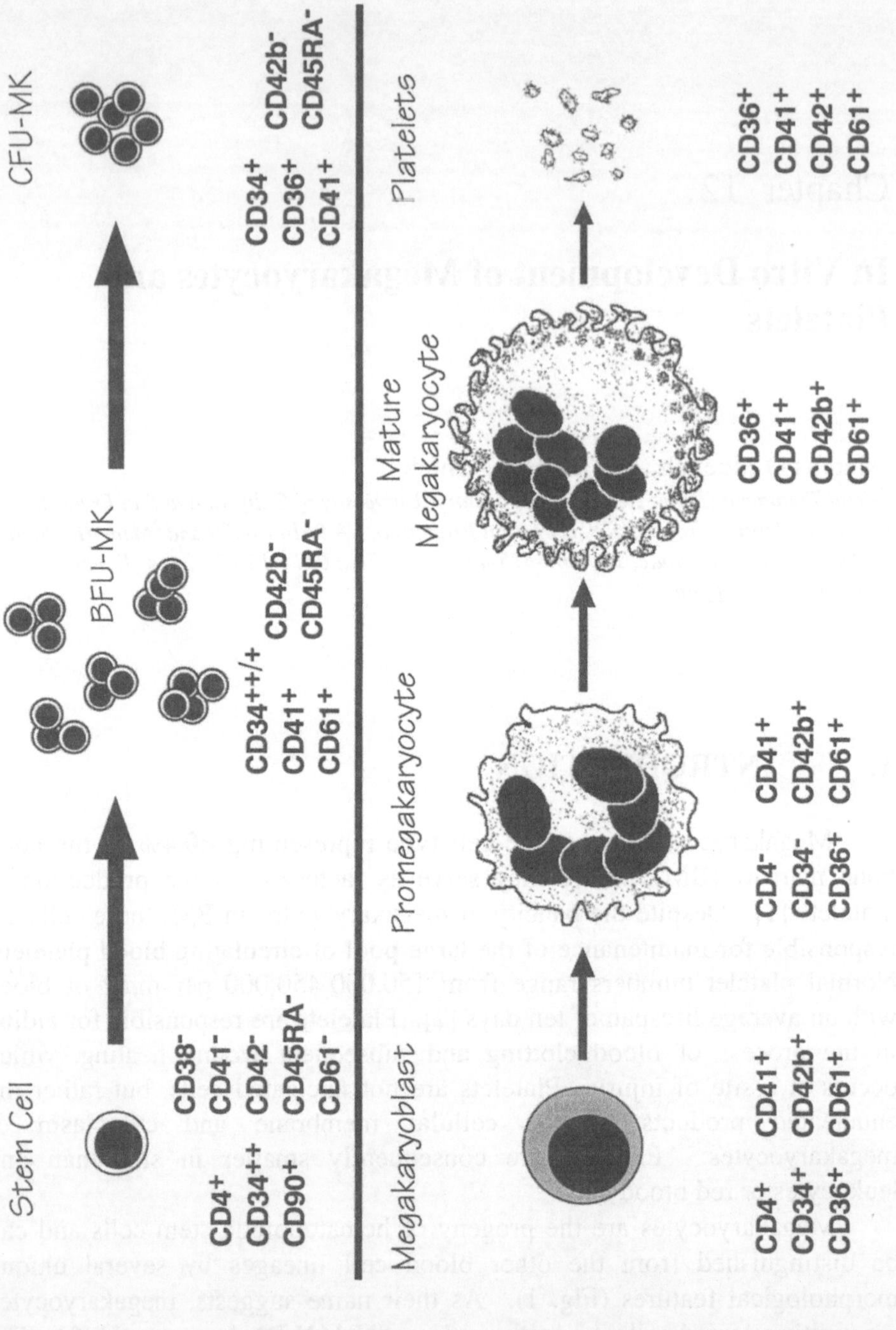

Figure 1. Overview of megakaryocytopoiesis and megakaryocyte maturation. The presence or absence of various cell surface markers on stem cells and megakaryocyte progenitors are indicated in the top half of the figure. The relationship between the expression of cell surface markers and megakaryocyte ploidy during the process of megakaryocyte maturation are shown in the lower half of the figure. The proposed scheme is based on the following studies: [47, 53, 56, 65-67, 69, 70, 72, 83-85].

Megakaryocytes have been generated in culture from various progenitor fractions. The growth of murine megakaryocyte colonies *in vitro*, as judged by the murine megakaryocyte-specific marker acetylcholinesterase, was first published by Metcalf et al. in 1975 [7]. Unfortunately, human megakaryocytes do not contain acetylcholinesterase. Using morphological criteria, Vainchenker et al. reported growth of human megakaryocyte colony formation from human BM cells as well as fetal, neonatal and adult peripheral blood (PB) [8, 9]. However, the methodology used by Vainchenker et al. to identify megakaryocytes was imprecise and tedious. Because human platelet glycoproteins are also expressed on human megakaryocytes [10], immunofluorescent staining using anti-CD41 and anti-CD42b monoclonal antibodies (mAb) was later used to identify megakaryocyte colonies grown in plasma clot cultures in the presence of heat-inactivated human AB plasma [11]. In addition to plasma clot based cultures, a variety of other clonal assay systems have been used to detect human megakaryocyte progenitors, including methylcellulose-, agar- and agarose-based culture systems [8, 11-17].

Understanding the different elements which regulate the growth and differentiation of stem cells into megakaryocytes requires culture systems consisting of defined components. Thus, most recent efforts at culturing megakaryocyte progenitors utilize media not supplemented with serum or plasma. A number of serum-deprived media formulations have been reported [18, 19], and several commercial vendors are marketing defined media for the growth of hematopoietic progenitors. We prefer the term serum-deprived instead of serum-free since most defined media use products isolated from serum, which can introduce various unknown serum contaminants into the medium. Commercially available media have the added disadvantage that most vendors do not disclose the constituents of their media. Nonetheless, defined media represent a significant advance in the study of megakaryocytopoiesis. Indeed, it has been shown that normal serum suppresses human megakaryocyte colony formation [20], and that a clonal assay for megakaryocyte progenitors requires plasma devoid of platelet products [16, 17, 21]. This may be due to the fact that normal serum contains platelet-derived products such as transforming growth factor-β (TGF-β) [22], platelet-derived growth factor (PDGF) [23], connective tissue-activating factor (CTAF) [24] and platelet factor 4 (PF4) [25]. These factors, either alone or in combination, inhibit the growth of megakaryocyte progenitors. In addition, accessory cells, which are supported by serum, also release cytokines such as interferon (IFN)-α, IFN-γ [26-28] and tumor necrosis factor (TNF) [29]. Therefore, media without serum are superior to serum-replete media for study of the effects of various regulatory factors on megakaryocytopoiesis.

A pivotal event in understanding the mechanism whereby the growth of megakaryocyte progenitors and the production of platelets is regulated came with the recent identification of *c-mpl* ligand (ML). The discovery of ML was the culmination of efforts by many groups of investigators seeking to identify the primary cytokine responsible for platelet production. The cDNA cloning and sequencing of the ML gene was reported in 1994 by four teams of researchers, resulting in four different names reflecting the biological activities of this hematopoietic growth factor (for review see [30, 31]): thrombopoietin (Tpo), megakaryocyte growth and development factor (MGDF), megapoietin, and, the name which we have chosen to use, ML. ML has been shown to be a major regulator of megakaryocyte growth, maturation and platelet production. In the midst of this discovery, ML was identified as a ligand for the cytokine-receptor *c-mpl* [32-34], which is expressed on human megakaryocyte progenitors [35]. The administration of ML to mice has shown this cytokine to be a potent thrombopoietin [32, 33], and ML has been shown to be present, naturally, in the sera of a number of mammalian species where its levels correlate inversely with PB platelet counts [32, 36-38]. The thrombopoietic effects of ML are the result of its ability to promote the growth and maturation of megakaryocytes as indicated by ML-stimulated increases in megakaryocyte numbers and ploidy [4, 5, 32, 34, 39, 40]. Consistent with these effects of ML, *c-mpl*$^{-/-}$ mice were found to have greatly reduced platelet and megakaryocyte numbers [41], demonstrating that ML and *c-mpl* are critical in the regulation of platelet homeostasis. In addition to the effects of ML on the megakaryocyte lineage, ML has also been shown to stimulate the growth of late erythroid progenitors, early-multipotent progenitors and stem cells [42-48].

Before the discovery of ML, a number of other cytokines were identified to have a role in the regulation of human megakaryocyte growth. Interleukin (IL)-3, IL-6, IL-11 and *c-kit* ligand (KL) have been shown to support the growth of megakaryocyte progenitors [5, 12, 49, 50]. IL-3, and to a lesser degree KL, have been shown to alone stimulate the growth of megakaryocyte progenitors, whereas IL-6 and IL-11 had no growth stimulating activity alone, but synergize with KL and IL-3 to promote growth. In addition to growth promoting activities of IL-3, IL-6 and KL, these cytokines have been reported to support the endoduplication of megakaryocytes and the acquisition of CD42b and CD62b antigens, markers of mature megakarocytes [12, 49]. However, ML has been shown to be the most potent cytokine in stimulating the maturation of megakaryocytes [5]. Indeed, the complexity of the regulation of megakaryocytopoiesis is in part indicated by the finding that IL-3 has a dual role in megakaryocyte growth and differentiation. Although IL-3 alone supports the growth, and to some degree, the differentiation of megakaryocytes, IL-3 also suppresses the endoduplication of megakaryocytes supported by ML [49]. Much has been learned about the growth factors required for the development of mature platelet-producing megakaryocytes from early hematopoietic progenitors. Nonetheless, a complete understanding of the many factors which regulate the number and ploidy of megakaryocytes remains elusive.

2. TISSUE PROCUREMENT AND PROCESSING

Megakaryocyte progenitors can be obtained from a number of different hematopoietic tissues including BM, mobilized peripheral blood (mPB) progenitors, umbilical cord blood (CB) and fetal hematopoietic tissues. Megakaryocyte progenitors can be isolated from these tissues by any of a number of methods used for the isolation of hematopoietic progenitors, depending on the population of progenitors being isolated. We describe below three protocols which we have used for the isolation of neonatal and fetal progenitors from CB, fetal BM (FBM) and fetal liver. These protocols can be readily adapted for the isolation of progenitors from adult tissues as well.

2.1 Isolation of Lineage$^-$ Light-Density CB Cells

Several techniques for the isolation of progenitors from CB have been reported. Most protocols begin with the removal of red cells and various mature leukocyte populations by separating low-density cells from high-density cells by centrifugation over a layer of high-density medium, such as Ficoll-Hypaque (d = 1.077, Histopaque; Sigma Chemical Company, St. Louis, MI) or Lymphoprep (d = 1.077, Life Technologies, Grand Island, NY) [51-53]. Alternatively, efficient removal of unwanted cells by sedimentation using gelatin has been reported [52]. There has been some disagreement among investigators as to the best method to enrich CB progenitors using these techniques. The success or failure of any one of these techniques in the hands of different investigators likely depends on subtle differences in the isolation protocols. In our experience, CB progenitors can be successfully isolated using the following protocol:

1. CB is harvested from the umbilical vein using 1-3 60ml syringes containing heparin. The greatest yields are possible when CB is harvested while the placenta is still *in utero* after delivery of the infant.
2. Mix the fresh CB with two volumes of room temperature (RT) phosphate-buffered saline without added Ca^{2+} or Mg^{2+} (PBS). The blood and PBS can be mixed in either 50 ml conical tubes or in a sterilized glass beaker, depending on the volume of the harvest.

3. Prepare a set of 15 ml or 50 ml polypropylene conical tubes with 5 ml or 15 ml, respectively, of Lymphoprep (density = 1.077 g/ml; Life Technologies).
4. Slowly layer the diluted CB on top of the Lymphoprep while holding the tubes at a 45° angle. Fill the tubes to no more than 3-5 ml below capacity. Centrifuge for 25 minutes at 800 *g* at RT with no brake.
5. After centrifugation, remove the plasma layer with a 5 ml pipette. Using a new pipette, transfer the mononuclear cell (MNC) layer found at the interface between the diluted blood and the Lymphoprep to new 50 ml conical tubes. The top two-thirds of the Lymphoprep solution can be harvested along with the interface.
6. Dilute the light-density CB cells with at least three volumes of PBS and centrifuge 10 minutes at 300 *g* at RT. Aspirate the supernatant and wash the cells two more times in PBS with 0.5% bovine serum albumin (BSA) (product #100 362; Boehringer Mannheim Biochemicals, Indianapolis, IN) and 50 μg/ml gentamicin sulfate (PBS/BSA), centrifuging the cells 7 minutes at 300 *g* at 4 °C. PBS containing 2% fetal bovine serum (FBS) can be used in place of PBS/BSA, however, the use of BSA is recommended if the cells are to be cultured in serum-deprived medium.
7. Resuspend the cells in PBS/BSA, count cells and determine viability by trypan blue (0.2%) exclusion. Alternatively, if cells are to be further purified by negative selection and cell sorting, it is recommended that the cells are resuspended in 2 ml of a blocking buffer consisting of 5% normal mouse serum (NMS), 0.5% human gamma-globulins, and 0.01% NaN_3 in PBS. The blocking buffer is used to prevent non-specific binding of mAb. The use of 15 ml conical polystyrene tubes is recommended throughout the negative selection procedure.
8. The light-density fraction of CB cells still contains only a low frequency of progenitors. These progenitors can be further enriched by negative depletion of mature blood cells and purified by fluorescence-activated cell sorting (FACS). Details of common procedures for the isolation of progenitors by FACS are described below for all three tissues. The depletion of mature blood cells prior to cell sorting serves several purposes. The purity of progenitors isolated by FACS is increased when the frequency of mature blood cells is reduced prior to cell sorting. The time required to complete the sorting of depleted cells is greatly reduced from the time required to sort undepleted cells. For many laboratories, the extra cost of depleting cells prior to cell sorting is more than compensated for by the decreased cost and time of the sorting procedure itself. Mature blood cells are depleted by immunomagnetic bead depletion. The cells are sensitized using fluorescein isothiocyanate (FITC)-labeled mAb against glycophorin A (GPA; nucleated red blood cells), CD3 (T-cells), CD14 (monocytes), CD19 (B-cells), CD20 (B-

cells), and CD56 (NK cells). mAb against these lineage (Lin)-specific antigens can be purchased from a number of different vendors, including Caltag Laboratories (South San Francisco, CA), Coulter-Immunotech (Miami, FL) or Becton Dickinson (San Jose, CA). An excellent FITC-labeled CD56 mAb (clone C5.9) can be obtained from the Exalpha Corporation (Boston, MA). It is recommended that the amount of mAb required for maximum labeling be determined for each of the lineage-specific mAb. The light-density CB cells are sensitized with saturating amounts of mAb for 20-30 minutes. The cells should be kept at 4 °C throughout the purification procedure, and be shielded from prolonged exposure to light. An aliquot of light-density CB cells should be saved from before the addition of labeled mAb. These cells are to be used for controls to adjust the voltage and compensation settings of the flow cytometer.

9. Excess mAb is removed by washing the cells twice with 14 ml of PBS/BSA (centrifuge for 7 minutes at 300 *g* at 4 °C) and resuspending the sensitized cells in 5 ml of the same solution.
10. The Lin^+ cells are removed using sheep anti-mouse coated Dynabeads (Dynal Inc., Lake Success, NY) at a ratio of five beads per cell. Prior to their use, wash the Dynabeads with PBS as recommended by the manufacturer. Combine the beads and cells and incubate the tube for 15 minutes at 4 °C with constant gentle mixing.
11. After binding of the Dynabeads to Lin^+ cells, place the tube next to a magnet (Dynal Inc.) and allow the particles to separate for 2 minutes. Recover the unbound cells in the supernatant. These Lin^- light-density CB cells can be placed in a second 15 ml tube and subjected to a second round of magnetic depletion. Recover the unbound cells and concentrate by centrifugation (7 minutes at 300 *g* at 4 °C). Resuspend the cells in 1-2 ml of blocking buffer and determine the number of viable cells. These cells can be stained with a number of different labeled mAb for the isolation of progenitor subpopulations by FACS as described below.

2.2 Isolation of Lineage⁻ Light-Density Fetal Liver Cells

Fetal livers at 16-24 weeks of gestational age are a rich source of hematopoietic progenitors. Isolating these progenitors requires, however, methods unlike those used for any other hematopoietic tissue. This is due to the predominance of erythropoiesis in the fetal liver which results in a high frequency of immature nucleated red cells [54]. These red cells can not be removed by density centrifugation, chemical lysis or through the use of sedimentation techniques commonly used with other hematopoietic tissues. These nucleated red cells can, however, be depleted using immunomagnetic beads to deplete GPA^+ cells [55, 56]. The following protocol can be used to prepare Lin^- light-density fetal liver cells for FACS:

1. In the United States, most fetal livers are obtained from elective abortions by cervical dilatation and extraction, which usually results in some disruption and microbial contamination of the liver [57]. Consequently, intact specimens are rare, but large pieces of liver tissue can often be obtained. The gestational age of the abortuses is approximated from the foot length of the fetus. Livers are placed in a 50 ml conical tube in a transportation medium consisting of PBS/BSA. Alternatively, this single antibiotic solution can be substituted by a more potent combination of anti-microbial drugs consisting of erythromycin lactobionate (100 μg/ml), gentamicin sulfate (50 μg/ml), vancomycin hydrochloride (100 μg/ml), and amphotericin B (2.5 μg/ml). It should be noted that although unprocessed fetal liver is most often contaminated, after processing and cell sorting, very few tissues show any evidence of microbial contamination [58]. Livers are transported on ice and processing should begin within four hours of abortion.
2. Place the liver, with about 20 ml of medium, into a cell-dissociation cup with a 50 (or 60) mesh screen (Sigma Chemical Company, #CD-1). Grind the liver through the screen using a glass pestle. Use vertical motions with the pestle as much as possible. The single-cell suspension can be recycled through the mesh to help wash cells through. Collect the single-cell suspension and place it in a 50 ml tube. Continue to grind the remaining fragments of liver and wash with small volumes (7-10 ml) of PBS/BSA. Attempt to obtain a final volume of 50 ml (one tube), unless the liver is particularly large in which case two tubes can be used. Dissociation of the liver can be considered finished when most of the red pulp has been recovered leaving behind a white-colored tissue.
3. The total number of viable cells, including red cells, is determined to quantify the amount of mAb required for depletion of the red cells. Typically, most livers between the gestational ages of 16-24 weeks contain approximately 2×10^9 cells. However, some of the older livers, when nearly intact, can contain over 1×10^{10} cells.
4. Add sufficient anti-GPA mAb for the number of cells obtained from the liver. We recommend the use of clone 10F7MN, an anti-GPA mAb available from the American Type Culture Collection (ATCC; Rockville, MD). Incubate for 30 minutes on ice.
5. Centrifuge the cells (7 minutes at 300 *g* at 4 °C), remove the supernatant and resuspend the cells in 50 ml. Wash the cells by centrifugation again and then resuspend the washed cells in 30 ml of PBS/BSA. If a large number of cell clumps have developed, the cells can be filtered using

either a cell dissociation cup with a #40 mesh screen or using a disposable 70 μm mesh cell strainer (Falcon #2350, Becton-Dickinson).

6. Remove the GPA^+ cells using BioMag goat anti-mouse IgG magnetic particles (PerSeptive Biosystems, Framinham, MA) at a ratio of five particles per cell. Up to 50 particles per cell can be used. However, the number of particles used has been reduced to a minimum sufficient to achieve a good level of depletion at a low cost. Prior to use, wash the magnetic particles twice in a T25 tissue culture flask against a magnet using PBS. Allow several minutes for these beads to be attracted towards the magnet. Add the liver cell suspension to the beads in the T25 flask. Mix the beads and cells and incubate on ice for 15 minutes. Every few minutes mix the flask. After 15 minutes, place the flask against the magnet and allow the beads to be attracted towards the magnet for at least 5 minutes. Remove the supernatant containing the mostly GPA^- fetal liver cells and place these cells into a 50 ml tube. Without removing the flask from the magnet, add 10-20 ml of PBS/BSA to the flask in order to help rinse out the remaining cells. Try to minimize disturbing the mass of magnetic particles and cells. Combine the rinse volume of PBS/BSA with cells in the 50 ml tube. Any contamination of the GPA^- fetal liver cells with magnetic particles or particle-coated cells is of little concern since these cells will be removed by density centrifugation in the next step of the protocol.
7. Spin down the cells (7 minutes at 300 *g* at 4 °C) in order to concentrate them. Resuspend in about 5 ml of PBS/BSA. Add 7 ml of 1.077 g/ml Nycoprep (Life Technologies) to a 15 ml polypropylene tube. Slowly layer the 5 ml of GPA^- liver cells on top of the Nycoprep. Centrifuge for 25 minutes at 800 *g* at RT with the brake off. Collect the light-density fraction and add these light-density fetal liver cells (LDFL) to a 15 ml tube. Discard the pelleted cells. Fill the tube containing the LDFL with PBS/BSA, invert the tube several times and centrifuge (7 minutes at 300*g* at 4 °C). Remove the supernatant down to 0.1 ml and add 1.9 ml of blocking buffer. Count the number of cells recovered. Place on ice for several minutes. Save an aliquot of these cells for use as controls for setting up the flow cytometer.
8. Add mAb for the depletion of Lin^+ cells and incubate on ice for 30 minutes. Use a saturating quantity of GPA-FITC, CD3-FITC, CD14-FITC, CD19-FITC, CD20-FITC, and CD56-FITC mAbs for the Lin panel.
9. Wash the cells twice (7 minutes at 300 *g* at 4 °C), resuspend the cells in about 7-10 ml of PBS/BSA and deplete using sheep anti-mouse coated Dynabeads at a ratio of five beads per cell. Prior to their use, wash the Dynabeads as was done with the BioMag particles above. Handle the Dynabeads in 15 ml polystyrene tubes instead of the T25 flask used

above. Mix the beads and cells and incubate the tube for 15 minutes with constant mixing. After binding of the Dynabeads to Lin^+ cells, place the tube next to a magnet and allow the particles to separate for 2 minutes. Recover unbound cells in the supernatant. These Lin^- LDFL can be placed in a second 15 ml tube and subjected to a second round of magnetic depletion. Recover unbound cells and concentrate by centrifugation (7 minutes at 300 *g* at 4 °C). Resuspend cells in 1-2 ml of blocking buffer. The Lin^- LDFL are now ready to be stained for cell sorting.

2.3 Isolation of Lineage⁻ Light-Density FBM Cells

FBM cells are another source of hematopoietic progenitors for the study of human megakaryocytopoiesis. BM is found primarily in the long bones of fetuses between the gestational ages of 18-24 weeks. The isolation of Lin^- light-density FBM cells follows a similar protocol as those described for CB and fetal liver cells [59].

1. Fetal limbs are obtained from elective abortions by cervical dilatation and extraction. The gestational age is approximated from the foot length of the fetus. Unlike fetal liver, intact fetal limbs are often found and the FBM, shielded within the marrow cavity, is sterile. The limbs are transported in PBS/BSA on ice.
2. FBM is harvested by first cleaning the long bones free of muscle and skin tissues. This is easily performed in a sterile Petri dish using a scalpel and forceps to hold the tissues. The cleaned bones are removed to a new Petri dish containing approximately 20 ml of PBS/BSA. Using a clean scalpel and forceps, FBM is dispersed by cutting the bones into small fragments. The scalpel blade is used to cut bones into pieces as small as possible, or to otherwise scrape FBM from the marrow cavities. A 10 ml pipette is then used to further rinse the FBM away from the bone fragments using the PBS/BSA solution in the Petri dish. This PBS/BSA solution is then transferred to another clean Petri dish where it is filtered through a cell-dissociation cup with a #50- or #60-mesh screen. The remaining bone fragments are also transferred to the cell-dissociation cup and rinsed with fresh PBS/BSA. A glass pestle is used to further break up any remaining large bone fragments. The filtered FBM suspension is transferred to a 50 ml conical tube. The bone fragments are rinsed with fresh PBS/BSA until they become white in appearance. The disappearance of red cells from the bone fragments is a good indicator that most of the FBM has been harvested from the bones.

3. The cells are centrifuged (7 minutes at 300 *g* at 4 °C) in order to concentrate them, and the light-density fraction is isolated using density centrifugation as described in step #7 of the fetal liver protocol. After isolating the light-density fraction of FBM cells, the cells are washed twice by centrifugation. The cells are resuspended in 2 ml of blocking buffer and the number of cells recovered is counted. An aliquot of these cells is saved for use as controls in setting up the flow cytometer.
4. The light-density FBM cells are depleted of Lin^+ cells by immunomagnetic bead depletion as described for fetal liver cells in steps #8 and #9 above. The Lin^- light-density FBM cells recovered are ready for staining with mAb followed by FACS to isolate progenitors.

2.4 Preparation of Cells for FACS

Lin^- progenitors can be stained in a number of different ways for isolation of different megakaryocyte progenitors and candidate hematopoietic stem cells. By far the most common antigen used to identify hematopoietic progenitors is CD34 [60, 61], although recent reports suggests that some stem cells may be contained among the $CD34^-$ fraction of hematopoietic cells [62, 63]. Staining with mAb against CD34 reveals a spectrum of CD34 expression. The level of CD34 expression can be used to differentiate the more immature progenitors/stem cells, which express high levels of CD34 ($CD34^{++}$), from the more mature progenitors which express lower levels of CD34 ($CD34^+$) (Fig. 1 and Fig. 2). Many other cell surface markers can be used to further differentiate different progenitor compartments from among the pool of $CD34^{++/+}$ cells. Investigators interested in the isolation of candidate hematopoietic stem cell populations have used CD90 (Thy-1) and CD4 to positively identify candidate stem cells from among the $CD34^{++}$ fraction of cells [53, 64-66]. Another marker used commonly to identify candidate stem cells is CD38. Data suggest that stem cells express little or no CD38, whereas progenitors express CD38 [56, 67, 68]. Staining hematopoietic cells with CD34 and CD38 can be, therefore, used to resolve several subpopulations of progenitors/stem cells (Fig. 2).

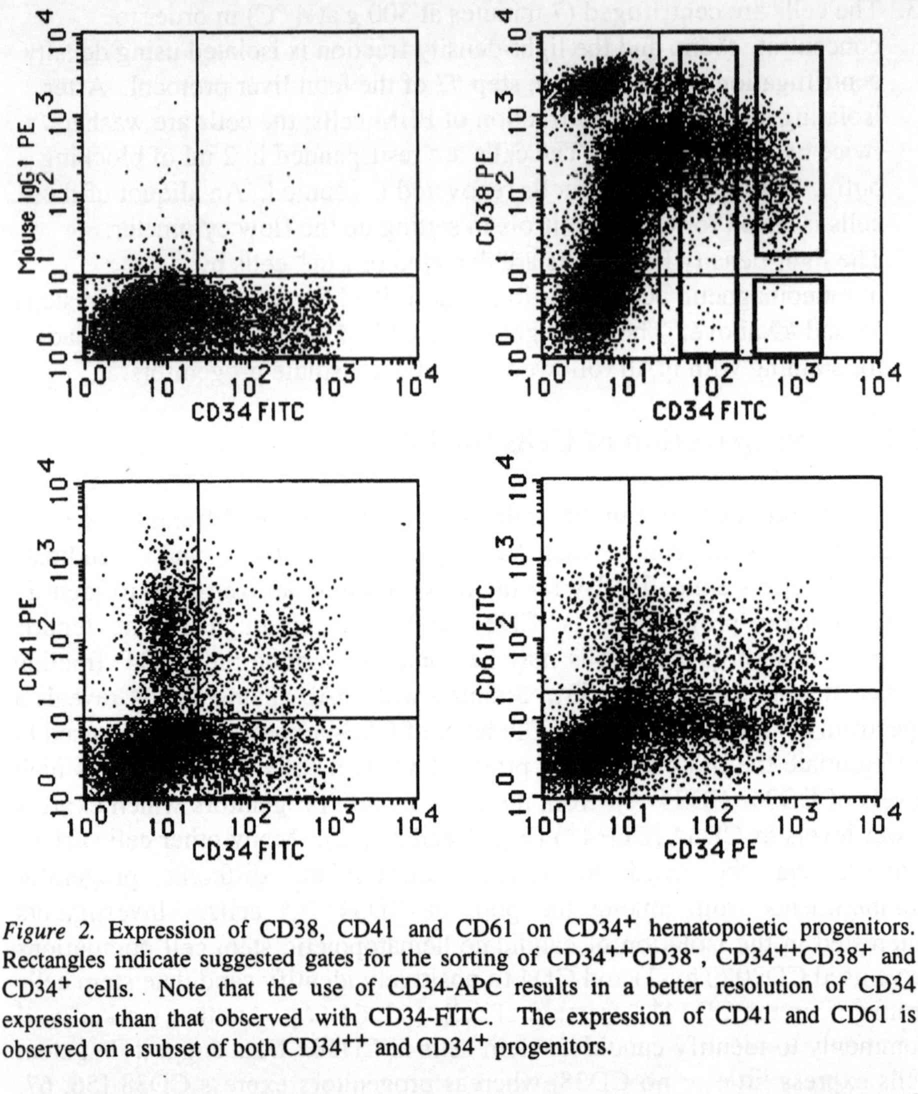

Figure 2. Expression of CD38, CD41 and CD61 on CD34$^+$ hematopoietic progenitors. Rectangles indicate suggested gates for the sorting of CD34^{++}CD38$^-$, CD34^{++}CD38$^+$ and CD34$^+$ cells. Note that the use of CD34-APC results in a better resolution of CD34 expression than that observed with CD34-FITC. The expression of CD41 and CD61 is observed on a subset of both CD34^{++} and CD34$^+$ progenitors.

Another cell surface marker employed in the study of megakaryocytopoiesis is CD41, which defines a subset of CD34$^{++/+}$ cells enriched in megakaryocyte progenitors (Fig. 2) [47, 69, 70]. CD41 expression is further observed on mature megakaryocytes and platelets where it serves as a receptor for fibrinogen or von Willebrand's factor. CD41 is also known as integrin αIIb or platelet gpIIb. CD41 forms a heterodimer with CD61 (also known as integrin β3 or platelet gpIIIa). Consequently, the pattern of CD61 expression on CD34$^+$ cells is similar to the pattern observed with CD41 (Fig. 2). mAbs recognizing CD41 are sometimes segregated into two classes, CD41a and CD41b. mAbs against CD41a recognize the gpIIb/gpIIIa heterodimer, whereas CD41b mAbs recognize gpIIb. Both CD41a and CD41b have been used to identify progenitors isolated from FBM [70]. The $CD41^{+}CD34^{++/+}$ population does not define progenitors solely committed to the megakaryocytic lineage. $CD41^{+}CD34^{++/+}$ cells contain a number of different progenitor types in addition to megakaryocyte progenitors, including myeloid-committed, erythroid-committed and mixed-lineage progenitors [47, 70]. Despite the presence of multilineage progenitors among the $CD41^{+}CD34^{++/+}$ fraction, stem cells are thought to be contained among the $CD41^{-}$ fraction [70].

The following protocol can used for the isolation of cells based on their expression of CD34 and either CD38 or CD41:

1. Lin^{-} cells are stained with saturating amounts of CD34-allophycocyanin (APC) (clone 581, Coulter-Immunotech) and either CD38-PE (Becton-Dickinson) or CD41-PE (Caltag Laboratories) for 30 minutes on ice.
2. The cells are washed twice by centrifugation in PBS/BSA for 7 minutes at 300 *g* at 4 °C. The washed cells are resuspended in PBS with 0.5% BSA and 1 µg/ml propidium iodide (PI; Molecular Probes, Eugene, OR) at a concentration of 1×10^6 to 5×10^6 cells/ml.
3. The cells are filtered through a 40 µm nylon cell strainer (Falcon #2340, Becton-Dickinson) into 12 x 75 mm tubes (Falcon brand tubes are recommended for flow cytometers manufactured by Becton-Dickinson). These cells should be kept in the dark and at 4 °C until cell sorting.
4. Controls for setting up the flow cytometer can be prepared using an aliquot of GPA^{-} light-density cells saved for this purpose. Use about 2×10^5 cells per staining. These cells can be stained using the same protocol used for phenotype analyses described below. Typically, the following controls are needed: unstained cells, cells stained with isotype-matched control antibodies and PI, cells stained with the panel of FITC-conjugated Lin mAbs and PI, cells stained with CD38-PE and PI and cells stained with CD34-APC and PI.

Live cells are isolated based on the lack of PI staining. An electronic gate is also set to include cells with a lymphoid-blastoid light scatter profile. This gate can be set by gating on $CD34^{+/++}$ cells and examining the forward and side scatter properties of these cells, most of which will have a low side scatter and an intermediate forward scatter profile. Any remaining Lin^{+} cells are removed by isolating only Lin-$FITC^{-}$ cells. Finally, electronic gates are set to isolate progenitors based on their expression of CD34-APC and either CD38-PE or CD41-PE. As shown in Fig. 2, we have isolated three subpopulations defined as $CD34^{++}CD38^{-}Lin^{-}$, $CD34^{++}CD38^{+}Lin^{-}$ and $CD34^{+}Lin^{-}$ using the combination of CD34 and CD38 mAbs.

3. ASSAY TECHNIQUES

3.1 Flow Cytometric Analysis of Cell Surface Markers

Flow cytometry has been successfully used to measure the production of megakaryocytes in liquid cultures. The generation of megakaryocytes is indicated by the acquisition of cell surface markers specific for megakaryocytes and platelets. Cell surface markers associated with differentiation along the megakaryocytic lineage include CD41, CD42 (a-d), CD61 and CD62b [47, 69, 71, 72]. Staining using mAbs against CD41 and CD42b have been widely used because these antigens can be used to distinguish both immature and mature megakaryocytes (Fig. 1) [72]. The levels of CD41 expression range from a low level of expression on progenitors and immature megakaryocytes (Fig. 2) to high levels of expression found on mature megakaryocytes (Fig. 3). Since CD61 forms part of the heterodimer complex with CD41, the expression of CD61 on megakaryocytes is identical to that of CD41. However, CD61 can also be associated with CD51 (integrin α_V), forming the vitronectin receptor. Expression of the CD51/CD61 complex is not restricted to the megakaryocytic lineage. Mast cells generated from cultured progenitors also express CD51/CD61 and may thus be confused for megakaryocytes if CD61 is the only marker used to identify megakaryocytes [73]. The following protocol can be used to stain megakaryocytes for analysis by flow cytometry:

1. Cultured or freshly harvested cells are washed and resuspended in blocking buffer at about 1 x 10^7 cells/ml. Blocking buffer consists of 5% NMS, 0.5% human gamma-globulins, 0.01% NaN_3 in PBS w/o Ca^{++} and Mg^{++}. The cells are held in blocking buffer for at least 5 minutes on ice before the addition of mAb. The blocking buffer is used to prevent non-specific binding of the labeled mAb.
2. Add 2 x 10^5 cells/well into 96-well V-bottom plates. As few as 5 x 10^4 cells can be used, or a higher cell number (up to 1 x 10^6 cells), can be used if >5 x 10^4 events are to be collected.
3. Add labeled mAb for the identification of megakaryocytes, e.g. CD41-FITC and CD42b-PE. All mAb should be titrated before use and used at the concentration which results in optimal staining.
4. Incubate the cells together with the labeled mAb on ice for 20-30 minutes.
5. Add 200 µl of washing buffer and centrifuge for 3 minutes at 300 *g* at 4 °C to pellet the cells. Washing buffer consists of 0.5% BSA, 0.01% NaN_3 in PBS w/o Ca^{++} and Mg^{++}. After centrifugation, remove the

supernatant by quickly inverting the plate. Repeat the washing procedure with another 250 µl of washing buffer. Alternatively, a single wash can be used if working with low numbers of cells.

6. Resuspend the samples in 200 µl of either washing buffer or washing buffer with 2 µg/ml propidium iodide (PI). PI is used to stain dead cells for their removal from the analyses. Dispense the cells into 1.2 ml polypropylene mini tubes (#AS-2015, Applied Scientific, So. San Francisco, CA) for analysis. These mini tubes can be placed inside the larger 12 x 75 mm tubes which fit the sample ports of flow cytometers.

3.2 Flow Cytometric Analysis of DNA Content

Polyploidy is another hallmark of megakaryocytes which can be measured by flow cytometry (Fig. 4). The DNA content of megakaryocytes can be determined by staining the cells with PI after permeabilization of their membranes. The emission spectrum of PI is broad and prevents the co-staining of cells with PE-labeled mAb. Cells can, however, be stained with a FITC-labeled megakaryocyte marker such as CD41 using the protocol described above. After cell surface staining, cells can be fixed and the DNA stained using the following protocol:

1. Wash the stained cells once with cold PBS to remove excess protein from the cell suspension. Resuspend the cells in 200 µl of PBS with 1% paraformaldehyde. The paraformaldehyde solution should be made fresh and can be used for one month. It should be stored in the dark at 4 °C. Cells should be kept in paraformaldehyde solution for at least 8 hours at 4 °C. The cells should also be kept in the dark until the time of analysis.
2. Add Triton X-100 to a final concentration of 0.1% and incubate for 40 minutes at 4 °C.
3. Pellet the cells by centrifugation (3 minutes at 300 *g* at 4 °C), remove the supernatant and wash twice with cold PBS with 0.5% BSA and 0.1% Triton X-100.
4. Resuspend the cells in PBS with 0.5% BSA, 0.1% Triton X-100, 50 µg/ml PI and 100 µg/ml DNAse-free RNAse (Boehringer Mannheim). Incubate for 1 hour at 37 °C in the dark.

Filter the cells through a 70 µm nylon mesh and run the samples on a flow cytometer. Note that the PI staining of DNA is best resolved on a linear scale.

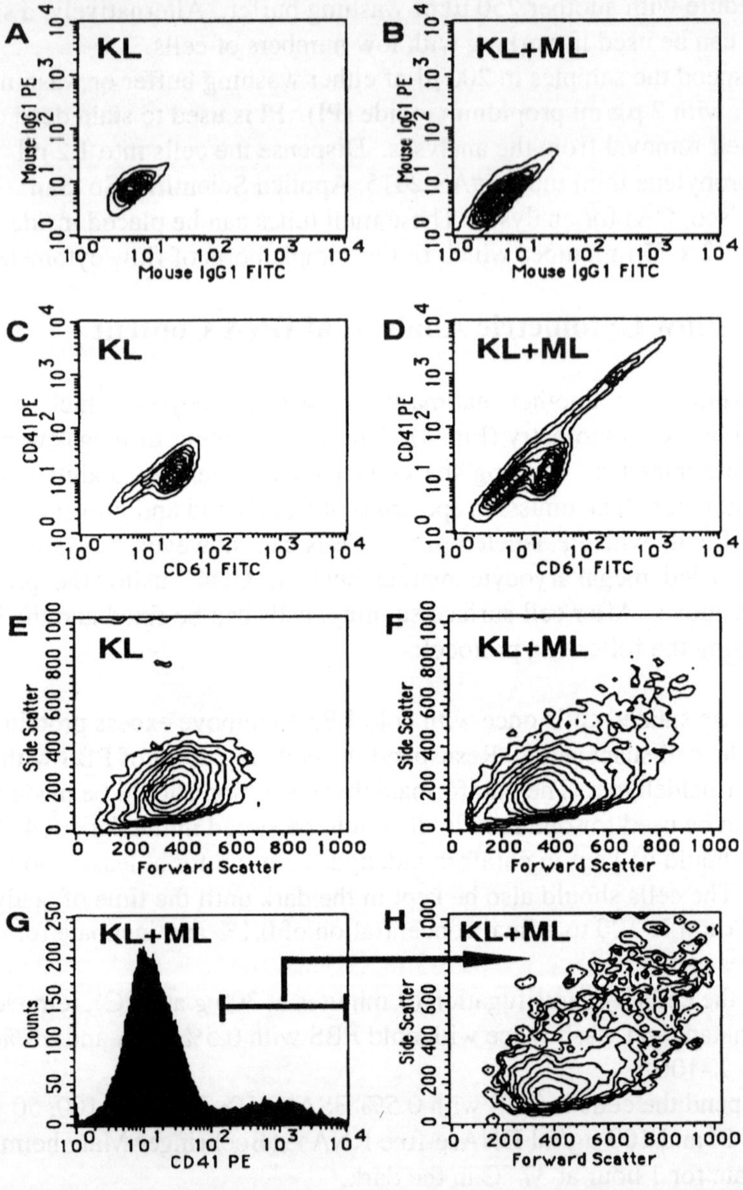

Figure 3. Expression of CD41 and CD61 on megakaryocytes generated from the culture of CD34^{++}CD38^{+}Lin^{-} fetal liver cells. Fetal progenitors were cultured for two weeks in either KL or KL+ML as indicated. Control stainings are shown in (A) and (B). Megakaryocytes expressing the CD41/CD61 complex were observed in cultures containing ML (D), but not in cultures containing KL alone (C). Note the presence of CD61^{dull} cells which do not express CD41 and are not megakaryocytes. Megakaryocytes are on average larger and have a higher side scatter than the other myeloid cells generated in culture (E-H).

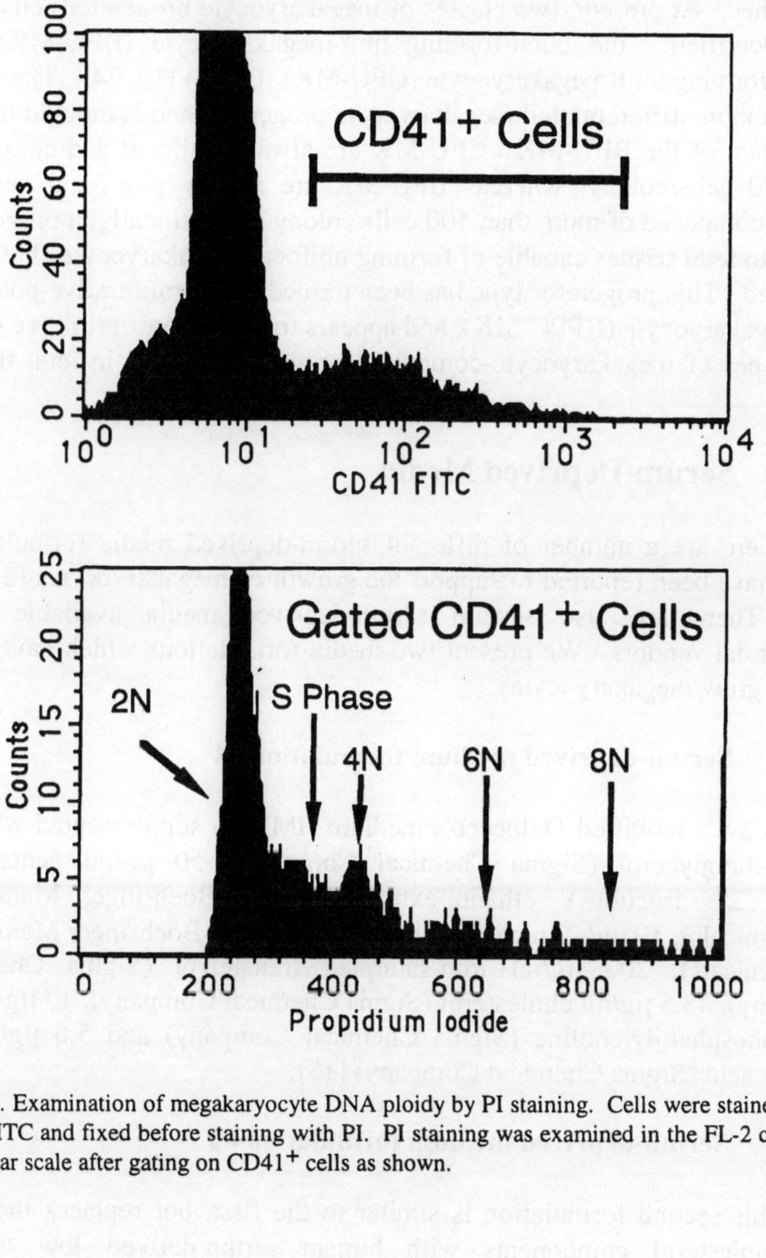

Figure 4. Examination of megakaryocyte DNA ploidy by PI staining. Cells were stained with CD41-FITC and fixed before staining with PI. PI staining was examined in the FL-2 channel on a linear scale after gating on $CD41^+$ cells as shown.

4. CULTURE TECHNIQUES

In vitro, megakaryocyte progenitors possess the ability to form colonies composed exclusively of megakaryocytes. Based on both the time of appearance and the cellular composition of colonies derived from such progenitor cells, the hierarchy of megakaryocyte progenitor cells was established. At present, two classes of megakaryocyte progenitor cells have been identified: the burst-forming unit-megakaryocyte (BFU-MK) and colony-forming unit-megakaryocyte (CFU-MK) (Fig. 1) [15, 74]. The CFU-MK is a more differentiated megakaryocyte progenitor and is thought to be a descendant of the BFU-MK. CFU-MK are always unifocal and composed of 3-100 cells/colony, whereas BFU-MK are mainly plurifocal and are usually composed of more than 100 cells/colony. Additionally, a progenitor unique to fetal tissues capable of forming unifocal megakaryocytes has been identified. This progenitor type has been termed high proliferative-potential cell-megakaryocyte (HPPC-MK), and appears to be the most primitive of the three types of megakaryocyte-committed progenitors found in fetal tissues [75].

4.1 Serum-Deprived Media

There are a number of different serum-deprived media formulations which have been reported to support the growth of megakaryocytes [22, 49, 76]. There are also several serum-deprived media available from commercial vendors. We present two media formulations which have been used to grow megakaryocytes:

4.1.1 Serum-deprived medium formulation #1

Iscove's modified Dulbecco's medium (IMDM) supplemented with 75 μM α-thioglycerol (Sigma Chemical Company), 50 μg/ml gentamicin sulfate, 2% fraction-V ethanol-extracted BSA (Boehringer Mannheim Biochemicals), 10 μg/ml recombinant human insulin (Boehringer Mannheim Biochemicals), 200 μg/ml iron-saturated transferrin (Sigma Chemical Company), 38.5 μg/ml cholesterol (Sigma Chemical Company), 10 μg/ml L-alpha-phosphatidylcholine (Sigma Chemical Company) and 5.6 μg/ml of linoleic acid (Sigma Chemical Company) [18].

4.1.2 Serum-deprived medium formulation #2

This second formulation is similar to the first, but replaces the lipid and cholesterol components with human serum-derived low density lipoproteins [19, 76, 77]. The medium consists of IMDM supplemented with 75 μM α-thioglycerol, 50 μg/ml gentamicin, 2% fraction-V ethanol-extracted BSA, 200 μg/ml human iron-saturated transferrin, 10 μg/ml recombinant human insulin, and 20 μg protein/ml human low density lipoprotein (Sigma Chemical Company).

These serum-deprived media can be used for either the growth of megakaryocyte progenitors in liquid cultures or for the generation of megakaryocyte colonies in semi-solid cultures. Recently, a commercial kit (MegaCult-C) to assay megakaryocyte progenitor cells under serum-deprived conditions has also been made available from StemCell Technologies, Inc. (Vancouver, BC, Canada). We present a similar protocol using avidin-biotin complex (ABC) immunochemical staining (Pierce, Rockford, IL) to detect megakaryocyte colonies grown in agarose-based cultures [78]:

4.2 Colony-Forming Unit Assay

1. Colony assays are performed in 35 mm tissue culture treated petri-dishes. At least three replicate plates should be made for each test group. The cultures consist of two layers, and a final volume of 1.5 ml. The lower layer has a final agarose concentration of 0.5%, and the upper layer, containing the progenitors, has a final concentration of 0.36% agarose. Stock solutions of 5% and 3.6% SeaPlaque agarose, a low gelling temperature form of agarose (FMC BioProducts, Rockland, ME), are made by dissolving the agarose in cell culture grade water and autoclaving for 5 minutes. At the time of use, the agarose is brought to a boil in a microwave oven and then cooled to approximately 40 °C. The lower layer of each culture consists of a final volume of 1 ml. Cytokines used to support the growth of megakaryocyte progenitors are added to the lower layer of the cultures. The growth factors used depend on the aim of each individual experiment. Nonetheless, efficient growth of megakaryocyte progenitors can be achieved with a combination of 100 ng each of either KL+ML or KL+IL-3+IL-6+IL-11. Cultures stimulated with KL+G-CSF can be used as negative controls for immunostaining of megakaryocytes. Before mixing with warm agarose, the culture medium containing cytokines is pre-warmed in a 37 °C water bath. Otherwise, addition of the agarose to cold medium results in a rapid solidification of the agarose. The final solution for preparing the lower layers consists of nine parts serum-deprived medium and cytokines and one part 5% agarose. A 1-2 ml excess of culture medium should be prepared to assure that exactly 1 ml can be dispensed into each culture dish. After adding 1 ml of medium to the culture dishes, each dish is checked to make sure that the medium covers the entire surface of the culture dish, and that no large bubbles have formed. Solidification of the underlayer is achieved by placing the culture dishes at 4 °C on a level surface for 5-10 minutes.

2. The upper layer of the cultures has a final volume of 0.5 ml and consists of nine parts serum-deprived medium and cells and one part 3.6% agarose. As with the lower layer, the culture medium and cells are pre-warmed before adding liquid agarose which has been cooled to near 37 °C. The number of cells to be used for each plate must be determined individually for each cell population being tested. Exactly 0.5 ml of medium is then added on top of the solidified lower layers and the culture dishes are again placed at 4 °C to solidify the upper layers.
3. The cultures are grown for 10-21 days at 37 °C in a fully humidified incubator containing ambient air supplemented with 5% CO_2. The length of culture depends on the immaturity of the progenitors and the source of tissue [75].
4. At the end of culture, the agarose plates are fixed with 2 ml of fixing solution (75% acetic acid and 25% methanol) for 5 min, washed twice with 2 ml of deionized water, and air-dried. Agarose plates are fragile during the fixation procedure and should be very gently handled. Specimens can be stored in a –20 °C freezer for more than 1 month prior to immunostaining.
5. To begin the ABC staining procedure, add 1 ml of ABC-blocking buffer to each culture dish for 10 minutes, and then aspirate the buffer. ABC-blocking buffer consists of PBS with 20 mM Tris, pH 7.5 (TBS) supplemented with 1.35% FBS.
6. Add 0.7 ml of primary mAb solution/plate for 30 minutes and wash four times with 1.5 ml of TBS supplemented with 0.1% BSA. The primary mAb solution consists of 970 μl TBS, 10 μl CD41b, and 10 μl of each of two anti-CD42b mAb. CD41b (#N42539M) and CD42b (#N42409M) mAb can be obtained from Biodesign International (Kennebunk, ME). The second anti-CD42b mAb (#M719) can be obtained from Dako Corporation (Carpinteria, CA).
7. Add 0.7 ml secondary antibody solution/plate for 30 minutes and then wash four times with 1.5 ml of TBS supplemented with 0.1% BSA. The secondary antibody solution is part of the ImmunoPure ABC alkaline phosphatase mouse IgG staining kit (#32044) available from Pierce.
8. Add 1 ml ABC complex solution/plate for 30 minutes and wash four times with 1.5 ml of TBS supplemented with 0.1% BSA.
9. Add 1 ml of substrate solution/plate, observe the reaction under a microscope and stop the reaction by washing with deionized water. Fast-red alkaline phosphatase substrate kit (#4034) is also available from Pierce. Cover the plates with water before viewing. The megakaryocyte colonies identified by fast-red staining can be easily scored using a regular light microscope (Fig. 5). Samples can be stored for more than two years.

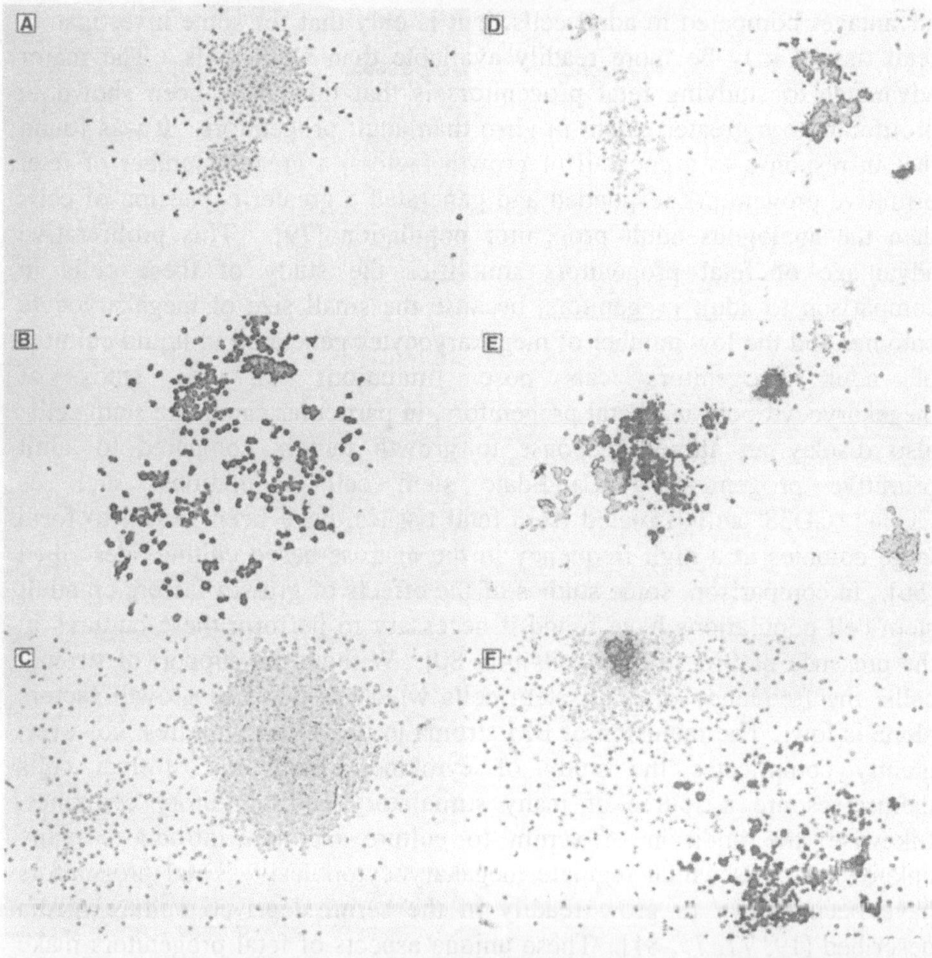

Figure 5. Expression of CD41b and CD42b antigens in colonies derived from CD34$^{+/++}$ cells grown under various cytokine conditions. The agarose plates were stained for the expression of CD41b and CD42b antigens. (B) Cultures with rhML exclusively formed megakaryocyte colonies as indicated by positive staining with Fast Red dye. Control cultures supported by (A) rhG-CSF or (C) rhKL+rhG-CSF did not demonstrate any staining. Cultures supported by (D) rhKL+rhML, (E) rhML+rhG-CSF, or (F) rhKL+rhML+rhG-CSF formed both megakaryocytic colonies and non-megakaryocytic colonies.

5. UTILITY OF SYSTEMS

There are several sources of hematopoietic progenitors available for study or transplantation. In this chapter, we have presented protocols for the isolation of megakaryocyte progenitors from fetal tissues. Small modifications to these protocols can be made to adapt them to the isolation of adult progenitors as well. The study of fetal progenitors offers some advantages compared to adult cells, if it is only that for some investigators fetal tissues may be more readily available than adult cells. The major advantage to studying fetal progenitors is that they have been shown to proliferate to a greater extent *in vitro* than adult progenitors. It was found that in response to a cocktail of growth factors, a greater number of fetal primitive progenitors responded and generated a greater expansion of cells than the analogous adult progenitor population [79]. This proliferative advantage of fetal progenitors simplifies the study of these cells in comparison to adult progenitors, because the small size of megakaryocyte colonies and the low number of megakaryocytes generated in liquid cultures of adult progenitors can pose limitations on the study of megakaryocytopoiesis. Fetal progenitors, in particular candidate stem cells, also display an altered response to growth factors compared to adult primitive progenitors. Candidate stem cell populations, such as $CD34^{++}CD38^{-}$ cells isolated from fetal tissues, have been shown to form large colonies at a high frequency in the agarose-based cultures described [56]. In comparison, some studies of the effects of growth factors on adult stem cell populations have found it necessary to perform these cultures in the presence of BM stromal cells [46, 80]. Without the support of stromal cells, the frequency of adult stem cells which respond to growth factors alone is low. The inclusion of BM stroma in progenitor cultures, however, greatly complicates the study of cytokines, since the stromal cells themselves are a source of many stimulatory and inhibitory cytokines. Likewise, the addition of serum to culture medium introduces many unknown factors which regulate megakaryocytopoiesis. Fetal progenitors have been shown to grow readily in the serum-deprived culture media described [19, 77, 79, 81]. These unique aspects of fetal progenitors make them ideal targets for the study of the growth and differentiation of stem cells into mature megakaryocytes.

Differences in the growth of fetal versus adult progenitors *in vitro* reflects differences in how fetal and adult hematopoiesis are regulated. Little is known about the regulation of fetal megakaryocytopoiesis and the mechanisms which account for the general differences in the growth properties of fetal and adult progenitors. The purpose of the HPPC-MK, the unique fetal megakaryocyte progenitor in fetal megakaryocytopoiesis, and the reason for its absence from adult BM, are questions that as of yet have no answers [75]. It is hoped that the protocols presented for the purification of fetal progenitors and the assay systems described may help in answering these and other questions relating to human megakaryocytopoiesis.

The clonal assay system described is best suited for analyzing different progenitor subpopulations for the presence of megakaryocyte progenitors. The effects of different growth and inhibitory factors on the recruitment and proliferation of megakaryocyte progenitors can be determined by counting the number and determining the size of megakaryocyte colonies. An agarose-based clonal assay has advantages over fibrin clot-based assays because it has been observed that MNC secrete proteases which can dissolve the fibrin clot. Methylcellulose-based assays are also not well suited for megakaryocyte progenitor assays which rely on immunochemical staining for the identification of megakaryocytes, because they are too delicate to withstand the staining procedure. The immunochemical staining procedure presented is also advantageous over protocols which use immunofluorescent staining, which requires that the megakaryocyte colonies be scored on fluorescence microscopes. Colonies stained using Fast Red can also be stored for up to 2 years, something which is not possible using fluorescent stains.

Clonal assays are the best method for studying the proliferation of megakaryocyte progenitors. However, clonal cultures are not well suited for the study of the maturation of megakaryocytes. Progenitors cultured in serum-deprived medium can be harvested and characterized using flow cytometry. Liquid cultures are, therefore, best for studying the differentiation of progenitors and maturation of megakaryocytes based on the expression of cell surface molecules and the degree of endomitosis. It should be noted that large mature megakaryocytes are delicate cells, and care should be taken not to disrupt these cells when staining with mAb or PI for DNA analysis. Liquid cultures tailored for the generation of megakaryocytes also present a method for generating large numbers of committed progenitors or mature megakaryocytes which can be used to study intracellular events associated with megakaryocyte differentiation and maturation [76].

6. CONCLUDING REMARKS

The discovery of ML as the major regulator of platelet production, megakaryocyte growth and megakaryocyte maturation has meant the culmination of the search for the growth factor activities known as thrombopoietin and megakaryocyte colony-stimulating factor. Indeed, not only was ML found to have a broad range of activities on the megakaryocyte lineage, but ML was also found to be capable of supporting the growth of erythroid progenitors, primitive progenitors and stem cells. These findings suggest that ML may have clinical value in stimulating the recovery of megakaryocytopoiesis and platelet production in patients given cytoablative therapy or a stem cell transplant. A further understanding of the role of ML and other cytokines in the growth of megakaryocytes from stem cells will also aid efforts to expand megakaryocyte cells *ex vivo* as a means to aid platelet recovery after transplantation of these cells [82]. These clinical endeavors will be supported by further research into the extracellular and intracellular mechanisms which regulate the growth and differentiation of stem cells into platelet-producing megakaryocytes.

LITERATURE CITED

1. Levine RF (1980) Isolation and characterization of normal human megakaryocytes. *Br. J. Haematol.* 45:487-497.
2. Mustard JF, Rowsell HC and Murphy EA (1966) Platelet economy (platelet survival and turnover). *Br. J. Haematol.* 12:1-24.
3. Williams N and Levine RF (1982) The origin, development and regulation of megakaryocytes. *Br. J. Haematol.* 52:173-180.
4. Kaushansky K, Lok S, Holly RD, Broudy VC, Lin N, Bailey MC, Forstrom JW, Buddle MM, Oort PJ, Hagen FS, Roth GJ, Papayannopoulou T and Foster DC (1994) Promotion of megakaryocyte progenitor expansion and differentiation by the c-Mpl ligand thrombopoietin. *Nature* 369:568-571.
5. Debili N, Wendling F, Katz A, Guichard J, Breton-Gorius J, Hunt P and Vainchenker W (1995) The Mpl-ligand or thrombopoietin or megakaryocyte growth and differentiative factor has both direct proliferative and differentiative activities on human megakaryocyte progenitors. *Blood* 86:2516-2525.
6. Levine RF, Hazzard KC and Lamberg JD (1982) The significance of megakaryocyte size. *Blood* 60:1122-1131.
7. Metcalf D, MacDonald HR, Odartchenko N and Sordat B (1975) Growth of mouse megakaryocyte colonies in vitro. *Proc. Natl. Acad. Sci. USA* 72:1744-1748.
8. Vainchenker W, Bouguet J, Guichard J and Breton-Gorius J (1979) Megakaryocyte colony formation from human BM precursors. *Blood* 54:940-945.
9. Vainchenker W, Guichard J and Breton-Gorius J (1979) Growth of human megakaryocyte colonies in culture from fetal, neonatal, and adult PB cells: ultrastructural analysis. *Blood Cells* 5:25-42.
10. Rabellino EM, Nachman RL, Williams N, Winchester RJ and Ross GD (1979) Human megakaryocytes. I. Characterization of the membrane and cytoplasmic components of isolated marrow megakaryocytes. *J. Exp. Med.* 149:1273-1287.
11. Mazur EM, Hoffman R, Chasis J, Marchesi S and Bruno E (1981) Immunofluorescent identification of human megakaryocyte colonies using an antiplatelet glycoprotein antiserum. *Blood* 57:277-286.
12. Debili N, Masse JM, Katz A, Guichard J, Breton-Gorius J and Vainchenker W (1993) Effects of the recombinant hematopoietic growth factors interleukin-3, interleukin-6, stem cell factor, and leukemia inhibitory factor on the megakaryocytic differentiation of CD34+ cells. *Blood* 82:84-95.
13. Levene RB, Williams NT, Lamaziere JM and Rabellino EM (1987) Human megakaryocytes. IV. Growth and characterization of clonable megakaryocyte progenitors in agar. *Exp. Hematol.* 15:181-189.
14. Bruno E, Briddell R and Hoffman R (1988) Effect of recombinant and purified hematopoietic growth factors on human megakaryocyte colony formation. *Exp. Hematol.* 16:371-377.

15. Long MW, Gragowski LL, Heffner CH and Boxer LA (1985) Phorbol diesters stimulate the development of an early murine progenitor cell. The burst-forming unit-megakaryocyte. *J. Clin. Invest.* 76:431-438.
16. Messner HA, Jamal N and Izaguirre C (1982) The growth of large megakaryocyte colonies from human BM. *J. Cell. Physiol.* Suppl. 1:45-51.
17. Kimura H, Burstein SA, Thorning D, Powell JS, Harker LA, Fialkow PJ and Adamson JW (1984) Human megakaryocytic progenitors (CFU-M) assayed in methylcellulose: physical characteristics and requirements for growth. *J. Cell. Physiol.* 118:87-96.
18. Iscove NN (1984) Culture of lymphocytes and hematopoietic cells in serum-free medium, in *Methods for serum-free culture of neuronal and lymphoid cells*, Alan R Liss Inc.
19. Valtieri M, Gabbianelli M, Pelosi E, Bassano E, Petti S, Russo G, Testa U and Peschle C (1989) Erythropoietin alone induces burst formation by human embryonic but not adult BFU-E in unicellular serum-free culture. *Blood* 74:460-470.
20. Vainchenker W, Chapman J, Deschamps JF, Vinci G, Bouguet J, Titeux M and Breton-Gorius J (1982) Normal human serum contains a factor(s) capable of inhibiting megakaryocyte colony formation. *Exp. Hematol.* 10:650-660.
21. Solberg LA, Jr., Jamal N and Messner HA (1985) Characterization of human megakaryocytic colony formation in human plasma. *J. Cell. Physiol.* 124:67-74.
22. Mitjavila MT, Vinci G, Villeval JL, Kieffer N, Henri A, Testa U, Breton-Gorius J and Vainchenker W (1988) Human platelet alpha granules contain a nonspecific inhibitor of megakaryocyte colony formation: its relationship to type beta transforming growth factor (TGF-beta). *J. Cell. Physiol.* 134:93-100.
23. Komatsu N, Suda T, Sakata Y, Eguchi M, Kaji K, Saito M and Miura Y (1986) Megakaryocytopoiesis in vitro of patients with essential thrombocythaemia: effect of plasma and serum on megakaryocytic colony formation. *Br. J. Haematol.* 64:241-252.
24. Han ZC, Bellucci S, Tenza D and Caen JP (1990) Negative regulation of human megakaryocytopoiesis by human platelet factor 4 and beta thromboglobulin: comparative analysis in BM cultures from normal individuals and patients with essential thrombocythaemia and immune thrombocytopenic purpura. *Br. J. Haematol.* 74:395-401.
25. Han ZC, Sensebe L, Abgrall JF and Briere J (1990) Platelet factor 4 inhibits human megakaryocytopoiesis in vitro. *Blood* 75:1234-1239.
26. Han ZC, Briere J, Abgrall JF, Le Gall G, Parent D and Sencebe L (1987) Effects of recombinant human interferon alpha on human megakaryocyte and fibroblast colony formation. *J. Biol. Regul. Homeost. Agents* 1:195-200.
27. Carlo-Stella C, Cazzola M, Gasner A, Barosi G, Dezza L, Meloni F, Pedrazzoli P, Hoelzer D and Ascari E (1987) Effects of recombinant alpha and gamma interferons on the in vitro growth of circulating hematopoietic progenitor cells (CFU-GEMM, CFU-Mk, BFU-E, and CFU-GM) from patients with myelofibrosis with myeloid metaplasia. *Blood* 70:1014-1019.
28. Ganser A, Carlo-Stella C, Greher J, Voelkers B and Hoelzer D (1987) Effect of recombinant interferons alpha and gamma on human BM-derived megakaryocytic progenitor cells. *Blood* 70:1173-1179.
29. Murase T, Hotta T, Saito H and Ohno R (1987) Effect of recombinant human tumor necrosis factor on the growth of human leukemia progenitor cells and normal hematopoietic progenitor cells. *Blood* 69:467-472.
30. Kaushansky K (1995) Thrombopoietin: The primary regulator of platelet production. *Blood* 86:419-431.
31. Eaton DL and de Sauvage FJ (1997) Thrombopoietin: The primary regulator of megakaryocytopoiesis and thrombopoiesis. *Exp. Hematol.* 25:1-7.

32. de Sauvage FJ, Hass PE, Spencer SD, Malloy BE, Gurney AL, Spencer SA, Darbonne WC, Henzel WJ, Wong SC, Kuang W-J, Oles KJ, Hultgren B, Solberg Jr. LA, Goeddel DV and Eaton DL (1994) Stimulation of megakaryocytopoiesis and thrombopoiesis by the c-Mpl ligand. *Nature* 369:533-538.
33. Lok S, Kaushansky K, Holly RD, Kuijper JL, Lofton-Day CE, Oort PJ, Grant FJ, Heipel MD, Burkhead SK, Kramer JM, Bell LA, Sprecher CA, Blumberg H, Johnson R, Prunkard D, Ching AFT, Mathewes SL, Bailey MC, Forstrom JW, Buddle MM, Osborn SG, Evans SJ, Sheppard PO, Presnell SR, O'Hara PJ, Hagen FS, Roth GJ and Foster DC (1994) Cloning and expression of murine thrombopoietin cDNA and stimulation of platelet production in vivo. *Nature* 369:565-568.
34. Bartley TD, Bogenberger J, Hunt P, Li YS, Lu HS, Martin F, Chang MS, Samal B, Nichol JL, Swift S, Johnson MJ, Hsu R-Y, Parker VP, Suggs S, Skrine JD, Merewether LA, Clogston C, Hsu E, Hokom MM, Hornkohl A, Choi E, Pangelinan M, Sun Y, Mar V, McNinch J, Simonet L, Jacobsen F, Xie C, Shutter J, Chute H, Basu R, Selander L, Trollinger D, Sieu L, Padilla D, Trail G, Elliott G, Izumi R, Covey T, Crouse J, Garcia A, Xu W, Del Castillo J, Biron J, Cole S, Hu MC-T, Pacifici R, Ponting I, Saris C, Wen D, Yung YP, Lin H and Bosselman RA (1994) Identification and cloning of a megakaryocyte growth and development factor that is a ligand for the cytokine receptor Mpl. *Cell* 77:1117-1124.
35. Debili N, Wendling F, Cosman D, Titeux M, Florindo C, Dusanter-Fourt I, Schooley K, Methia N, Charon M, Nador R, Bettaieb A and Vainchenker W (1995) The Mpl receptor is expressed in the megakaryocytic lineage from late progenitors to platelets. *Blood* 85:391-401.
36. Wendling F, Maraskovsky E, Debili N, Florindo C, Teepe M, Titeux M, Methia N, Breton-Gorius J, Cosman D and Vainchenker W (1994) cMpl ligand is a humoral regulator of megakaryocytopoiesis. *Nature* 369:571-574.
37. Kuter DJ, Beeler DL and Rosenberg RD (1994) The purification of megapoietin: A physiological regulator of megakaryocyte growth and platelet production. *Proc. Natl. Acad. Sci. USA* 91:11104-11108.
38. Hunt P, Li Y-S, Nichol JL, Hokom MM, Bogenberger JM, Swift SE, Skrine JD, Hornkohl AC, Lu H, Clogston C, Merewether LA, Johnson MJ, Parker V, Knudten A, Farese A, Hsu RY, Garcia A, Stead R, Bosselman RA and Bartley TD (1995) Purification and biologic characterization of plasma-derived megakaryocyte growth and development factor. *Blood* 86:540-547.
39. Broudy VC, Lin NL and Kaushansky K (1995) Thrombopoietin (c-mpl ligand) acts synergistically with erythropoietin, stem cell factor, and interleukin-11 to enhance murine megakaryocyte colony growth and increases megakaryocyte ploidy in vitro. *Blood* 85:1719-1726.
40. Banu N, Wang JF, Deng B, Groopman JE and Avraham H (1995) Modulation of megakaryocytopoiesis by thrombopoietin: The c-Mpl ligand. *Blood* 86:1331-1338.
41. Gurney AL, Carver-Moore K, de Sauvage FJ and Moore MW (1994) Thrombocytopenia in c-mpl-deficient mice. *Science* 265:1445-1447.
42. Zeigler FC, de Sauvage F, Widmer HR, Keller GA, Donahue C, Schreiber RD, Malloy B, Hass P, Eaton D and Matthews W (1994) In vitro megakaryocytopoietic and thrombopoietic activity of c-mpl ligand (TPO) on purified murine hematopoietic stem cells. *Blood* 84:4045-4052.
43. Kobayashi M, Laver J, Kato T and Ogawa M (1996) Thrombopoietin supports proliferation of human primitive hematopoietic cells in synergy with steel factor and/or interleukin-3. *Blood* 88:429-436.

44. Sitnicka E, Lin N, Priestly GV, Fox N, Broudy VC, Wolf NS and Kaushansky K (1996) The effect of thrombopoietin on the proliferation and differentiation of murine hematopoietic stem cells. *Blood* 87:4998-5005.
45. Ku H, Yonemura Y, Kaushansky K and Ogawa M (1996) Thrombopoietin, the ligand for the Mpl receptor, synergizes with steel factor and other early acting cytokines in supporting proliferation of primitive hematopoietic progenitors in mice. *Blood* 87:4544-4551.
46. Young JC, Bruno E, Luens KM, Wu S, Backer M and Murray LJ (1996) Thrombopoietin stimulates megakaryocytopoiesis, myelopoiesis, and expansion of $CD34^+$ progenitor cells from single $CD34^+Thy\text{-}1^+Lin^-$ primitive progenitors. *Blood* 88:1619-1631.
47. Papayannopoulou T, Brice M, Farrer D and Kaushansky K (1996) Insight into the cellular mechansims of erythropoietin-thrombopoietin synergy. *Exp. Hematol.* 24:660-669.
48. Petzer AL, Zandstra PW, Piret JM and Eaves CJ (1996) Differential cytokine effects on primitive ($CD34^+CD38^-$) human hematopoietic cells: Novel responses to Flt3-ligand and thrombopoietin. *J. Exp. Med.* 183:2551-2558.
49. Dolzhanskiy A, Basch RS and Karpatkin S (1997) The development of human megakaryocytes: III. Development of mature megakaryocytes from highly purified committed progenitors in synthetic culture media and inhibition of thrombopoietin-induced polyploidization by interleukin-3. *Blood* 89:426-434.
50. Yonemura Y, Kawakita M, Masuda T, Fujimoto K, Kato K and Takatsuki K (1992) Synergistic effects of interleukin 3 and interleukin 11 on murine megakaryopoiesis in serum-free culture. *Exp. Hematol.* 20:1011-1016.
51. Harris DT, Schumacher MJ, Rychlik S, Booth A, Acevedo A, Rubinstein P, Bard J and Boyse EA (1994) Collection, separation and cryopreservation of umbilical cord blood for use in transplantation. *BM Transplant.* 13:135-143.
52. Denning-Kendall P, Donaldson C, Nicol A, Bradley B and Hows J (1996) Optimal processing of human umbilical cord blood for clinical banking. *Exp. Hematol.* 24:1394-1401.
53. Muench MO, Roncarolo MG and Namikawa R (1997) Phenotypic and functional evidence for the expression of CD4 by hematopoietic stem cells isolated from human fetal liver. *Blood* 89:1364-1375.
54. Rowley PT, Ohlsson-Wilhelm BM and Farley BA (1978) Erythroid colony formation from human fetal liver. *Proc. Natl. Acad. Sci. USA* 75:984-988.
55. Bárcena A, Muench MO, Galy AHM, Cupp J, Roncarolo M-G, Phillips JH and Spits H (1993) Phenotypic and functional analysis of T-cell precursors in the human fetal liver and thymus: CD7 expression in the early stages of T- and myeloid-cell development. *Blood* 82:3401-3414.
56. Muench MO, Cupp J, Polakoff J and Roncarolo MG (1994) Expression of CD33, CD38 and HLA-DR on CD34+ human fetal liver progenitors with a high proliferative potential. *Blood* 83:3170-3181.
57. Rice HE, Hedrick MH, Flake AW, Donegan E and Harrison MR (1993) Bacterial and fungal contamination of human fetal liver collected transvaginally for hematopoietic stem cell transplantation. *Fetal Diagn. Ther.* 8:74-78.
58. Mychaliska GB, Muench MO, Rice HE, Leavitt AD, Cruz J and Harrison MR (1998) The biology and ethics of banking fetal liver hematopoietic stem cells for *in utero* transplantation. *J. Pediatr. Surg.* 33:394-399.
59. Namikawa R, Muench MO, de Vries JE and Roncarolo M-G (1996) The FLK2/FLT3 ligand synergizes with interleukin-7 in promoting stromal-cell-independent expansion and differentiation of human fetal pro-B cells in vitro. *Blood* 87:1881-1890.

60. Civin CI, Strauss LC, Brovall C, Fackler MJ, Schwartz JF and Shaper JH (1984) Antigenic analysis of hematopoiesis. III. A hematopoietic progenitor cell surface antigen defined by a monoclonal antibody raised against KG-1a cells. *J. Immunol.* 133:157-165.
61. Di Giusto D, Chen S, Combs J, Webb S, Namikawa R, Tsukamoto A, Chen BP and Galy AHM (1994) Human fetal BM early progenitors for T, B, and myeloid cells are found exclusively in the population expressing high levels of CD34. *Blood* 84:421-432.
62. Goodell MA, Rosenzweig M, Kim H, Marks DF, DeMaria M, Paradis G, Grupp SA, Sieff CA, Mulligan RC and Johnson RP (1997) Dye efflux studies suggest that hematopoietic stem cells expressing low or undetectable levels of CD34 antigen exist in multiple species. *Nat. Med.* 3:1337-1345.
63. Zanjani ED, Almeida-Porada G, Livingston AG, Flake AW and Ogawa M (1998) Human BM $CD34^-$ cells engraft in vivo and undergo multilineage expression that includes giving rise to $CD34^+$ cells. *Exp. Hematol.* 26:353-360.
64. Baum CM, Weissman IL, Tsukamoto AS, Buckle AM and Peault B (1992) Isolation of a candidate human hematopoietic stem-cell population. *Proc. Natl. Acad. Sci. USA* 89:2804-2808.
65. Louache F, Debili N, Marandin A, Coulombel L and Vainchenker W (1994) Expression of CD4 by human hematopoietic progenitors. *Blood* 84:3344-3355.
66. Zauli G, Furlini G, Vitale M, Re M-C, Gibellini D, Zamai L, Visani G, Borgatti P, Capitani S and La Placa M (1994) A subset of human $CD34^+$ hematopoietic progenitors express low levels of CD4, the high-affinity receptor for human immunodeficiency virus-type 1. *Blood* 84:1896-1905.
67. Terstappen LWMM, Huang S, Safford M, Lansdorp PM and Loken MR (1991) Sequential generation of hematopoietic colonies derived from single nonlineage-committed CD34+CD38- progenitor cells. *Blood* 77:1218-1227.
68. Hao QL, Shah AJ, Thiemann FT, Smogorzewska EM and Crooks GM (1995) A functional comparison of $CD34^+CD38^-$ cells in cord blood and bone marrow. *Blood* 86:3745-3753.
69. Dolzhanskiy A, Basch RS and Karpatkin S (1996) Development of human megakaryocytes: I. Hematopoietic progenitors ($CD34^+$ bone marrow cells) are enriched with megakaryocytes expressing CD4. *Blood* 87:1353-1360.
70. Murray LJ, Mandich D, Bruno E, DiGiusto RK, Fu WC, Sutherland DR, Hoffman R and Tsukamoto A (1996) Fetal BM $CD34^+CD41^+$ cells are enriched for multipotent hematopoietic progenitors, but not for pluripotent stem cells. *Exp. Hematol.* 24:236-245.
71. Dercksen MW, Weimar IS, Richel DJ, Breton-Gorius J, Vainchenker W, Slaper-Cortenbach CM, Pinedo HM, von dem Borne AE, Gerritsen WR and van der Schoot CE (1995) The value of flow cytometric analysis of platelet glycoprotein expression of CD34+ cells measured under conditions that prevent P- selectin-mediated binding of platelets. *Blood* 86:3771-3782.
72. Higuchi T, Koike K, Sawai N and Koike T (1997) Proliferative and differentiative potential of thrombopoietin- responsive precursors: expression of megakaryocytic and erythroid lineages. *Exp. Hematol.* 25:463-470.
73. Shimizu Y, Irani AM, Brown EJ, Ashman LK and Schwartz LB (1995) Human mast cells derived from fetal liver cells cultured with stem cell factor express a functional CD51/CD61 (alpha v beta 3) integrin. *Blood* 86:930-939.
74. Hoffman R, Briddell R and Bruno E (1990) Numerous growth factors can influence in vitro megakaryocytopoiesis. *Yale J. Biol. Med.* 63:411-418.
75. Bruno E, Murray LJ, DiGiusto R, Mandich D, Tsukamoto A and Hoffman R (1996) Detection of a primitive megakaryocyte progenitor cell in human fetal BM. *Exp. Hematol.* 24:552-558.

76. Guerriero R, Testa U, Gabbianelli M, Mattia G, Montesoro E, Macioce G, Pace A, Ziegler B, Hassan HJ and Peschle C (1995) Unilineage megakaryocytic proliferation and differentiation of purified hematopoietic progenitors in serum-free liquid culture. *Blood* 86:3725-3736.
77. Muench MO, Roncarolo MG, Menon S, Xu Y, Kastelein R, Zurawski S, Hannum CH, Culpepper J, Lee F and Namikawa R (1995) FLK-2/FLT-3 ligand (FL) regulates the growth of early myeloid progenitors isolated from human fetal liver. *Blood* 85:963-972.
78. Shieh J-H, Chen Y-F, Briddell RA, Stoney G and McNiece IK (1995) Megakaryocyte growth and development factor (c-mpl ligand) acts in combination with stem cell factor and granulocyte colony-stimulating factor to expand myeloid progenitors and megakaryocytes in vitro. *Exp. Hematol.* 23:Abstract #579.
79. Lansdorp PM, Dragowska W and Mayani H (1993) Ontogeny-related changes in proliferative potential of human hematopoietic cells. *J Exp Med* 178:787-791.
80. Shah AJ, Smogorzewska EM, Hannum C and Crooks GM (1996) Flt3 ligand induces proliferation of quiescent human BM $CD34^{+}CD38^{-}$ cells and maintains progenitor cells in vitro. *Blood* 87:3563-3570.
81. Muench MO, Roncarolo M-G, Rosnet O, Birnbaum D and Namikawa R (1997) Colony-forming cells expressing high levels of CD34 are the main targets for granulocyte colony-stimulating factor and macrophage colony-stimulating factor in the human fetal liver. *Exp. Hematol.* 25:277-287.
82. Bertolini F, Battaglia M, Pedrazzoli P, Da Prada GA, Lanza A, Soligo D, Caneva L, Sarina B, Murphy S, Thomas T and della Cuna GR (1997) Megakaryocytic progenitors can be generated ex vivo and safely administered to autologous PB progenitor cell transplant recipients. *Blood* 89:2679-2688.
83. Craig W, Kay R, Cutler RL and Lansdorp PM (1993) Expression of Thy-1 on human hematopoietic progenitor cells. *J. Exp. Med.* 177:1331-1342.
84. Lansdorp PM, Sutherland HJ and Eaves CJ (1990) Selective expression of CD45 isoforms on functional subpopulations of CD34+ hemopoietic cells from human BM. *J. Exp. Med.* 172:363-366.
85. Basch RS, Dolzhanskiy A, Zhang XM and Karpatkin S (1996) The development of human megakaryocytes. II. CD4 expression occurs during haemopoietic differentiation and is an early step in megakaryocyte maturation. *Br. J. Haematol.* 94:433-442.

Chapter 13

Perspectives, Ethics and Clinical Issues in the Use of Primary Human Cells

Mary Pat Moyer
INCELL™ Corporation, LLC, Suite B-200, 12000 Network Boulevard, San Antonio, TX 78249.
Tel: 001-800-364-1765; Fax: 001-210-694-4452; E-mail: mpmoyer@incell.com

1. INTRODUCTION

For many centuries, the idea of transplanting organs, tissues, cells and genetic material, combined with the collection and storage of human body parts, have conjured up visions of monsters and medical creations that have often blurred science-fact with science-fiction. Interestingly, the fantasies proposed in science-fiction in some ways mirror those parts of scientific endeavor that go beyond fact. The fantasies reflect fear of the unknown with imagination, dreams, forward-thinking and risk-taking that can wreak havoc on human-kind, or can stimulate scientific discoveries and clinical applications that would have been considered miracles by our ancestors. Thus, the biomedical community has an awesome responsibility to quell what has been termed the Frankenstein Myth [1] by maintaining high ethical standards, regulatory policies, and quality assurance procedures that optimize public trust, safety, and efficacy.

Historical events and the enactment of laws are indirectly related to ethics. Historical events are what happen, law is the product of politics, and biomedical ethics is the philosophical reflection on science and the technology of healing [e.g., 2]. Laws may imply ethical values. If we disagree, we can work to change the laws. Alternatively, we can act on our own ethical values, but we must accept the consequences if they transcend the law of the land. Thus, a consideration of history and law are integral to a discussion of ethics. This must be done with careful consideration of the historical events and legal framework. Or, restated: "Just because we can do something technically, do we have a right or responsibility to do it?"

2. HISTORICAL AND LEGAL PERSPECTIVES

After World War II, the United States shifted research from wartime efforts to the needs at home. Lessons learned on the battlefield and in the clinic stimulated research in hematology, surgery, infectious disease, and many other areas. Over the past four decades, as the nation focused on peacetime research and development activities, major discoveries were made in basic research and clinical applications relevant to cell-based therapies in the overall context of transplantation (Table 1). These have led to new ethical, political, and legal issues which apply to cells alone, as carriers of genetic information, as tools for gene therapy and vaccine production, or as components of tissues or whole organs. For brevity, this chapter primarily focuses on the events and policies in the United States of America, with scientific and clinical contributions that are relevant to global ethical responsibility to the human condition. Although not intended as a review or to be historically inclusive, this chapter provides a starting point for thoughtful discussion and a perspective on ethical issues pertaining to applications of cell-based therapies now, and as we enter the 21st century. Reviews and texts which discuss a variety of historical, legal, and ethical issues should be consulted for more details [e.g., 3-9].

2.1 Whole Organ Transplantation

Although there are references to the transplantation of human organs at earlier times, whole organ transplantation only became a common medical practice with strong public support after the landmark successful heart transplantation of Dr. Christian Barnard in 1967 and the kidney and liver transplants of Dr. Thomas Starzl in the mid-to-late 1960s. In the 1970s, organ transplantation became accepted medical practice and continued success led to political support and new laws. In the 1980s, technological advances increased the potential for survival, funding was available to those in need, and the ethical, legal and public policy decisions that might limit transplantation were reviewed [e.g. see 3,7,10]. Although new tools and technologies have emerged in the 1990s, cost containment issues have overshadowed many important aspects of medical ethics and technology.

Table 1. Historical highlights of basic science discoveries and clinical applications in the context of ethical, political and legal issues

	Basic science discoveries	Clinical applications	Ethics, politics, law and public sentiment
1940s to 1950s	Blood cells and components Discovery of antibiotics Cell culture technology advances DNA is the genetic code	Blood transfusion methods Use of antibiotics Polio vaccine Different look at genetic diseases	Strong post-WWII public support for these efforts Ethics of doctors and industry not questioned Medical breakthroughs
1960s	Discovery of SV40; other viruses Biological and chemical defense "War on Cancer" bill provides funding	Search for human cancer-causing viruses Organ transplantation: kidney, heart, liver	Space exploration helps fuel support for technology UNOS and other agencies "War on Cancer" bill
1970s	Studies of anti-cancer agents Restriction endonucleases, DNA cloning and the new era of biotechnology DNA used for gene transfection studies Molecular analysis of viruses and other microbial pathogens Monoclonal antibodies Fetal and other tissue studies contribute to understanding disease	New treatments for cancer Bone marrow transplantation (BMT) New diagnostic and therapeutic tools Advances made in obstetrics and developmental biology Many new medical devices developed	Vietnam War; questions of government ethics raised Asilomar conference to promote ethical responsibility for scientists The Recombinant Advisory Committee (RAC) is created Strong public and government support Abortion choice becomes available: Roe vs. Wade
1980s	New products from recombinant technology & other molecular advances New media and methods to grow human and other cells *in vitro* Immunology advances: cyclosporin, blood cell populations Discovery of oncogenes, protooncogenes and tumor suppressor genes HIV molecular attributes quickly defined Cytokines and growth control Blood progenitors in bone marrow (BM) and umbilical cord blood (CB) Isolation of neural origin cells Human Genome Project Major advances in reproductive technology: embryo transfer, freezing, endocrinology Transgenic animals developed	New diagnostic and therapeutic tools Improved transplant success BMT adds use of cytokines AIDS epidemic; increased concern over safety of blood and its products Early treatment of patients with Parkinson's and other neural disorders BM banking for high dose chemotherapy Transplantation of CB cells New information on the molecular basis of diseases Skin tissue transplantation BMT banks established	Strong public and government support NOTA, UAGA Ban on use of fetal tissue for NIH funded research War on AIDS; education emphasis New quality assurance required for human biologicals Issues of privacy, sexual preference predominate 1987: OTA assessment on cell ownership 1987: FDA Points to consider…re: Cell Lines RAC oversight of new gene (and cell) therapies Babies born to older mothers; multiple births Surrogate parenting and artificial insemination Issues of ownership, parental definitions and other legal rights begin to emerge Greater acceptance of new models of "family" Concerns about insurability, privacy New baby conceived as source of BM for sick sibling Patent on whole animal transgenic allowed

1990s	Less government funding	Cost containment issues	Budget deficit; medical insurance issues
	available	New diagnostic and	Reduction of military force; increased
	Polymerase chain reaction (PCR)	therapeutic tools for	research
	& other molecular advances	transplantation:	Partnering of academic-industry-government
	CB banking	autologous, heterologous	more common and accepted
	Cellular and/or molecular basis of	New diagnostic and	Increased public trust for research and
	AIDS becoming better defined	therapeutic tools: longer	technology
	New biomaterials, scaffolds and	patient survival	New approaches to make healthcare more
	better cell culture technology;	New approaches to	available
	differentiation and tissue	transplantation	1996: FDA Regulations defined for biologics
	engineering	new therapies with	license application (BLA)
	Mammalian cloning	cultured cells and tissues	1997: FDA Rule 21 CFR Part 1270 to
	Cytokines, cell surface molecules	Clinical use of cytokines	regulate banked human tissue
	and cell signaling	derived from	Ban lifted on use of fetal tissues for research
	Genetic basis of cancer and other	recombinant DNA	New accreditation for cell, tissue and gene
	diseases	research	therapies and banking of human tissue
	Emerging pathogens, drug	Cell and gene therapy to	Discovery of human cancer genes
	resistance	treat human disease	Concerns about privacy, insurability, etc
	Development of specialized cell	Drug discovery; many	Enthusiastic support of the public for new
	& gene banks	clinical trials	technology
	Cellular and/or molecular basis of	New discovery,	Bill preventing human cloning
	some human diseases discovered	diagnostic and	High public expectations of technology
	Rapid throughput screening	therapeutic tools	Patenting of human genes and engineered
	methods with emphasis on	Improved record-keeping	cells
	molecular mechanisms	New diagnostic and	Patenting of whole animals
	Fluorescence/luminescence	therapeutic tools	Cloning of whole animals
	technologies	involving computerized	
	Computer and information	technology	
	technologies; new equipment	Use of kits and rapid	
	developments	testing methods for	
		diagnostics	
		Changes in medical	
		training and practice	

2.1.1 The National Organ Transplant Act

In 1984, the National Organ Transplant Act (NOTA) of the United States (Public Law No. 98-507), which grew out of the committee spearheaded by Senator Al Gore in 1983, was enacted. It began the current policy framework for organ transplantation (i.e., it was enacted with an ethical view that "new hope" would be provided to patients with disease that would otherwise "inevitably lead to total disability and death"). NOTA was not formally regulatory in character except for two provisions: (1) sale of organs was banned; and (2) the responsibilities for procurement and distribution should be in the private sector, not with the government. NOTA provided funding for the Organ Procurement Agencies (OPA) via the Organ Procurement and Transplantation Network (OPTN) as a vehicle for effective implementation and quality assurance (42 U.S.C. § 274). The pre-existing United Network for Organ Sharing (UNOS) organization, a central computer registry of potential kidney recipients, was designated by DHHS as the OPTN. The legislative history is sparse, but the Senate committee report stated that individuals and organizations should not profit by the sale of human organs. Furthermore, it distinguished the sale of blood and blood derivatives which can be replenished and do not compromise the donor's health. It also dealt mainly with cadaveric donor organs and not a single duplicate organ, such as a kidney, or more recently, other cells and tissues, such as part of a liver, bone marrow (BM), or cord blood (CB) from living related donors.

Under the terms of the NOTA, a Task Force was convened to clarify the issues and make recommendations. In 1986, it transmitted its Final Report (US DHHS 1986) with a strong recommendation for increasing numbers of available organs through hospital policies that would make organ donation an obligation. Implementation of this recommendation was achieved by 1986 legislation that required hospitals with Medicare or Medicaid participation to request organ donation inquiries from family members of potential organ donors. It also led to a quasi-governmental regulatory system and hospital membership qualifications and compliance for transplantation under the control of UNOS and the Health Care Financing Administration (HCFA). Hospitals with transplant programs must necessarily meet UNOS criteria or risk loss of Medicare and Medicaid payment for all services, not just transplantation. Some [e.g., 11] have argued that this authority might conflict with anti-trust laws and is not necessarily consistent with evolving health policies or the donor's preferences. In the 1990s, some limits were placed on organ transplantation in Oregon, where state legislative initiatives sought better distribution of shrinking medical support resources to the general population and questioned the cost per transplant recipient compared to the overall needs of the public.

2.1.2 The Uniform Determination of Death Act

According to the 1981 Uniform Determination of Death Act (UDDA): "An individual who has sustained either (1) irreversible cessation of circulatory or respiratory functions, or (2) irreversible cessation of all functions of the entire brain, including the brain stem, is dead. A determination of death must be made in accordance with accepted medical standards." New questions and ethical discussions have been raised for potential organ, tissue and cell donors who are outside of the UDDA definition. In particular, these include anencephalic newborns who have no hope for survival beyond a few days and certain other non-heart beating donors [12]. Special issues relevant to fetal donors have been discussed in detail [12-14], but the National Commission for the Protection of Human Subjects of Biomedical and Behavioral Research [15] defined a dead fetus as: "a fetus *ex utero* which exhibits neither heart beat, spontaneous respiratory activity, spontaneous movement of voluntary muscles, or pulsation of the umbilical cord (if still attached)." Generally, some organs, tissues, and cells (referred to collectively as "fetal tissues") remain alive for varying periods of time after the total organism is dead.

2.1.3 The Uniform Anatomical Gift Act

The donor's preferences can be made prior to death through the Uniform Anatomical Gift Act (UAGA), which was initially promulgated in 1968 and adopted in some form by all states by 1974. A driver's license signature stating "yes" or "no" to donation and other methods are in place in many states, and the UAGA is a legal means for people to sign donor cards that indicate disposition of their organs upon death. Nevertheless, the organ transplant community prefers independent approval from family members rather than reliance on the contractual validity of the donor card. Thus, the NOTA's mandatory requests for organ donation are done when families are under severe emotional stress. Even though the procurement specialists may be sensitive in presenting the altruistic and ideological elements of donation, the time and place of this decision-making are questionable. There are several reasons for this, including our litigious society and the argument that the donor is no longer alive and controlling his/her own body. Thus, the body becomes the property of the "estate" immediately upon death, and the beneficiaries must make the decision about its disposition.

With that legal premise, it would seem that an individual could include a "Body Donation" addendum to his/her will (much like a "Living Will" is done now). It would require that the estate dispose of the body and its organs according to the wishes of the donor, or upon decision of his/her designated representative. A variety of options might be considered and included in the Body Donation addendum to a will. This would allow for the necessary processing and distribution of organs and other potentially useful tissues for direct clinical applications and, optionally, for research use. Many scenarios can be envisioned for this to work well, but it might include a pre-death contractual arrangement with an established procurement organization or within the context of overall funeral planning. As an option, these arrangements might include specified financial inducements (not direct purchase of the organs) that would become part of the estate of the donor.

Examples might include reimbursing some or all of the costs for medical expenses and the burial or cremation of the body. This would preferably be done by the donor prior to death, but would have mechanisms in place to minimize bidding wars that might exacerbate the emotional stress of the potential donor's family if such decisions were made during the last hours of the donor's life. A Body Donation addendum to a will is simply a proposal. It would work only for a percentage of older donors, but would probably not apply to most of the normal organ donors who are usually younger and would be less likely to have a will.

As described in detail by Hansmann [16], another approach might be to have an optional health insurance premium reduction or other plans that allow individuals who elect to be organ donors and transfer the rights to harvest this property (i.e., the organs) to an insurance company or designated organ procurement agency. This would require some reformation of national and state laws with judicious regulation of procurement and distribution. Such proposals are not outside of existing legal precedent, since state laws have already been developed within the notion of property protection for an individual's body parts [17], and several courts have recognized the family's quasi-property right in the corpse [18,19]. Indeed, it is noteworthy that when the family agrees to donation, it is actually serving as the donor, whereas the decedent is the source of the donated organs or other body parts. Furthermore, many other countries, particularly in Europe, have presumed consent policies in which it is presumed that consent is authorized unless the family protests. There are obvious major ethical implications to these policies. Examples include the possibility that the family may not approve, or that organs might be harvested more zealously by the clinicians, accelerating the determination of death. Finally, it is of interest that United States laws authorizing the removal of corneas on the basis of presumed consent alone have persisted and survived constitutional challenges.

2.2 Tissue, Cell and Gene Transplantation

Although key elements of tissue, cell, and gene transplantation are reviewed in the following sections, these are presented as an overview and are not intended to detail all of the relevant components of this explosive, exciting area of discovery and its promising clinical applications.

2.2.1 Blood and bone marrow

It is of interest that the collection, banking and distribution of blood and its products was specifically noted as outside of the purview of NOTA, which was in the domain of surgery. Blood and its products were already being successfully managed by hematologists and pathologists. And, since the transplantation of BM from living related donors had achieved reasonable clinical success, blood and whole organ transplantation generally evolved along separate, albeit interweaving, pathways (Table 1).

Hematology research and its clinical applications have made enormous strides since the 1950s, when cellular morphology and cytochemistry were the major available tools [20]. Kinetic studies dominated the literature of the 1960s, with the validation of stem cell renewal and migration (concepts first proposed in the early 1900s!) combined with early cryopreservation studies. The 1970s heralded better methods for cell separation and identification of histocompatibility types. BM transplantation (BMT) became a recognized therapy for treating aplastic anemia and various malignant lymphoproliferative disorders of the blood. Living related donors were the most common source, but the establishment of more uniform standards of banking, typing and matching, and the creation of a network for sharing banked BM, improved the success in identifying more compatible matches with unrelated donors. This consequently led to the establishment of Bone Marrow Transplant Registries. In the 1980s, defining the differentiation pathways and subpopulations of blood cells from a common stem cell in the context of a plethora of cytokines and growth factors dominated many basic science and clinical efforts. Mobilization of stem cells by patient pre-treatment with specific recombinant DNA technology-generated cytokines, such as colony-stimulating factors, interleukins, and others, followed by the collection of peripheral blood (PB), has become more common standard medical practice in the 1990s. Additional areas of major progress at the end of the 1980s and through the 1990s have been a movement toward molecular typing and matching methods for transplantation, stem cell transplantation, and gene therapy [21-24].

Human blood stem cells can now be routinely isolated from allogeneic and autologous sources. These include mobilized peripheral blood (mPB) from normal and sick patients, CB, and BM. Products of these efforts are the stem cells and progenitors which can now be *ex vivo* expanded for adoptive immunotherapy and to treat immune deficiency, cancer, infectious disease, and autoimmune disease. These and other types of stem cells are being actively investigated as tools for ongoing and potential applications in gene therapy, *in utero* therapy, and graft/tissue engineering. Many new and promising cell therapy protocols have been proposed or are in progress. These include non-myeloablative stem cell therapies as an alternative to BMT [22,25-27], *in vitro* generation of specific types of blood cells (e.g., dendritic cells [28]), and adoptive immunotherapies [29,30]. Autologous banking of mPB has become a common medical procedure in the 1990s for patients requiring hematopoietic reconstitution after receiving high-dose chemotherapy. An interest in banking stem cells (mPB and CB) for autologous transplantation, in addition to donation or general banking for altruistic reasons, has surfaced as a new commercial outcome of these capabilities.

2.2.2 Non-blood tissues, and cell and gene therapy

Established practices for blood and whole organ transplantation, skin grafting, and the engineering of medical devices have provided the standard basis for cell and gene therapy research and applications, which have technically evolved along multiple lines of work (e.g., Table 1). As reviewed elsewhere [7,31-33], historic basic science work was developed by cell, tissue and organ culture pioneers whose technological contributions created the cellular tools for discovery. These were complemented with the development of molecular tools from the 1950s to the present. Although the first successful *ex vivo* culture of vertebrate (chick embryo) cells was done in the early 1900s, cell culture was not an active scientific area until cells were used as substrates for vaccine production in the 1950s. This beginning of the modern cell culture era and its commercialization led to many important discoveries in virology, vaccines for infectious disease and the production of monoclonal antibodies (mAb). It also led to the more recent efforts in tissue engineering [31,34] and cell-based therapies for life-threatening and chronic diseases, including cardiovascular disease, cystic fibrosis, diabetes, neural disorders, burns, chronic pain, and cancer. For example, stable cell cultures have been used to obtain high quality products, and cells genetically modified *ex vivo* have been used to enhance therapeutic targeting. These potential cell and tissue therapy applications have been heralded as the breakthroughs of the 1990s, and public sentiment has been generally supportive. Although some unexpected problems remain, significant progress has been made [e.g., 23,35], and is expected to continue. Some types of tissue therapies may be integrated with gene therapies and standard clinical intervention strategies. Recently reviewed examples include treatments for angiogenesis [36] and kidney disease [37-38].

Historically, work in molecular biology with biotechnology-derived products and, more recently, the promise of gene therapy, are linked to understanding both cells and genetic vectors (e.g., viruses and plasmids) at the molecular level. The discovery that DNA was the genetic material of the cell fueled interest in understanding molecular mechanisms. Clinically, in the 1950s, the polio epidemic led to funding for work to develop viral vaccines. When simian virus 40 (SV40) was found as a contaminant in the monkey kidney cells being used to prepare polio vaccine, an explosion of work investigated the safety of these and other cell systems from animals as well as from human tumors and normal cells [39]. There was an overriding concern that unknown viruses might play a role in the development of human cancers, and that caution must be exercised when using any type of human cell, malignant or normal. SV40 and other viruses that were discovered in the 1960s and 1970s were extensively characterized and evaluated for their oncogenic potential in animals and cell culture. This progress was an important prelude towards defining some of the issues for the use of cell lines, as well as freshly obtained cells, in the commercial development of biologicals. "Points to Consider" and other documents were prepared as guidelines [40-44]. Other work during that time, and continuing into the 1980s, led to increased assurances that DNA injected *in vivo* (even if it contained subgenomic "transforming regions" or oncogenes of a virus alone, or as part of a vector containing another gene of interest) did not necessarily pose a threat of inducing tumors (e.g., [33,45]). This development paved the way for gene therapy and DNA vaccines.

In the early to mid 1970s, it was realized that the tools of recombinant DNA technology might pose a threat if used inappropriately. Thus, the Asilomar Conference [46] was held to consider a moratorium in light of the ethical implications of the technology, not only to medicine, but to other sciences and the environment, and indeed, the planet. From this meeting, the NIH Recombinant Advisory Committee (RAC) was born. Members of the committee worked hard with other advisory groups and regulatory agencies to develop levels of containment for laboratory workers based on the types of studies being done. They gave careful consideration to the risks, benefits, and overall public safety based on the expected use of "new life forms" created by protocols involving recombinant DNA technologies. The daunting task of RAC was its initial oversight of everything from clinical applications to genetically engineering new plants which would be placed in the environment. Some of the critics of gene technology during the 1970s and 1980s worked very hard to thwart anything relevant to genetic engineering. Many projects were delayed or ruined, but this enthusiasm served the purpose of reminding scientists and their funding sources that the public had watchdogs. However, as the benefits of genetic engineering (i.e., molecular technology) to scientific discovery and its applications became more evident, and it was clear that the scientists and lawmakers were taking their ethical responsibilities seriously, the quixotic efforts of many self-proclaimed crusaders diminished. The explosive numbers of protocols and applications of molecular technology far exceeded even the greatest imaginations. The magnitude and rapid growth of discovery, combined with regulatory needs, using these new tools were awesome. In 1981, the first recombinant protein (human growth hormone) heralded the new age of genetically-engineered production of pharmaceuticals, and in 1988, the first gene therapy protocol was approved by RAC. In 1991, FDA issued a "Points to Consider" document for human gene transfer studies, and under the leadership of Dr. Phillip Noguchi, created the Division of Cellular and Gene Therapies within the Center for Biologic Evaluation and Research of FDA. Statutory authority for regulation of somatic cell and gene therapies was published [40,47], and at a 1994 meeting, it was agreed that RAC would cede its role in regulatory issues to FDA and the NIH Office of Recombinant DNA Activities (ORDA). As authority shifted to FDA, the 1997 RAC Charter indicated that it would continue its service through May, 1999 unless re-chartered [e.g., 48-51 and the FDA, ORDA and NIH Websites on the Internet]. Also in 1996, Epstein [52] published addenda to the "Points to Consider" in human somatic cell and gene therapy.

3. ETHICAL ISSUES AND RESPONSIBILITIES

Detailed reviews with considerations of important ethical concerns relevant to organ, tissue, cell, and gene transplantation have focused on a wide range of issues ranging from medical capabilities, to spiritual reflections, to a global agenda for bioethics [e.g., 2-3,34,52-59]. The major issues of ethics and responsibility are embedded in the four general moral principles of biomedical ethics:

1. *Respect*--for persons, including their autonomous choices and actions;
2. *Beneficence*--including the obligation to benefit others and maximize the good consequences;
3. *Nonmaleficence*--the obligation not to inflict harm; and
4. *Justice*--the principle of fair and equitable distribution of benefits and burdens.

The ethical questions in the context of legal and historical considerations are fueled by the serious gap between the need for organs and the supply of organs, combined with new cell and gene technologies.

3.1 Acquisition and Distribution

Acquisition of human organs, tissues, and cells occurs by: *donation* (express or presumed), *expropriation*, *abandonment*, and *sale*. The questions of ownership and providing human organs, tissues, and cells as a commodity are frequently raised, although providing processed or renewable cells or tissues as a service (with an associated remuneration for the final product) has been generally accepted. Also, the use of one's own cells or tissues has been considered as a right of the patient. This includes "standard medical practice" sources such as blood, BM and skin grafts, as well as "experimental" sources (e.g., engineered tissues from autologous cells such as bone, cartilage, blood vessels).

Donation, which can be requested by a potential donor or on behalf of the donor or the donor's family, has already been considered in some detail above as an outcome of NOTA and the UAGA. *Expropriation* of human organs or tissues upon autopsy of a cadaver, or in preparation of the decedent for burial, without approval from the decedent or the family, is an unlikely source because it generally would not be justified in the current context of legal and ethical guidelines. *Abandonment* or the failure to claim bodies and their parts is another mode of transfer of human organs or tissues. *Sale* results in the transfer of a commodity for some type of financial remuneration, consideration or benefit. Donation has been dealt with throughout this chapter, so the last three modes of acquisition are considered in more detail in the following paragraphs.

3.1.1 Expropriation

This is the most feared of the means of acquisition since those in want of the cells or tissues might kill or facilitate the donor's death by whatever means necessary. This issue received some notoriety in 1998 when a sting operation revealed a conspiracy to take organs from executed Chinese prisoners and sell them in the United States. This same issue is relevant to the selling of fetal tissues by Russian brokers. When there is a disregard for human life, the probability of expropriation may increase in some countries or among some groups [54,58,60], although it is not a justified retrieval method in our society. Nevertheless, arguments have been posed that presumed consent might be considered a type of expropriation. The debates continue within the context of all of the four general moral principles of biomedical ethics.

3.1.2 Abandonment

Although acquisition by abandonment is common when unwanted tissues are removed by a variety of surgical procedures, it is unlikely to increase the supply of organs for transplantation. However, it is a major source of normal and diseased tissues and cells for research. Studies with some of these human organs, tissues, or cells may eventually lead to commercialization, in which the abandoned human organs, tissues, or cells provide a new and useful commodity. Potential commercial benefit is usually unknown at the time of acquisition since it results from the research outcomes. However, not informing patients about this possibility has been raised as an important consent issue as we get closer to new applications in cell and gene therapies. Furthermore, it became quite controversial in the case where a cell line derived from a patient's discarded tumor cells produced an important cytokine and the cell line was commercialized [17]. The problem with that case was that it was more complicated than simply cell line development and commercialization from an abandoned tissue source. The patient, who was eventually cured of his cancer, was asked to repeatedly return to donate more cells and was not told why this was being done or informed about the possible commercial potential of the donation. The conduct of his physicians during the discovery process, rather than the initially unexpected outcome of a commercially useful cell line, was the major breach of ethical expectations.

At this time, all abandoned human organs, tissues, or cells should not be considered to have commercial potential XE "human tissue acquisition:commercial potential". On the other hand, if the patient consents to use of the discarded human organs, tissues, or cells for research, the consent could include wording that commercialization might result, but is unlikely. However, if the human organs, tissues, or cells are being specifically obtained for a known commercial application at the time of collection [*e.g.*, 17], the individual should be informed about the commercial potential. This is a fair and ethical way to deal with the untenable notion that every discarded specimen would require continual follow-up with the patient and his/her heirs in the unlikely event that a commercial product might result from the research. Such a policy gets quite complicated. For example, in the context of fetal tissues and their use, the tissues are considered to be abandoned by the mother upon termination of a pregnancy. They may be processed for specific tissues, cells or products, potentially for transplantation or other applications, including research, under appropriate Institutional Review Board (IRB) guidelines and biomedical ethics. However, even if potential transplant recipient(s), a research program, or the general public will immediately or eventually gain benefit from the fetal tissues, current IRB policy does not allow contact between the mother and anyone who would influence a decision to terminate pregnancy by suggesting that fetal tissue donation is a gift in the same context of organs donated under the UAGA. Also, since the decision process may not include the father (if he is or is not known) or other family members, later complications could result from questions of ownership, genetic rights, and other issues.

3.1.3 Sale

Although the NOTA made selling organs for human transplantation unlawful, it is of interest that the original UAGA did not prohibit the sales of organs; in fact, the Uniform Commissioners on State Laws "believed that it was improper to include an absolute bar to commercial relationships and concluded that this would best be handled at the local level by the medical community". Education and research were not specifically included in the NOTA statute, and sales by living donors are not specifically precluded by UAGA or NOTA. As mentioned above, sales imply a commodity, but the collection and processing of blood, cells, tissues, and secreted or excreted products have been variably sold or prepared as a service. Sometimes, payments are given as a consideration for time spent rather than a specific payment for the human organs, tissues, or cells. With regard to this issue, transfer of ownership would occur upon the payment for the services transaction, such as occurs with a blood donation. Internationally, there have been efforts to regulate use of human organs for transplantation and to establish methods to assure equity, value for human life and new policies through approved guidelines, the most extensive of these being the 1991 World Health Organization "Guiding Principles". Critical to these considerations is vulnerability, which is relevant to bioethical policy issues and fundamental human rights. For example, in India and other poor countries, it is not uncommon for someone to be a living donor and donate one of his/her kidneys as a source of income (about $10,000). Also, the trading of dead Chinese prisoners' organs as commodities recently received public attention [*e.g.*, 61]. These practices have been raised as a human rights issue by many groups, and recently the number of transplants has dropped.

Greenberg and Kamin [62] show that in the rare case when a prospective value can be placed on cells or tissues [*e.g.*, 17], a one-time payment is sufficient to grant a property right and transfer ownership. Since the majority of discarded or other research tissues are of unknown value at collection, the views on their donation have varied. In cases where cell and tissue collection and distribution services are done for clinical applications or research through a non-profit service organization or a for-profit company, it has become common to provide a procurement payment. Within proper guidelines, this should not pose an ethical dilemma or conflict with current standards of practice. If follow-up research leads to new discovery upon use of a human source of cells or tissue (i.e., the raw materials), then the costs/benefits to society should be weighed. For example, proposals for cumbersome ownership tracking and transactions costs with the granting of full property rights would be detrimental to areas such as tissue engineering which may include cells from allogeneic, as well as autologous, source tissues.

3.2 Selection of Donors and Recipients

The criteria for selection of donors and recipients, as well as other aspects of transplantation, are well established for general blood and organ donations. Triage procedures are in place, donors and recipients are tested for histocompatibility matching and infectious agents, and there are local, national and international networks that help match donors and recipients.

Since organ recipients are selected based on tissue matching criteria between a potential donor and recipient, one of the critical problems of organ transplantation is the decreased probability that ethnic minorities will be recipients. Several factors have contributed to this problem, but most notably, there are fewer organ donors among minorities and a lower probability for access to a regional transplant center due to costs for travel, family care, work and other factors. Once family approval is received for organ or tissue donation, the allocation system essentially transfers property rights to the hospital and/or the organ procurement agency by a "rule of capture". The individual organs or tissues do not have an associated cost *per se*, but there are procurement costs which are eventually added to the cost of a transplant operation. Also, it is usually in the best interests of the local hospital or procurement group to maximize the potential to perform the transplant within their own institution, and thus gain additional remuneration from the operation. Because local transplantation enhances the probability of maintaining higher organ viability, and thus a successful outcome, the financial benefits do not necessarily compromise ethical considerations. There are many other selection issues. For example, in the United States, no more than 10% of transplants can be given to non-resident aliens [3].

Although it has many issues in common with adults, transplantation to and from pediatric patients has some notable exceptions. The greatest of these is the intrinsic lack of autonomy [63]. Thus, any decision to be made is on behalf of the child by informed consent of the parents or guardians. This is true for transplant recipients, organ or tissue donations from pediatric cadavers, and sibling donors of blood stem cells. CB transplantation is in the purview of BMT with regard to the science and clinical practice of stem cell transplantation, but the bioethics are different. Although BMT has been confirmed an ethical practice, there have been intense debates on "conceiving a child to save a child" or a "child conceived to give life." Others have argued against *in utero* HLA typing when a termination of pregnancy would occur if the fetus were incompatible with the sick patient. In addition to ownership issues discussed below, a variety of other advantages and disadvantages to the donor and recipient have been described [13,63]. Most notably, since the CB cells are in limited quantity and are donated to a sick sibling (or other autologous recipient), they are not available to the original owner (i.e., the donor) if a disorder that required autotransplantation later developed. Furthermore, issues relevant to cell banking, commercialization, and costs and profits increase the ethical concerns. Examples include needs for proper consent, high level quality assurance standards, follow-up and privacy issues, cost/benefit considerations, and equality in the availability of new therapy regardless of ability to pay. For-profit versus non-profit banking has been analyzed with some suggesting that CB cells, like other human body parts, should not be commercialized, even for possible autologous donation [*e.g.*, 64-66]. In the 1997 report of the Working Group on Ethical Issues in Umbilical Cord Blood Banking [67], the major conclusions were summarized as follows: (1) CB technology is promising but has several investigational aspects; (2) during this investigational phase, secure linkage to CB donor identity should be maintained; (3) CB banking for autologous use has greater uncertainty than for allogeneic use; (4) marketing practices for CB banking in the private sector need close attention; (5) more data are needed to ensure that recruitment for CB banking and use are equitable; and (6) the process of obtaining informed consent for collection of CB should begin before labor and delivery. Except for items 2 and 3, these same issues are relevant to most fetal cell-based therapies, for which standards and regulatory issues must be developed in concert with banking [*e.g.*, 68-69].

Recent advances in embryoscopy and genetics allow early prenatal diagnosis [70], *in utero* cell transplantation [71] and gene therapy [72]. This is very exciting for early genetic or cellular therapies that may be administered *in utero*. However, these new technical capabilities also raise the ethical issues of life and death that may result from treatment which could terminate the pregnancy, or, when an irreparable defect is found, whether pregnancy should continue to term. For example, it has been argued that newborns with specific conditions such as anencephaly should be immediately used as a source of organs or tissues for transplantation or research since the baby will not live more than a few days after birth. However, the use of fetuses or newborns as sources of organs, tissues and cells has historically been intimately integrated with the moral and religious issues of life, death and abortion [73]. The moral relevance of the humanity and personhood of the nonviable fetus have been central to such discussions, and changing the criteria of brain death or even creating exceptions to brain death might be threatening and lead to undesirable outcomes. However, in recent years controversial medical issues such as euthanasia and abortion, when performed in a medically appropriate setting and under ethical guidelines, have gained increasing (sometimes silent) public support, even though they are volatile subjects with major political implications. Anti-abortion arguments led to a ban on the use of human fetal tissue for US government-sponsored research, even though the intent of that work was within the context of the main ethical criteria for the procurement and distribution of human cells, tissues, and organs. That ban was recently lifted, but the ethical questions and stigma against using fetal tissue sources still remain, even though the potential benefits for therapy and research are substantial [73]. Since fetal cells are "nature's modeling clay" for the tissues that will comprise the whole organism and they have a greater growth potential than adult cells, they have been proposed as donors for many cell and tissue-based therapeutic applications. Indeed, early-stage pluripotent embryonic stem cells which result after *in vitro* fertilization and the first embryonic population doublings might be a better choice for certain applications. Nevertheless, the substantial benefits of these tissues do not diminish the ethical battlefields of abortion and the creation of new life in the laboratory followed by terminating that life to harvest cells or tissues. Such ethical questions in cell and gene therapy have fueled international debates in the struggle to deal fairly with the issues [49,74].

3.3 Genetic Manipulation

The role of human gene therapy in the treatment of human genetic diseases, including cancer, have been reviewed [51,56,75]. These reviews realistically present the current technical limitations, safety and efficacy issues and ethical considerations that are controversial or remain to be resolved. Viral vectors (reviewed by Robbins [76]) have been extensively integrated into the overall approach to gene therapy for three main reasons: (1) viruses and recombinant DNA technology have been intimately linked historically; (2) the parasite-host (virus-cell) interactions that have naturally evolved have provided viruses with cell entry and gene delivery function systems that are exquisitely able to work in host cells; and (3) gene delivery by viruses is more efficient than delivery by laboratory methods. Although there have been important technological advances, viral vectors have specific advantages and disadvantages [*e.g.,* 76] that require their use to be weighed from an ethical as well as a technical perspective.

The cloning of mammals from somatic cells [77] is a landmark study. That work, and the mapping and sequencing of the human genome, transgenics, cloning of mammals, development of recombinant viral and other vectors for use in humans, and genetic manipulation are some of the technical capabilities that have generated excitement while creating new ethical issues. Genetic screening, gene therapy, issues of self-determination, eugenics, germ-line therapy and confidentiality are intertwined in the excitement of discovery and the need for policies and laws that will protect individuals and society. Concerns range from the confidentiality of DNA stored in banks to the legal and ethical issues of cloning humans and generating a "Master Race." Patenting and other intellectual property issues relevant to gene research remains controversial internationally. The Question: "Who owns the human genome?" is a continuous debate. The NIH patent on basic techniques covering all *ex vivo* gene therapy [78] was followed by the out-licensing of gene therapy by Novartis, Inc. to avoid the retributions of creating a monopoly [79]. Fears and questions have been raised regarding transgenic humans created like transgenic animals, using viruses that might pose a long-term biohazard, or creating vectors that might irrevocably change normal genetic progression. As pointed out by Sade [75], scientists have a responsibility to educate the public to secure the acceptance of new genetic technology which could be threatened by those who are anti-science and anti-technology. In an essay, Juengst [80] proposed that the FDA "Points to Consider" documents will ultimately lead us to approach the moral limits of gene therapy in the context of professional policy and the goals of medicine, rather than as a social policy question about the public good. The debates continue as we integrate molecular, cell, and tissue technologies, not only in the United States, but internationally [*e.g.*, 81].

4. HARVESTING, PREPARATION AND BANKING ISSUES

4.1 Safety and Efficacy of the Products

The ethical use of any cell-based or tissue product requires validation of safety and efficacy. Safety includes quality control, the development of Standard Operating Procedures (SOPs), and staff training. Efficacy includes pre-clinical and clinical validation. Some of this will come with the newly established FDA committees for "Tissue Engineered Medical Products." FDA will not demand the impossible but will require testing for safety, sterility, potency, purity and identity. There is inherent biological variability of human cells and tissues and many procedures are still considered investigational and have not yet been approved for "standard clinical care." Thus, *ex vivo* handling procedures that would precede therapeutic use require attention to the details of quality control and assurance. An FDA-approved Investigational New Drug (IND) Application followed by a Biologics License Application (BLA) and the implementation of current Quality System Regulation (QSR) may be required [*e.g.*, 82]. It is important to remember that cells are a biological product, but materials such as cell separators, substrates (processed from tissues or engineered), culture media, and other products may be classified as a device. All of these factors have implications for the FDA approval process (Table 2).

Table 2. FDA or other agency (OA) regulation of organs, tissues and cells

Description	Prior FDA Approval	Type of Source
Unmanipulated	None; OA	Allogeneic or syngeneic organs
	None; OA or CBER	Allogeneic or syngeneic cells/tissues
	CBER and OA (eg USDA)	Xenogeneic organs
Manipulated Cells/Tissues	Biologicals	Autologous
		Allogeneic or syngeneic
		Encapsulated
		Transgenic
Nonliving Tissues	Minimal	Human
		Animal
Mechanical	Device	Organs/tissues

CBER: Center for Biologics Evaluation and Research; USDA: United States Department of Agriculture

Examples of some of the key issues relevant to establishing the manufacturing process and FDA approval level include: (1) the amount of *ex vivo* manipulation (i.e., whether the cells will be cultured *in vitro*, cryopreserved, treated with specific cytokines or other factors, or if they will receive a gene or gene product which may be in a viral or other vector); (2) specific features of the end-use product and its validation for suitability; and (3) tracking from the source material to the final product for distribution. It is important that cells and tissues be appropriately processed even if they are being returned to their autologous donor. Current guidelines indicate that whole organs and minimally manipulated blood and blood products would continue to follow standard organ transplantation and clinical hematology practices, respectively. Tissue procurement facilities accredited by the American Association of Tissue Banks (AATB) have defined tissue manufacturing protocols which include patient selection, tissue harvesting, tissue testing, and criteria for tissue discard. Furthermore, the tissues are stringently tested for hepatitis core antibody, hepatitis B surface antigen, human immunodeficiency virus (HIV) antibody and antigen, human T-cell leukemia virus (HTLV)-1, hepatitis C, VDRL, ABA blood group and Rh factor. However, additional guidelines are being developed through the Association for Standards Materials (ASTM) in concert with FDA. Certification from AATB or other regulatory groups, within the context of FDA Biologics regulatory approval, will usually be required for manipulated cells and tissues (i.e., the product). Some of the self-governing guidelines for new applications and approvals of cell, tissue and gene therapies are currently being developed. As these guidelines evolve, it is anticipated that high ethical standards in marketing and distribution, combined with quality assurance, will not only be important for public safety, but will enhance the potential success of the products.

Concerns were originally posed in the 1950s and 1960s about unknown adventitious agents in animal cells used for vaccines. These debates were re-fueled with active protocols for xenotransplantation of animal cells and organs into humans, and *ex vivo* therapies using animal cells for intermediate therapy of transplant patients awaiting organs [83]. Important ethical arguments have been raised for the possibility of introducing a new infectious agent from animals to man as has been strongly implicated in the epidemiology of a primate origin for HIV. Although the animals used for production of therapeutic cells and tissues are well-maintained and tested for all known infectious agents under FDA, USDA and other regulatory guidelines, there is still some concern that such testing is only as good as having a probe, and what can be used as a probe for an unknown? Other ethical issues raised with these efforts have been the use of animals, the cost-effectiveness of the approach, and the differences in human and animal physiology that might lead to new, unexpected problems, such as enhanced immune sensitization against the needed organ and a greater probability for rejection upon transplantation. The counter-arguments that this can potentially reduce the morbidity and mortality of those who await the ever-diminishing source of organs is also a good one. The resolution of this ethical dilemma may not be realized for decades, or it may be resolved in the next few years. We can only hope that the Pandora's box arguments are wrong, and that heretofore undescribed, and possibly catastrophic, zoonoses will not be released upon entry into human hosts from the animal donors.

4.2 Ownership and Cells, Tissues and Organs as Commodities

Technological developments in transplantation and reproduction have led to new legal and ethical or moral questions. In particular: Who owns human body parts? What are the moral limits on body parts as property within a philosophical and legal context? Since payment has routinely been made for replenishable body parts such as blood, sperm, skin, and hair [19,84], ownership by the seller of the parts is implied. However, the extension to payment of unrelated living donors for supplying duplicate organs such as kidneys, or other tissues and cells, is fraught with controversy and ethical considerations of many types, including whether or not these are commodities [16,18]. Of particular concern has been payments for fetal tissues, a problem which surfaced when USSR brokers were exporting fetuses as a commodity. The complexities of ownership issues are varied and not clear-cut. These new ownership and privacy questions are exemplified by recent decisions on distribution of frozen human embryos for research purposes or as an asset (e.g., upon a couple's divorce), the technological advances and new laws that have allowed human genes to be cloned and patented, and the development of gene banks for studying normal functions and specific diseases, such as cancer or diabetes. The new and unanswered questions support the realization that new legal and ethical decisions will be made as technology continues to advance. However, successful mammalian cloning technology has led to a general consensus and adoption of new legislation in the United States and other countries to prohibit human cloning.

As described above, issues on the ownership of human cells, tissues, and organs must be considered in the context of acquisition, and the historical and legal considerations that have resulted in our present laws and policies. Thus, solid organs are usually donated, blood may be donated or sold, semen is usually sold, etc. Since placentas and umbilical cords are temporary, short-lived structures, they have been considered discard tissues under IRB guidelines. However, since the placenta is owned by the neonate, it is now standard practice to provide an enlightened informed consent to the mother. CB donation is given as an option in the context of its utility, donating it to a bank that might use it for transplantation or research, or the purported insurance (with its associated costs for long-term storage, and without guarantee of success) that it might provide if an autologous transplant were needed. The relatively recent notion that CB may have commercial value has added a complication to the decision-making and altruism previously associated with donation.

5. SUMMARY

Within the realm of social responsibility and ethical concerns, it is prudent to periodically re-evaluate principles and criteria pertaining to patients' rights, standards of treatment and health care delivery, and the use of community resources for health care needs. We are fortunate to live in exciting times, and to have new tools available that will allow us to establish cell, tissue and gene therapies as standard medical practice. The known and to-be-discovered technical and conceptual advances present us with choices and opportunities to make a difference to many people in need. Towards that end, we have a responsibility to make decisions based on strong ethical and moral values. This goes beyond obtaining appropriate regulatory approvals, or tempering caution with enthusiasm for novel and pioneering clinical therapies. It goes to the very core of medical ethics, which is defined as *subjecting moral dilemmas to systematic rational analysis*. It goes to the heart of people who need the help, those who donate the organs, cells and tissues, and the psychosocial well-being of all. It means doing the right thing after being well-informed about technical and historical elements of the technology, (i.e., risks *vs.* benefits, costs *vs.* benefits, and political *vs.* moral/ethical issues). It means being fair in a realm of difficult choices. Not only is this true for clinicians and their patients, it is also important for those of us who use human cells, tissues, and organs to perform biomedical research studies. In all stages of the work, we have a responsibility to handle human cells, tissues, and organs (and their molecular components) with reverence, emphasizing respect and appreciation for the donor source. Similar respect and responsible decision-making should also be afforded to the animals that contribute to the development of these therapeutic and research efforts. This should be a conscious part of our everyday work through our interactions with patients, donors, colleagues, staff, trainees, and the general public.

In summary, biomedical researchers and clinicians will continue to be afforded opportunities to make decisions with far-reaching effects. Our intent is to help those who are in need now, but the outcomes are a legacy to the generations that follow. This was more succinctly stated by American playwright Tennessee Williams: "We're all of us just guinea pigs in the laboratory of God. Humanity is just a work in progress." So, as we continue on our journey to pursue new clinical applications and technologies, and spark support for our vision and passion, we must temper our enthusiasm in light of our ethical responsibility for the beautiful gift of life.

REFERENCES

1. Dresselhaus MS (1998). What scientists can do to fight the Frankenstein myth. The Scientist, Philadelphia.
2. Keyes CD and Wiest WE (eds.) (1991) *New Harvest: Transplanting Body Parts and Reaping the Benefits*. Humana Press, Clifton.
3. Mathieu D (1988) *Organ Substitution Technology: Ethical, Legal, and Public Policy Issues*. Westview Press, Boulder & London.
4. Palmer JG (1994) Human gene therapy: suggestions for avoiding liability. *Ann. NY Acad. Sci.* 716:294-305. {discussion} 305-306.
5. Demmy TL and Magovern GA (1991) History of transplantation and future trends, in CD Keyes and WE Wiest (eds.), *New Harvest: Transplanting Body Parts and Reaping the Benefits,* Humana Press, Clifton.
6. Kletzel M, Morgan E, and Frader J (1998) Editorial: Ethical issues in stem cell transplantation. *J. Hematotherapy* 7:197-204.
7. Swazey JP (1992) Spare parts: Organ replacement in American Society. New York: Oxford University Press.
8. Wailoo K (1997) *Drawing Blood: Technology and Disease in Twentieth Century America*. Johns Hopkins University Press, Baltimore.

9. Report of the Task Force on Organ Transplantation (1986). U.S. Dept of Health and Human Services, Washington DC
10. Blumstein JF and Sloan FA (eds.), (1988). *Organ Transplantation Policy: Issues and Prospects*, Duke University Press, Durham and London.
11. Blumstein JF (1988) Government's role in organ transplantation policy, in JF Blumstein and FA Sloan (eds.), *Organ Transplantation Policy: Issues and Prospects*, Duke University Press, Durham and London.
12. Arnold RM, Youngner SJ, Schapiro R, and Spicer CM (1995) *Procuring Organs for Transplant: The Debate over Non-Heart-Beating Cadaver Protocols*, Johns Hopkins, Baltimore.
13. Bell NM (1994) Regulating transfer and use of fetal tissue in transplantation procedures: the ethical dimensions. *Am. J. Law Med.* 20:277-294.
14. Rojansky N and Schenker JG (1993) The use of fetal tissues for therapeutic application. *Int. J. Gynaecol. Obstet.* 41:233-240.
15. National Commission on the Protection of Human Subjects (1975) *"Fetal Viability and Death," Report and Recommendations: Research on the Fetus*, DHEW Publ. No. (05) 76-127 .
16. Hansmann H (1988) The economics and ethics of markets for human organs, in JF Blumstein and FA Sloan (eds.), *Organ Transplantation Policy: Issues and Prospects*, Duke University Press, Durham and London.
17. Moore vs. The University of California Regents. (1988) 249 Cal Regents (Cal App., 2nd District.
18. Office of Technology Assessment (1987) *New Developments in Biotechnology: Ownership of Human Tissues and Cells: Special Report,* US Govt. Printing Office, Washington DC.
19. Matthews P (1983) Whose body? People as property. *Curr. Leg. Probl.* 1983:193-239.
20. Korbling M and Fliedner TM (1996) The evolution of clinical PB stem cell transplantation. *BM Transplant.* 17:765-678.
21. Lee JH and Klein HG (1995) Cellular gene therapy. *Hematol. Oncol. Clinics of North America* 9:91-113.
22. Snyder EL (1995) Hematopoietic cell therapy and transfusion medicine--carpe diem? *Transfusion* 35:371-373.
23. Karlsson S (1997) Gene therapy of hematopoietic cells *J. Intern. Med.* 740:95-99.
24. Orkin SH and Motulsky AG (1995) Report and recommendations of the panel to assess the NIH investment in research on gene therapy. NIH, US Govt Printing Office, Washington DC.
25. Gale RP, Juttner CA, and Henon P (1994) *Blood Stem Cell Transplants,* Cambridge University Press, Cambridge.
26. Slavin S et al. (1998) Nonmyeloablative stem cell transplantation and cell therapy as an alternative to conventional BM transplantation with lethal cytoreduction for the treatment of malignant and nonmalignant hematologic diseases. *Blood* 91:756-763.
27. Yee GC (1996) The future of cell therapy. *Pharmacotherapy* 16:109S-115S.
28. Gluckman JC et al. (1997) In vitro generation of human dendritic cells and cell therapy. *Cytokines Cell Mol. Ther.* 3:187-196.
29. Riddell SR et al. (1995) Principles for adoptive T cell therapy of human viral diseases. *Ann. Rev. Immunol.* 13:545-586.
30. Yee C et al. (1997) Prospects for adoptive T cell therapy. *Curr. Opin. Immunol.* 9:702-708.
31. Dalemans W (1994). From in vitro to in vivo: Progress in the use of cultured cells for human therapy. *Cytotechnology* 16:189-194.

32. Freshney RI (1992) *Culture of Animal Cells: A Manual of Basic Technique*, IRL Press, New York.
33. Moyer MP (1989) Human cell models for genetic engineering. *Medical Progress Through Technology Series,* 15:83-100.
34. Lanza RP and Chick WL (1996) *Yearbook of Cell and Tissue Transplantation 1996/1997*, Kluwer Academic Publishers, Dordrecht and Boston.
35. Dunbar CE and Young NS (1996) Gene marking and gene therapy directed at primary hematopoietic cells. *Curr. Opin. Hemat.* 3:430-437.
36. Lewis BS, Flugelman MY, Weisz A, Keren-Tal I, and Schaper W (1997) Angiogenesis by gene therapy: a new horizon for myocardial revascularization? *Cardiovasc Res* 35:490-497.
37. Imai E and Isaka Y (1998). Strategies of gene transfer to the kidney. *Kidney Int.* 53:264-272.
38. Woods JD et al. (1997) Prospects for the bioartificial kidney. *Semin. Nephrology* 17:381-386.
39. Hayflick L, Plotkin S, Stevenson RE (1986) History of the acceptance of human diploid cell strains as substrates for human virus vaccine manufacture. *Develop. Biol. Standard.* 68:9-17.
40. US Food and Drug Administration (FDA) (1993) Application of current statutory authorities to human somatic cell therapy products and gene therapy products. *Federal Register Notice*, Oct. 14.
41. US Food and Drug Administration (FDA) (1985) Points to consider in the design and submission of human somatic cell gene therapy. *Recomb. DNA Tech. Bullet.* 8:181-186.
42. Office of Biologics Research and Review (1987) *Points to Consider in the Characterization of Cell Lines to Produce Biologicals*, FDA, Bethesda, MD.
43. WHO Study Group (1987) *Acceptability of Cell Substrates for Production of Biologicals*, Technical Report (No. 747) of a WHO Study Group, World Health Organization, Geneva.
44. International Association of Biological Standardization (IABS), (acting editors: L. Hayflick and W. Hennessen), (1989) *Continuous Cell Lines as Substrates for Biologicals, Developments in Biological Standardization*, Vol. 70, Karger, Basel.
45. Moyer MP and Moyer RC (1986) Potential risks of tumor virus subgenomes in the production of biologicals. Develop. Biol. Standard. 68:51-62.
46. Berg P, Baltimore D, Brenner S, Roblin RO, Singer MF (1975) Summary statement of the Asilomar conference on recombinant DNA molecules. *Proc. Natl. Acad. Sci. USA* 72:1981-1984.
47. Kessler DA, Siegel JP, Noguchi PD, et al. (1993) Regulation of somatic cell and gene therapy by the Food and Drug Administration. *N. Engl. J. Med.* 329:1169-1173.
48. Friedman T (1993) Milestones and events in the early development of human gene therapy. *Molec. Genet. Med.* 3:1-32.
49. Jenks S (1997) Genes to order: regulators struggle against future abuses. *JNCI* 89: 1478-1479.
50. Wolff JA and Lederberg J (1994) An early history of gene transfer and gene therapy. *Human Gene Therapy* 5:469-480.
51. Sandhu JS, Keating A, and Hozumi N (1997) Human Gene Therapy. *Crit. Rev. Biotechnol.* 17:307-326.
52. Epstein S (1996) Addendum to the points to consider in human somatic cell and gene therapy. *Human Gene Therapy* 7:1181-1190.
53. Anderson WF (1985) Human gene therapy: scientific and ethical considerations. *Recomb. DNA Tech. Bull.* 8:55-63.

54. Bankowski A and JH Bryant (eds) 1994. Poverty, vulnerability, and the value of human life: A global agenda for bioethics. Highlights and Papers of the XXVIIIth CIOMS Conference, CIOMS.
55. Childress JF (1989) Ethical criteria for procuring and distributing organs for transplantation. *J. Health Polit. Policy Law* 14:87-113.
56. Eisenberg VH and Schenker JG (1997) Genetic engineering: moral aspects and control of practice. *J. Assisted Reprod. Genet.* 14:297-316.
57. McGleenan T (1995) Human gene therapy and slippery slope arguments. *J. Med. Ethics* 21:350-355.
58. World Health Organization. Human Organ Transplantation: A Report on Developments under the Auspices of WHO (1987-1991). *Int. Dig. Health Legis.* 42:389-413.
59. President's Commission for the Study of Ethical Problems in Medicine and Biobehavioral Research (1981) *Defining Death*, US Govt. Printing Office, Washington DC.
60. Connor SS, and HL Fuenzalida-Puelma (eds.) (1990). Bioethics Issues and Perspectives. DAHO Scientific Publication No. 527, DAHO.
61. Goodman E (1998) The organ trade. The Boston Globe (3/8/98).
62. Greenberg W & Kamin D (1993) Property rights and payment to patients for cell lines derived from human tissues: an economic analysis. *Social Sci. Med.* 36:1071-1076.
63. Burgio GR and F Locatelli (1997) Transplant of BM and CB hematopoietic stem cells in pediatric practice, revisited according to the fundamental principles of bioethics, *BM Transplant.* 19:1163-1168.
64. Auerbach AD (1994) Umbilical CB transplants for genetic diseases: diagnostic and ethical issues in fetal studies. *Blood Cells* 20:303-309.
65. Lind SE (1994) Ethical considerations related to the collection and distribution of CB stem cells for transplantation to reconstitute hematopoietic function. *Transfusion* 34:828-834; {Comment in}-1995. *Transfusion* 35: 967-968.
66. Marshall E (1996) Clinical promise, ethical quandary. *Science* 271:586-588.
67. Sugarman J, Kaalund V, Kodish E, Marshall MF, Reisner EG, Wilfond BS, and Wolpe PR (1997) Ethical issues in umbilical CB banking. *JAMA* 278:938-943.
68. Newman-Gage H (1996) Fetal tissue banking: standards and regulatory issues. *J. Hematotherapy* 5:189-193.
69. O'Neil B (1996) Implementing a validation program in a CB bank. *J. Hematotherapy.* 5:139-143.
70. Reece EA (1997) Embryoscopy and early prenatal diagnosis. *Obstet. Gynecol. Clin. North Amer.* 24:111-121.
71. Flake AW and Zanjani EJ (1997) In utero hematopoietic stem cell transplantation. A status report. *JAMA* 278:932-937.
72. Pergament E et al. (1995) Prenatal gene therapy: prospects and issues. *Prenat. Diagn.* 15:1303-1311.
73. Vawter DE, Kearney W, Gervais KG, Caplan AL, Garry D, and Tauer C (1990) *The Use of Human Fetal Tissue: Scientific, Ethical, and Policy Concerns.* University of Minnesota, Minneapolis.
74. Tichtchenko P and Yudin B (1997) The moral status of fetuses in Russia. *Cambridge Quarterly Healthcare Ethics* 6:31-8.
75. Sade RM (1994) Issues of social policy and ethics in gene technology. *Methods Find. Exp. Clin. Pharmacol.* 16:477-489.
76. Robbins PD, Tahara H, and Ghivizzani SC (1998) Viral vectors for gene therapy. *Trends Biotechnol.* 16:35-40.
77. Wilmut I, Schnieke AE, McWhir J, Kind AJ, and Campbell KHS (1997) Viable offspring from fetal and adult mammalian cells. *Nature* 385:810-813.

78. Gavaghan H (1994) Fetal gene therapy under the microscope. *Nature* 372:490.
79. Glaser V (1997) Novartis licenses gene therapy to avoid monopoly. *Nature Biotechnol.* 15:118-119.
80. Juengst ET (1990) The NIH "Points to Consider" and the limits of human gene therapy. *Hum. Gene Therapy* 1:425-433.
81. Lindemann et al. (1995) Guidelines for the design and implementation of clinical studies in somatic cell therapy and gene therapy. The German Working Group for Gene Therapy. *J. Mol. Med.* 73:207-211.
82. duMoulin GC, Pitkin Z, Shen YJ, Conti E, Stewart JK, Charles C, and Hamilton D (1994) Overview of quality assurance/quality control compliance program consistent with FDA regulations and policies for somatic cell and gene therapies: a four year experience. *Cytotechnology* 15:365-72.
83. Institute of Medicine (1996) *Xenotransplantation: Science, Ethics, and Public Policy.* National Academy Press, Washington, DC.
84. Scott R (1981) *The Body as Property*, Viking Press, New York.